STRUCTURE OF COMPLEX NUCLEI

STRUKTURA SLOZHNYKH YADER

СТРУКТУРА СЛОЖНЫХ ЯДЕР

STRUCTURE OF COMPLEX NUCLEI

Lectures presented at an International Summer School for Physicists, Organized by the Joint Institute for Nuclear Research and Tiflis State University in Telavi, Georgian SSR

Edited by

Academician N. N. Bogolyubov

Academy of Sciences of the USSR

Translated from Russian

Springer Science+Business Media, LLC 1969

Library of Congress Catalog Card Number 69-12510

The original Russian text was published by Atomizdat in Moscow in 1966

ISBN 978-1-4899-4835-9 ISBN 978-1-4899-4833-5 (eBook)
DOI 10.1007/978-1-4899-4833-5

Originally published by Consultants Bureau in 1969.

PREFACE

The International Summer School on the Structure of Complex Nuclei was held from August 11 to 28, 1965, in Telavi, Georgian SSR. Organized by the Joint Institute for Nuclear Research together with the Tbilisi (Tiflis) State University, it was attended by 146 physicists from many different countries, including Poland, Rumania, Bulgaria, East Germany, North Vietnam, and Mongolia.

The Telavi Summer School dealt with one of the most important questions of physics – the study of atomic nuclei. In recent years, considerable progress has been recorded in the experimental study of the properties of atomic nuclei, and new favorable possibilities have appeared for determining the most important characteristics of the ground and excited states of both light and heavy nuclei. Significant advances have been made in the theory of atomic nuclei, mainly in the application of mathematical methods developed in the many-body problem. Work on the theory of nuclear matter has also been important. Appreciable progress has resulted from the application of mathematical methods developed in superconductivity theory to nuclear theory and in the development of the superfluidity model of the nucleus. Research on the theory of finite Fermi systems as applied to the nucleus has been interesting. The trend due to the application of group methods to the study of light nuclei has also proved important.

It was the aim of the Summer School to acquaint the audience with the principal achievements in the study of the structure of complex nuclei. The lectures given at the School covered three main sections: the theory of heavy nuclei (including many-body nuclear problems), the classification and analysis of experimental data on heavy nuclei, and the theory of light nuclei. The activities of the School included seminars, at which original communications were presented.

The School owes its success to the considerable assistance provided by our Georgian colleagues and the authorities of the Republic and of the town of Telavi. Those who took part in the School are grateful to the people of Telavi for their cordiality and hospitality.

Professor V. G. Solov'ev

Chairman of the School Organizing Committee

CONTENTS

ATOMIC NUCLEAR THEORY

I. M. Ulehla

Joint Institute for Nuclear Research

Introduction

The real atomic nuclear theory cannot be based on model concepts, but should start from the general laws of motion and interaction of the particles forming the atomic nucleus. In addition to explaining the fundamental physical properties of nuclei, binding energy, nucleon distribution, charge distribution, spin, magnetic, and quadrupole moments, etc., this theory ought to predict the spectrum of atomic nuclei and nuclear reactions. One of the main problems is also the explanation of one model or the other.

This is a complex problem and the physics of the atomic nucleus is at the beginning of its solution. In the course of the last 15 years, theory has been provided with a basis which starts with the following assumptions.

1. The nucleus may be regarded as a many-particle system of nucleons.
2. The laws of nonrelativistic quantum theory suffice to describe this system.
3. The interaction between nucleons in the nucleus has the character of a two-particle force which may be derived from the potential.

Each of the assumptions enumerated is open to criticism. From the point of view of field theory, to regard the nucleus as a system of nucleons is too limited. It is not clear how large is the relativistic effect of nuclear forces which vary rapidly with distance. It has not been demonstrated that the forces acting between the nucleons may be described by means of a two-particle potential, and that many-particle forces are not active between the nucleons.

However, the assumptions expressed are the only ones of their kind, and they already make it possible at the present time to formulate a nuclear theory. From the fundamental point of view, it is reasonable to verify such a theory also as a function of the results obtained or to develop it or to exclude certain basic assumptions and adopt others.

A relatively simple verification of nuclear theory may be given by the two extreme cases, the application of this theory to the examination of the properties of nuclear matter and to relatively light nuclei.

By nuclear matter is understood the infinite system of nucleons, between which electromagnetic forces do not act. This system is uniform and isotropic. The density of nuclear matter is defined by a finite quantity. Extrapolation of the data for heavy nuclei gives a binding energy of a nucleon of nuclear matter of the order of 16 MeV and $r_0 \sim 1f$, the radius characterizing the elementary volume $\frac{4}{3}\pi r_0^3$ of one nucleon. The advantage of nuclear matter, which may perhaps exist in heavy nuclei or in some astronomical objects, is defined by the statement that in nuclear matter the nucleons are described by plane waves, since the laws of translational invariance are then operative.

From the point of view of theory, all nuclei up to Ca^{40} are assumed to be light nuclei. They are suitable because the number of particles is not too large to prevent the system from being examined within the framework of modern mathematical methods.

As shown by the two extreme cases, the results of theory depend to a great extent on the form of the two-particle potential. It is well known that up to the present there has been no law of two-nucleon interaction, such as the Coulomb law or the law of gravitation. Experiment provides us with data which are not completely unambiguous and from which it is difficult to derive an analytical form of the two-nuclear potential. The most successful potentials which describe two-nuclear data up to energies of 300 MeV are the potentials of Gammel and Thaler [1], the Yale group potentials [2] and the potentials of Hamada and Johnson [3]. It is typical of these potentials that they have a hard core and different parameters for singlet and triplet states and for even and odd l.

If a complete spin of two nucleons is denoted by S, the isospin operator (scalar in isospace) by τ, having the value +1 for the isotriplet and −3 for the isosinglet, the orbital moment by $\mathbf{L}$, and the relative coordinates of the two nucleons by $\mathbf{r}$, the general two-nucleon potential then has the form

$$v = V_1 + V_2 s^2 + V_3(\mathbf{L}\mathbf{s}) + V_4(\mathbf{L}\mathbf{s})^2 + V_5 \frac{(\mathbf{r}\mathbf{s})^2}{r^2}, \tag{1}$$

where

$$V_i \equiv V_i\left(r, \frac{\partial}{\partial r}, \tau, J^2\right), \quad J^2 = j(j+1), \quad j = 0,1,2 \ldots .$$

If V_i increases unlimitedly in a certain finite region, we are dealing with a potential containing a hard core. If V_i contains the operator $\partial/\partial r$, we have a potential which is dependent on velocity.†

In the potentials [1-3], the V_i's are independent of the velocities. Potentials which are dependent on velocity and do not contain a hard core are used relatively little, and principally only for tentative calculations. Potentials containing a hard core have, in the outer region, the form:

$$V_i(r, \tau, J^2) = \sum_j a_{ij}(\tau, J^2) \frac{e^{-\mu_j r}}{r^{n_j}}.$$

Recently, potentials with a soft core are being used. These potentials in the vicinity of $r=0$ are repulsive and in the extreme case have a singular point at $r=0$. As shown by researches at the Joint Institute for Nuclear Research, these potentials also are not completely exact, and therefore the results obtained in calculations using them must be regarded critically.

The values of the isospin operator τ, spin operator s^2 and parity $\pi(l)$ are interdependent:

s^2		τ	$\pi(l)$
0	singlet	−3	−
0		1	+
2	triplet	−3	+
2		1	−

†Terms containing $(\mathbf{L}\mathbf{s})$, $(\mathbf{L}\mathbf{s})^2$ are also said to be terms dependent on velocity.

Here, in the singlet spin state, the operators $(\mathbf{Ls})$ (spin-orbital) and $(\mathbf{rs})^2/r^2$ (tensor) are equal to 0. In all the calculations in nuclear theory, based on the above-mentioned assumptions, the perturbation theory is used as the basic method. An important part of this theory is the determination of the two-particle reaction matrix from which, as shown by Watson [4], the optical potential may be derived.

Perturbation Theory

We shall examine the perturbation theory in the second-quantization concept. We have a system of fermions, described by the Hamiltonian

$$H = H_0 + V,$$

$$H_0 = \sum_n E_n \eta_n^* \eta_n, \quad V = \frac{1}{4} \sum_{n_1 n_2 n_3 n_4} (n_1 n_2 | v | n_3 n_4)\, \eta_{n_1}^* \eta_{n_2}^* \eta_{n_3} \eta_{n_4}. \tag{2}$$

E_n are the eigenvalues of the Hamiltonian of the single-particle problem. The subscript n symbolizes all the quantum numbers determining the corresponding single-particle state, $N = \{n_{x1} n_{y1} n_{z1} s_1 \tau\}$. The quantity $(n_1 n_2 | v | n_3 n_4)$ is defined as follows:

$$(n_1 n_2 | v | n_3 n_4) = \langle n_1 n_2 | v | n_3 n_4 \rangle - \langle n_1 n_2 | v | n_4 n_3 \rangle,$$

$$\langle n_1 n_2 | v | n_3 n_4 \rangle \equiv \int \varphi_{n_1}^*(x_1) \varphi_{n_2}^*(x_2)\, v(x_1 x_2)\, \varphi_{n_3}(x_1) \varphi_{n_4}(x_2)\, dx_1 dx_2,$$

where $\varphi_{n_i}(x_i)$ are the wave functions of the single-particle problem; $v(x_1 x_2)$ is the potential (1), acting between the first and second particles; η_n^*, η_n are the creation and annihilation operators of fermions in the n state satisfying the known commutation relations

$$\{\eta_{n_1}^*, \eta_{n_2}^*\} = \{\eta_{n_1}, \eta_{n_2}\} = 0, \quad \{\eta_{n_1}^*, \eta_{n_2}\} = \delta_{n_1 n_2}.$$

The system of fermions described by the Hamiltonian H_0 is regarded as the unperturbed system. We denote the vacuum of the unperturbed system by $|\rangle$. In the ground state of the unperturbed system $-|0\rangle$, which (as is assumed) is nondegenerate, all the states whose energy is less than the Fermi energy are occupied and the other states are free.

For the occupied states we use the subscript k, for the free states l. We determine the new operators

$$\eta_l = b_l, \quad \eta_l^* = b_l^*, \quad \eta_k = a_k^*, \quad \eta_k^* = a_k. \tag{3}$$

Evidently

$$b_l |0\rangle = 0, \quad a_k |0\rangle 0$$

because the operators b* create particles and operators b annihilate them, while the operators *a**, *a* create and annihilate holes.

The expression for the Hamiltonian (2) may be transformed by means of the operators *a*, b. For example,

$$H_0 = E_0 + \sum_l E_l b_l^* b_l - \sum_k E_k a_k^* a_k,$$

where

$$E_0 = \sum_k E_k$$

is the energy of the ground state of the unperturbed system. Similarly, V may be transformed so that all the products of the operators will be written in the form of normal products. This means that the creation operators will be on the left and the annihilation operators will be on the right.

A detailed deduction of the perturbation theory for a Hamiltonian of type (1) is given in [5, 6].

For the ground state $|\psi>$ of the Hamiltonian H we get

$$|\psi> = \left(1 + \frac{1}{E_0 - H_0} V + \frac{1}{E_0 - H_0} V \frac{1}{E_0 - H_0} V + \ldots\right)_L |0>. \tag{4}$$

The state $|\psi>$ is normalized, such that

$$<0|\psi> = 1.$$

The energy corresponding to this state is equal to

$$\left.\begin{aligned} E &= E_0 + \Delta E, \\ \Delta E &= <0|\left(V + V\frac{1}{E_0 - H_0} V + \ldots\right)_L |0>. \end{aligned}\right\} \tag{5}$$

The subscript L means that in all the expressions only those diagrams are considered which do not contain vacuum parts, i.e., parts having no external lines and which are not connected with other parts of the diagram. The permissible diagrams are called coupled diagrams. Since expressions (4), (5) contain only coupled diagrams, the denominators in these equations cannot be equal to zero.

Using the Wick theory, an exact expression for the energy in any approximation may be derived from expression (5). Putting

$$\Delta E = \sum_{i=1}^{\infty} \Delta E^{(i)},$$

the following expressions or the corresponding diagrams will be contained in the individual terms of the energy:

$$\Delta E^{(1)} = \frac{1}{2} \sum_{k_1 k_2} (k_1 k_2 | v | k_1 k_2) \tag{6a}$$

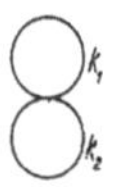

$$\Delta E^{(2)} = \sum_{k_1 k_2 k_3} \frac{(k_3 k_2 | v | k_3 l_1)(k_1 l_1 | v | k_1 k_2)}{E_{k_3} - E_{l_1}} + \frac{1}{4} \sum_{\substack{k_1 k_2 \\ l_1 l_2}} \frac{(k_1 k_2 | v | l_1 l_2)(l_1 l_2 | v | k_1 k_2)}{E_{k_1} + E_{k_2} - E_{l_1} - E_{l_2}} \tag{6b}$$

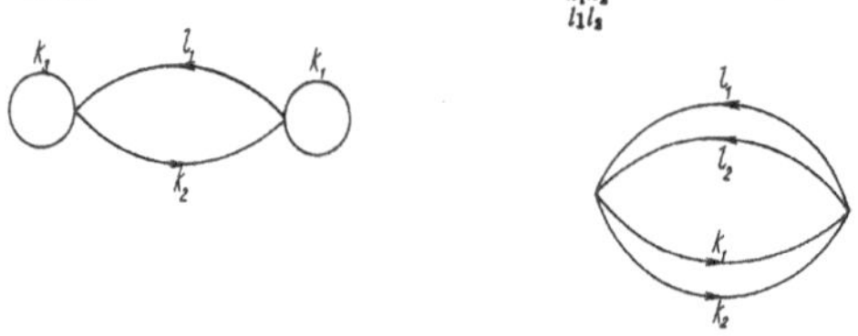

$\Delta E^{(3)}$ contains the term

$$-\frac{1}{8}\sum_{\substack{k_1k_2k_3k_4\\ l_1l_2}}\frac{(k_3k_4|v|l_1l_2)(k_1k_2|v|k_3k_4)(l_1l_2|v|k_1k_2)}{(E_{k_3}+E_{k_4}-E_{l_1}-E_{l_2})(E_{k_1}+E_{k_2}-E_{l_1}-E_{l_2})} \qquad (6c)$$

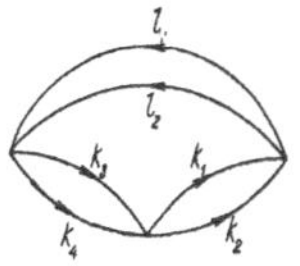

and 12 other terms which we shall not consider. Corresponding expressions may be connected to the individual diagrams on the basis of the following rules.

1. The diagram is drawn from right to left and the corresponding expression is read in the same way.

2. The vertex of the diagram corresponds to the matrix element of two-particle interaction.

3. Four lines intersect in each vertex.

4. The lines for the holes (denoted by k) go from left to right, and the lines for the particles go from right to left (the lines for the particles are denoted by l).

5. The closed lines correspond to holes.

6. The factor ½ corresponds to each pair of lines of the same direction, a closed line is disregarded.

7. The sign of the corresponding expression is determined as follows: The order in which the quantum numbers k and l of a given expression are written must be used for writing the corresponding creation and annihilation operators. These operators commute, such that operators having the same quantum numbers are situated together and creation operators are situated on the right. For an even number of permutations we put the plus sign, for an odd number, the minus sign. For example, in $\Delta E^{(2)}$

$$(k_3\, k_2|v|k_3l_1)(k_1l_1|v|k_1k_2)$$

$$a_{k_3}a_{k_2}\quad a^*_{k_3}b_{l_1}a_{k_1}b^*_{l_1}\quad a^*_{k_1}a^*_{k_2}\ldots +$$

8. The denominators are determined by the diagram being intersected downward after each vertex (on the left) (closed lines must not be intersected); for the intersected lines of particles we write E_k, for holes E_l, and the energies are summed.

For example,

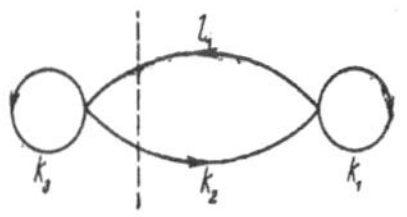

$$\rightarrow \frac{1}{E_{k_2}-E_{l_1}}.$$

9. Summation is taken over all the quantum numbers. Expressions (6) obtained for the energy may be used for calculating the energy of the ground state for all potentials which do not have a hard core. For potentials having a hard core, the matrix elements are meaningless. For this case, the perturbation theory must be modified so as to get rid of the divergences. We shall carry out this modification in the general form and show that the newly-derived perturbation theory may also be used for potentials having a hard core.

Let us assume that there is a vertex of some diagram at which four lines meet. Of these lines, one or more ought not to be lines for a particle

If the next order, there is a diagram for v which differs from the one considered in that, instead of the given vertex, it contains

In the next order, we find a diagram where, instead of the given vertex there is

and so forth. The sum of all these diagrams is determined by means of an expression for the initial diagram in which instead of the expression $(n_1n_2|v|n_3n_4)$ (n denotes both holes and particles), we have the expression $(n_1n_2|t(\omega)|n_3n_4)$. This matrix element is determined by the equation:

$$(n_1n_2|t(\omega)|n_3n_4) = (n_1n_2|v|n_3n_4) + \frac{1}{2}\sum_{l_1l_2}\frac{(n_1n_2|v|l_1l_2)(l_1l_2|t(\omega)|n_3n_4)}{\omega - E_{l_1} - E_{l_2}}. \tag{7}$$

The operator $t(\omega)$ is called the reaction matrix, and in some sense is a two-particle operator. If an exact solution of equation (7) is obtained, we say that we have found an exact solution of the two-particle problem. After this, in the expression for the energy, we take into account the influence of other particles on the pair under consideration by means of the perturbation theory. Actually, the influence of other particles is also taken into account in equation (7) as well, i.e., in the parameter ω which is equal to

$$\omega = E_{n_3} + E_{n_4} + \delta E,$$

δE corresponds to the excitation energy and is determined by intersecting the diagram downwardly between the vertex we are considering and the closest vertex on the right, and determining δE according to the eighth rule

$$\delta E = E_{k_1} + E_{k_2} + \ldots - E_{l_1} - E_{l_2} - \ldots$$

We employ the process described for all the vertices of all the diagrams. We thus replace the v interaction by the t interaction. This change must be made carefully so as not to take some diagrams into account several times. We therefore exclude beforehand the vertex

because this diagram with this vertex is added to some lower-order diagram. We denote vertices with a t interaction by a point

It should be noted that in equation (7) in the second term, summation is taken only over particles. This summation method is a consequence of the Pauli principle which permits a particle to enter some of the occupied places in the |0> state.

By this method, corrections are obtained to the energy defined by means of t. They differ from the corrections (6) and therefore we denote them by $\overline{\Delta E}^{(i)}$. The first correction

$$\overline{\Delta E}^{(1)} = \frac{1}{2} \sum_{k_1 k_2} (k_1 k_2 | t(E_{k_1} + E_{k_2}) | k_1 k_2) \qquad \text{(8a)}$$

consists of: $\Delta E^{(1)}$, a second term in $\Delta E^{(2)}$, and so forth. For it, $\delta E = 0$, because on the right-hand side of the corresponding diagram there are no lines.

The second correction has the form

$$\overline{\Delta E}^{(2)} = \sum_{\substack{k_1 k_2 k_3 \\ l_1}} \frac{(k_3 k_2 | t(E_{k_2} + E_{k_3}) | k_3 l_1)(k_1 l_1 | t(E_{k_1} + E_{k_2}) | k_1 k_2)}{E_{k_2} - E_{l_1}}. \qquad \text{(8b)}$$

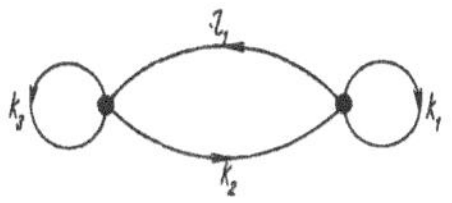

For third-order corrections we get the expression

$$\overline{\Delta E}^{(3)} = \frac{1}{8} \sum_{\substack{k_1 k_2 k_3 k_4 \\ l_1 l_2}} \frac{(k_3 k_4 | t(E_{k_3} + E_{k_4}) | l_1 l_2)(k_1 k_2 | t(E_{k_1} + E_{k_2} + E_{k_3} + E_{k_4} - E_{l_1} - E_{l_2}) | k_3 k_4)}{(E_{k_3} + E_{k_4} - E_{l_1} - E_{l_2})(E_{k_1} + E_{k_2} - E_{l_1} - E_{l_2})} \times \qquad \text{(8c)}$$

$$\times \; (l_1 l_2 | t(E_{k_1} + E_{k_2}) | k_1 k_2) + \ldots .$$

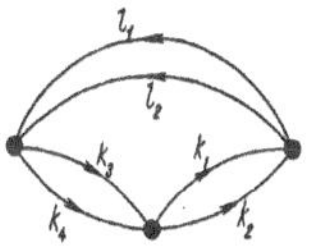

This correction arises out of the third-order term in expression (6) and terms of higher order. The sum of the other terms in the correction of this order gives

$$\frac{1}{2} \sum_{\substack{k_1 k_2 k_3 k_4 \\ l_1 l_2}} \frac{(k_1 k_4 | t(E_{k_1} + E_{k_4}) | l_1 l_2)(k_2 k_3 | t | (E_{k_1} + E_{k_2} + E_{k_3} + E_{k_4} - E_{l_1} - E_{l_2}) | k_3 k_4)}{(E_{k_1} + E_{k_4} - E_{l_1} - E_{l_2})(E_{k_1} + E_{k_2} - E_{l_1} - E_{l_2})} (l_1 l_2 | t |(E_{k_1} + E_{k_2}) | k_1 k_2) +$$

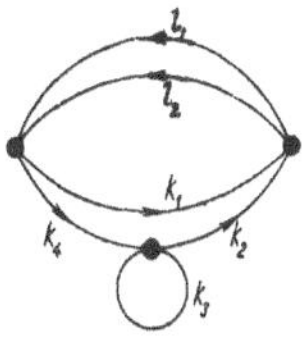

$$+ \sum_{\substack{k_1 k_2 k_3 \\ l_1 l_2 l_3}} \frac{(k_1 k_3 | t(E_{k_1} + E_{k_3}) | l_3 l_1)(k_2 l_3 | t(E_{k_1} + E_{k_2} + E_{k_3} - E_{l_1}) | k_3 l_2)}{(E_{k_1} + E_{k_1} - E_{l_1} - E_{l_3})(E_{k_1} + E_{k_2} - E_{l_1} - E_{l_2})} \quad (l_1 l_2 | t(E_{k_1} + E_{k_2}) | k_1 k_2) +$$

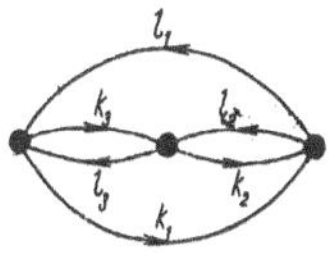

$$+\frac{1}{2}\sum_{\substack{k_1k_2k_3\\ l_1l_2l_3}}\frac{(k_1k_2|t(E_{k_1}+E_{k_2})|l_1l_3)(k_3l_1|t(E_{k_1}+E_{k_2}+E_{k_3}-E_{l_1})|k_1l_2)}{(E_{k_1}+E_{k_3}-E_{l_1}-E_{l_2})(E_{k_1}+E_{k_2}-E_{l_1}-E_{l_2})}(l_1l_2|t(E_{k_1}+E_{k_2})|k_1k_2)+$$

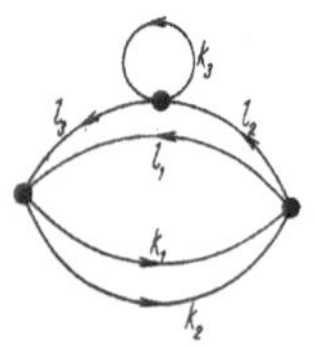

$$+\frac{1}{2}\sum_{\substack{k_1k_2k_3k_4\\ l_1l_2}}\frac{(k_3k_4|t(E_{k_3}+E_{k_4})|l_2l_1)\;(k_2l_2|t(E_{k_2}+E_{k_3}+E_{k_4}-E_{l_1})|k_3k_4)}{(E_{k_3}'+E_{k_4}-E_{l_1}-E_{l_2})(E_{k_3}-E_{l_1})}(k_1l_1|t(E_{k1}+E_{k_2})|k_1k_2)+$$

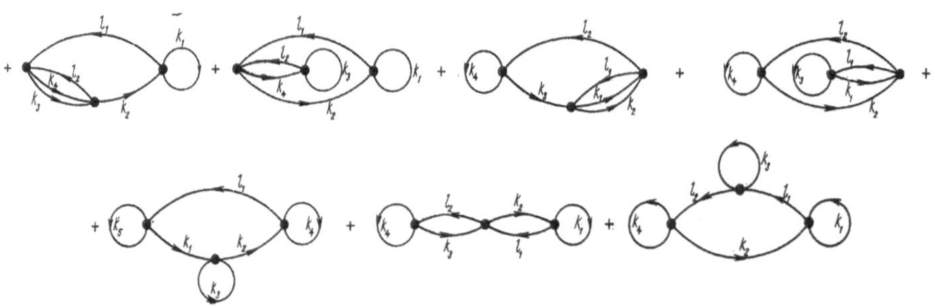

Nuclear Matter

We shall first of all examine the special case of a low-density Fermi gas. Here, the Hamiltonian of the single-particle problem has the form $\mathbf{p}^2/2m$ and its spectrum is continuous. In the $|0\rangle$ state, all levels with a momentum $|\mathbf{p}|<p_F$, where p_F is the Fermi momentum, are occupied. The Fermi momentum is a small quantity, since: v (total volume) = A (number of nucleons $^4/_3\pi r_0^3$,

$$\rho=\frac{A}{v}=\frac{3}{4\pi r_0^3}.$$

The mathematical significance of the momentum is easily determined

$$p_F=\left(\frac{9\pi}{8}\right)^{1/3}\frac{1}{r_0},$$

and therefore

$$\rho\sim p_F^3.$$

In expressions for the energy, integration over the entire volume of the Fermi sphere corresponds to each hole line. The result of integration is proportional to p_F^3, i.e., to the density. This means that the largest contribution to the energy is provided by those diagrams which contain the minimum number of hole lines. Such diagrams are diagrams having two hole-lines

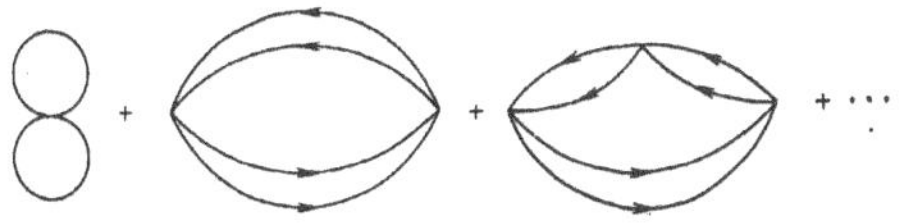

Altogether, they provide a contribution to the energy of

$$\Delta E = \frac{1}{2}\sum_{k_1 k_2}(k_1 k_2 | t | k_1 k_2), \tag{9}$$

where the t matrix is determined by equation (7) for $n_3 = k_1$, $n_1 = k_2$, $\omega = E_{k_1} + E_{k_2}$. Here $k \equiv \{\mathbf{p}, s_z\}$, $|\mathbf{p}| < p_F$; $l \equiv \{\mathbf{p}, s_z\}$, $|\mathbf{p}| > p_F$, and Σ denotes integration over continuous variables and summation over discrete variables.

In the case considered, equation (7) is an integral equation. It can be shown that this equation also has a solution for potentials having a hard core [7]. As will be seen from expressions (9) and (8a), the case of the low density of the Fermi gas is identical with the first approximation of the perturbation theory. Real nuclei and, probably, nuclear matter do not constitute a low-density Fermi gas. It is therefore clear that in the perturbation theory we can not confine ourselves to a first approximation. The question as to how far it is necessary to proceed in the perturbation theory is very important, because the calculation of higher-order corrections rapidly becomes complicated. The original reasoning of Brueckner and Bethe [8] showed that third-order corrections were small and that it was possible to be content with first- and second-order corrections. If this reasoning is correct it is advantageous because, for first- and second-order terms, calculation of the t reaction matrix for $\delta E = 0$ will suffice. Doubt has recently arisen, however, regarding the low value of the third-order correction, and it appears to be more advisable to calculate higher-order corrections.

Three possible ways of elucidating this question are indicated:

1. Calculate $\Delta\bar{E}^{(2)}$, $\Delta\bar{E}^{(3)}$, and so forth and study the convergence. It appears that at least third-order terms are important [9].

2. Calculate the energy contribution of all diagrams containing three, four, etc., hole-lines. This method, proposed by Bethe [10], is basically an expansion into a series of powers of density. Bethe considers that the series should converge rapidly. The difficulty is that we do not know how to accurately sum diagrams of this type. Bethe summed the diagrams containing three hole-lines by the method discussed by L. D. Fadeev [11], and obtained an integral equation which he solved approximately. His calculations show that the influence of three-hole corrections for the theory of nuclear matter is small compared with the influence of two-hole corrections.

In addition to hard-core potentials, Bethe made use of soft-core potentials and obtained a binding energy for them of the order of 13 MeV, which is a relatively good result.

There is not, however, a single case of more exact calculations having been performed. The only exact calculation has been done by the third method, which is fairly difficult to characterize from the point of view of the perturbation theory.

3. Brueckner theory. It may be a good approximation for densities which are characteristic of nuclear matter.

In this theory, as also for a low-density Fermi gas, ΔE is determined by the expression

$$\Delta E = \frac{1}{2} \sum_{k_1 k_2} (k_1 k_2 | t | k_1 k_2), \tag{10}$$

but the t matrix elements are not the solution of equation (7); they satisfy the equation originally derived by Brueckner [11, 8]:

$$(n_1 n_2 | t | n_3 n_4) = (n_1 n_2 | v | n_3 n_4) + \frac{1}{2} \sum_{l_1 l_2} \frac{(n_1 n_2 | v | l_1 l_2)(l_1 l_2 | t | n_3 n_4)}{\mathcal{E}_{n_3} + \mathcal{E}_{n_4} - \mathcal{E}_{l_1} - \mathcal{E}_{l_2}}, \tag{11}$$

where

$$\mathcal{E}_n = E_n + \sum_k (k_n | t | kn). \tag{12}$$

In the low-density approximation, summation is performed only over diagrams of the type

In the Brueckner theory, summation proceeds over all the diagrams formed from the derived diagrams by substitution of the lines for particles —← by the expression

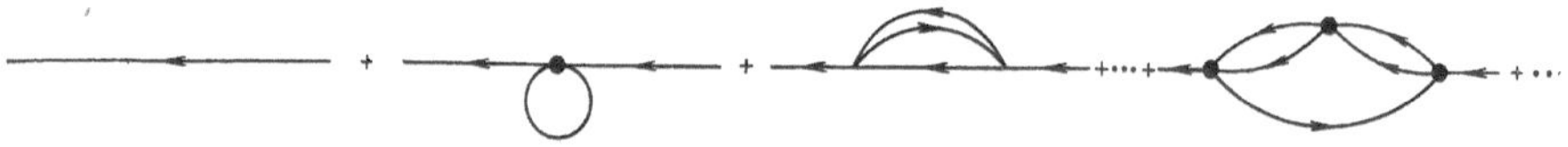

and hole lines —← by the expression

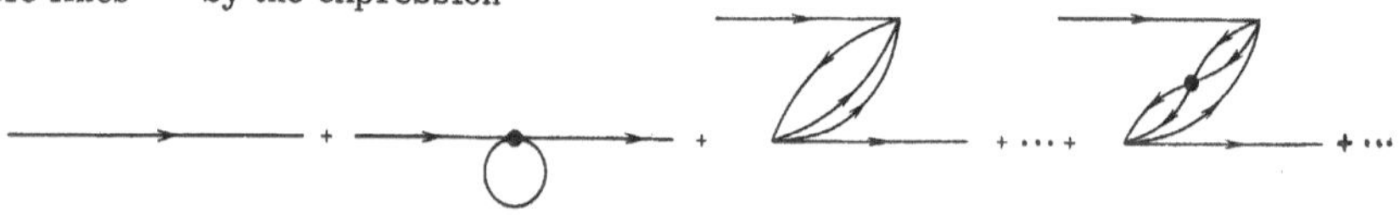

It is evident that in the Brueckner theory, many more diagrams are integrated than in the low-density approximation. It is, however, possible to show that this theory is also true for densities encountered in nuclear matter only by comparing the results of theory with experimental data, which is a fairly complicated problem.

A solution of equations (11) and (12) can be obtained for realistic potentials only by means of high-speed computers, with the simultaneous use of certain approximations.

The first calculation of this kind was made by Brueckner and Gammel [13]; they used some types of Gammel and Thaler potential for the two-nucleon interaction, and found the variation of binding energy of a nucleon with density of nuclear matter. From the extreme value of the energy they obtained $r_0 \sim 1f$ and a binding energy of 16 MeV. The calculation was performed by the self-consistent field method. The first selected the energy $\mathcal{E}_n$, solved the equation for t and after this recalculated the energy $\mathcal{E}_n$. All the calculations were performed in coordinate representation. They made the following approximation on the IBM-704 computer:

1. Exact calculation gives a complex expression for Green's function. In this function, the dependence on the angle between the relative momentum $\mathbf{p}$ and the vector $\mathbf{r}-\mathbf{r}'$ appears in two places. One type of dependence is in the argument of this function, the second follows from the Pauli principle. To eliminate this dependence, Brueckner and Gammel averaged the argument of Green's function over the angles, and also the function representing the Pauli principle.

2. The influence of the hard core was considered in the same way as it is taken into account in the ordinary scattering theory. This theory, however, is not identical with that under consideration.

3. The two-nuclear potentials were considered only in the s and d states; in all other cases, these authors used the potentials v = 0.

The latter approximation only may be justified. It is known that between nucleons one operates mainly in even s, d... states.

On the basis of the relatively good results of Brueckner and Gammel, it may be expected that potentials better describing the experimental data for two nucleons will give a still better result. The calculations have therefore been repeated with new potentials. A corresponding calculation with the Yale potential was made by Brueckner and Masterson [14] who used the first and second approximations of the preceding case. They took the potentials for the s-p-d states.

In addition, they used the following approximation for Green's function in equation (11)

$$\mathcal{E}_{n_3} + \mathcal{E}_{n_4} - \mathcal{E}_{l_1} - \mathcal{E}_{l_2} \approx 2[\,\mathcal{E}(p_{n_3n_4}) - \mathcal{E}(p_{l_1l_2})].$$

Here $p_{n_3n_4}$, $p_{l_1l_2}$ are the relative momenta. The approximation is a corollary of the assumption that the denominator in Green's function does not depend on the total momentum. The approximation is correct if the energy depends quadratically on the momentum, or if the relative momentum is fairly high compared with the total momentum. Taking these variations into account, they repeated the previous calculations on the CDC 1604 machine and obtained practically the same results (the discrepancies were less than 2%). With the Yale potential, however, an unexpected result was obtained: the binding energy for one nucleon was found to be equal to 8.3 MeV for $r_0 = 1.28f$.

A similar result was obtained by Razavy [15] who used the Hamada and Johnson potential, while completely ignoring the influence of the Pauli principle and the influence of the motion of the center of gravity, but taking into account the s-p-d states and approximately the influence of higher states. His result was almost identical with that of Brueckner and Masterson. For the binding energy he obtained 7.8 MeV for $r_0 = 1.35f$.

The new potentials gave a relatively low binding energy and relatively large values of r_0, i.e., insufficiently dense nuclear matter.

Perhaps the situation is even worse than it appears. Brown [16] showed that the approximations concerning the Pauli principle and also the hard core were relatively rough and increased the binding energy by approximately 3-5 MeV. If his estimate is correct, the Brueckner theory, using potentials with a hard core gives a binding energy ~4 MeV.

The incorrect result is perhaps due to the fact that hard-core potentials are unsuitable or the approximations are rough or summation by the Brueckner method does not take a sufficient number of diagrams into account.

The latter questions have so far not been examined in detail, with the exception of the potentials. The radius of the hard core is probably too high for all the potentials, and perhaps the potentials have too great a repulsive part in the p state.

Preliminary calculations by Wong [17] show this to be possible. For the nuclear matter theory in Brueckner's formulation, Wong used a soft-core potential and obtained an increase in the binding energy of approximately 5 MeV. Wong calculated the binding energy of nuclear matter only for two potentials and only for the s state. One of the potentials had a hard core, the other a soft one, representing a Yukawa potential. He performed the calculations using the self-consistent field method and took into account the influence of the Pauli principle. Up to the present, soft-core potentials have not been adequately studied and therefore his results must be approached critically.

The calculation of the properties of nuclear matter with different potentials is evidently one of the criteria in the choice of the correct two-nucleon potential.

Finite Nuclei

The perturbation theory may also be used for calculating the physical properties of finite nuclei. Concrete results with realistic two-nucleon potentials, however, have been obtained only by Brueckner and coworkers [8] and also by Eden and his group [19].

Brueckner used as basis for his investigations the solution of equations (11) and (12) by the self-consistent field method, and replaced the matrix of short-distance reactions by an expression known from his theory of nuclear matter. With his coworkers, he calculated the properties of O^{16}, Ca^{40}, Zr^{90}, using the Gammel and Thaler potential.

The calculated values of the nucleon binding energy in MeV are given in Table 1 (Theory 1), where they are compared with experimental data.

They then determined the density of nucleons in the nucleus. The calculations gave a value 6% lower for O^{16}, and 20% lower for Ca^{40} and Zr^{90} than in real nuclei. The calculations were made on the IBM 704 computer.

These calculations were performed later, with slight corrections, on the CDC 1604 computer. In addition to O, Ca, and Zr, the nucleus of Pb^{208} was also considered, for which it was necessary to use the CDC-1604 computer, which has better parameters than the IBM 704. The results are given in Table 1 (Theory 2). The values obtained for the density are similar to the original values. This follows from the calculated radius of the nucleus in R_{rms} given in Table 2.

Evidently, the energy, density and radius are small.

Eden and his coworkers based their calculations on the perturbation theory formulated by Goldstone, and calculated the energy from formulas (8a), (8b). In the calculations, they used the self-consistent field method, determined by the condition that the second-order correction for the energy ought to be equal to zero: $\overline{\Delta E}^{(2)} = 0$. This condition was only approximately satisfied. The Pauli principle was modified so that it was possible to ignore the motion of the center of gravity of the two particles.

Equation (7) was solved approximately for the tensor and spin-orbit potentials, and the potentials were considered only in the s, p and d states.

The computations were performed on the EDSAG II computer and were restricted to a calculation of the binding energy and nuclear radius of O^{16} for different potentials of Gammel and Thaler type.

TABLE 1

Nucleus	Theory 1	Experiment	Theory 2
O^{16}	4.41	7.98	—
Ca^{40}	6.12	8.55	6.55
Zr^{90}	5.80	8.67	—
Pb^{208}	—	7.87	6.86

TABLE 2

Nucleus	Theory 1	Experiment	Theory 2
O^{16}	2.41	2.57 ± 0.05	—
Ca^{40}	2.91	3.52 ± 0.07	3.00
Zr^{90}	3.56	~ 4.2	—
Pb^{208}	—	5.42 ± 0.11	4.62

Table 3 gives the results for different potentials and compares them with experimental data.

For the 4b and 5 potentials, Eden and his group obtained relatively good values for the energy, but a relatively large discrepency for the radius. In no case does the radius of the nucleus attain the experimental value.

It should be noted that the calculation was done by a fundamentally different method from that used in the Brueckner theory. In the work of Eden and his group, however, approximations and their influence were not evaluated. It is interesting that in both Brueckner's and Eden's calculations, the relatively correct values obtained were lower than the experimental values. If the discrepancies are not the result of approximations, this means that the limitation of the energy to first- and second-order corrections is too rough. In subsequent calculations, therefore, the effect of the different approximations will have to be estimated exactly.

The region of light nuclei is the only region in which the accuracy of the perturbation theory for calculating the properties of atomic nuclei can even now be verified. The physical constants and structure of light nuclei are well known, and at the same time, these nuclei are not very complex systems, so that it is possible to solve the many-body mathematical problem with relative accuracy.

Even in the case of light nuclei, the problem is mathematically very complex and electronic computers are absolutely necessary for exact calculation. This is clear from the matrix element $(n_1 n_2 | t(\omega) | n_3 n_4)$, in which each quantum number represents five quantum numbers (radial, orbital, magnetic, spin, and isospin). This means, for example, that for calculating third-order corrections it is necessary to know hundreds or even thousands of different matrix elements. For a first- and second-order correction, the argument in $t(\omega)$ does not contain the excitation energy, and therefore for these corrections several equations of type (7) are solved. To obtain a correction of higher order for each value of the excitation energy, a system of equations of type (7) must be solved. Actually, equation (7) is an infinite system.

We shall explain in simple form the idea of the calculation, which is described in more detail in [20].

TABLE 3

Potential	Binding energy, MeV	R_{rms}
2	256	1.5
3	160	1.7
4a	150	1.8
4b	123	1.9
5	118	2.0
6	155	2.0
Experiment	127.3	2.65

We write the Hamiltonian of a nucleon system A in the form

$$H = \sum_{i=1}^{A} T_i + \frac{1}{2} \sum_{\substack{i,j \\ i \neq j}} V'_{ij}, \tag{13}$$

where T_i is the kinetic energy of the i-nucleon, and V'_{ij} is the energy of interaction of the i, j nucleons. This Hamiltonian enables us to eliminate the motion of the center of gravity, which is described by a plane wave and has no effect on the internal energy of the nucleus; therefore, the Hamiltonian (13) may be replaced by another which describes the same system of A nucleons, but with a center of gravity situated in the potential well V_T of the harmonic oscillator. Such a variation is convenient for concrete calculations.

The Hamiltonian (13) is still unsuitable for calculation because the unperturbed Hamiltonian $\sum_i T_i$ represents the free motion of the nucleons. We therefore write expression (13) in the form

$$H=\sum_{i=1}^{A}(T_i+V_i)+\left(\frac{1}{2}\sum_{ij}V'_{ij}-\sum_i V_i+V_T\right), \tag{14}$$

where V_i is the single-particle potential. We shall consider the first term in expression (14) as the unperturbed Hamiltonian, and the second as the perturbed Hamiltonian. The potential V_i is selected as potential of the harmonic oscillator, because it is only in the field of this potential that it is pratically possible to perform the transition from a laboratory system of coordinates to a system of relative coordinates. The choice does not contradict the essence of the physical problem. It is well known from the shell model theory that the potential of the harmonic oscillator is suitable for light nuclei. The perturbation theory used may be applied only to the nondegenerate initial state. Therefore, only the nuclei He^4, O^{16}, Ca^{40}, etc., can be investigated by the above-mentioned method. For other nuclei different approximations must be used from the very start. The second term of expression (14) may be transformed into the form

$$\frac{1}{2}\sum_{ij}V'_{ij}-\sum_i V_i+V_T=\frac{1}{2}\sum_{ij}V_{ij},$$

where V_{ij} are the two-particle potentials, containing the potentials of the harmonic oscillator, in addition to the actual interaction. The value of V_{ij} depends solely on the relative coordinates. The eigenfunctions of the Hamiltonian T_i+V_i are the functions of the harmonic oscillator $\varphi_n(x_i)$, where the x_i's are the coordinates of the i-nucleon. We shall now write the equation for the t matrix in the space of the function of the harmonic oscillator, and for simplicity we shall not consider the dependence on spin and isospin, and the antisymmetrization of the matrix elements, and shall put in equation (7) $\delta E=0$ (the general case from the mathematical point of view does not differ from this special case).

$$(n_1n_2|t|k_1k_2)=(n_1n_2|v|k_1k_2)+\sum_{l_1l_2}(n_1n_2|v|l_1l_2)\,\frac{(l_1l_2|t|k_1k_2)}{E_{k_1}+E_{k_2}-E_{l_1}-E_{l_2}}\,. \tag{15}$$

We multiply this equation from the left by $\varphi_{n_1}(x_1)\,\varphi_{n_2}(x_2)$, and performing the summation over all the n_1, n_2's we get

$$t(x_1x_2)\,\varphi_{k_1}(x_1)\,\varphi_{k_2}(x_2)=v(x_1x_2)\,\varphi_{k_1}(x_1)\,\varphi_{k_2}(x_2)+\sum_{l_1l_2}v(x_1x_2)\varphi_{l_1}(x_1)\,\varphi_{l_2}(x_2)\,b_{l_1l_2}, \tag{16}$$

where

$$b_{l_1l_2}=\frac{1}{E_{k_1}+E_{k_2}-E_{l_1}-E_{l_2}}\int\varphi^*_{l_1}(x_1)\,\varphi^*_{l_2}(x_2)\,v(x_1x_2)\,\varphi_{k_1}(x_1)\,\varphi_{k_2}(x_2)\,dx_1dx_2.$$

Here x_1, x_2 are the coordinates of the first and second particles. We determine

$$t(x_1x_2)\,\varphi_{k_1}(x_1)\,\varphi_{k_2}(x_2)=v(x_1x_2)\,\psi_{k_1k_2}(x_1x_2) \tag{17}$$

and get

$$\psi_{k_1k_2}(x_1x_2) = \varphi_{k_1}(x_1)\,\varphi_{k_2}(x_2) + \sum_{l_1l_2}\varphi_{l_1}(x_1)\,\varphi_{l_2}(x_2)\int\frac{\varphi^*_{l_1}(x'_1)\,\varphi^*_{l_2}(x'_2)}{E_{k_1}+E_{k_2}-E_{l_1}-E_{l_2}}\,v(x'_1x'_2)\,\psi_{k_1k_2}(x'_1x'_2)\,dx'_1dx'_2. \quad (18)$$

Let $n_i = 0, 1, 2, 3, \ldots$; $k = 0, 1$; $l_i = 2, 3, 4\ldots$, then the pair of quantum numbers n_1, n_2 may be represented as a point in a plane (Fig. 1).

Summation over l_1l_2 is defined by the region of white circles; the black circles represent the quantum states forbidded by the Pauli principle. It is evident that the sum over all the quantum numbers n_1n_2 is equal to

$$\sum_{n_1n_2} = \sum_{k_1k_2} + \sum_{k_1l_2} + \sum_{k_2l_1} + \sum_{l_1l_2} = \sum_{n_1n_2}^{(0)} + \sum_{l_1l_2}.$$

The index (0) denotes the region in which the Pauli principle is operative.

If, now, we apply to equation (18) the operator $E_{k_1} + E_{k_2} - H_1 - H_2$, $(H_i = T_i + V_i)$, we get

$$v(x_1x_2)\,\psi_{k_1k_2}(x_1x_2) = (E_{k_1} + E_{k_2} - H_1 - H_2)\,\psi_{k_1k_2}(x_1x_2) + \sum_{n_1n_2}^{(0)} a^{(0)}_{n_1n_2}\varphi_{n_1}(x_1)\,\varphi_{n_2}(x_2), \quad (19)$$

where

$$a^{(0)}_{n_1n_2} \equiv (n_1n_2\,|\,t\,|\,k_1k_2) = \int\varphi^*_{n_1}(x_1)\,\varphi^*_{n_2}(x_2)\,v(x_1x_2)\,\psi_{k_1k_2}(x_1x_2)\,dx_1dx_2.$$

We determine

$$a_{l_1l_2} \equiv (l_1l_2\,|\,t\,|\,k_1k_2) = \int\varphi^*_{l_1}(x_1)\,\varphi^*_{l_2}(x_2)\,v(x_1x_2)\,\psi_{k_1k_2}(x_1x_2)\,dx_1dx_2,$$

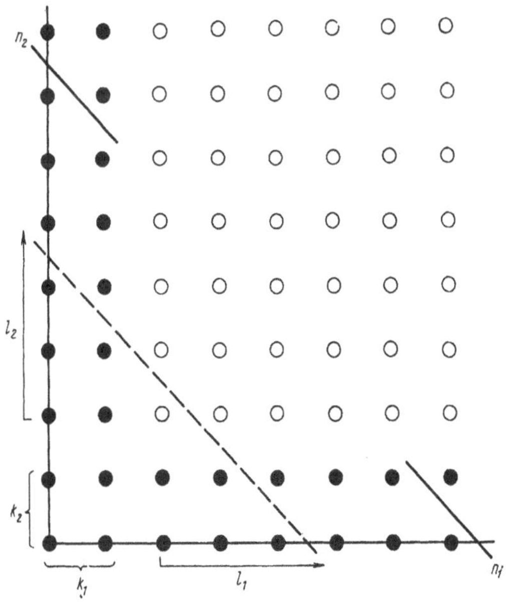

Fig. 1

substitute the expression $v(x_1x_2)\psi_{k_1k_2}(x_1x_2)$ from equation (9) and get

$$a_{l_1l_2} = (E_{k_1} + E_{k_2} - E_{l_1} - E_{l_2}) \int \varphi^*_{l_1}(x_2)\varphi^*_{l_2}(x_2)\psi_{k_1k_2}(x_1x_2)\,dx_1dx_2. \tag{20}$$

If we multiply equation (18) by the function $\varphi_{n_1}(x_1)\varphi_{n_2}(x_2)$ and integrate over the entire space, we get

$$\int \varphi^*_{n_1}(x_1)\varphi^*_{n_2}(x_2)\psi_{k_1k_2}(x_1x_2)\,dx_1dx_2 = \delta_{n_1k_1}\delta_{n_2k_2}, \tag{21}$$

when n_1n_2 belongs to the region denoted by the index (0).

If the solution of the differential equation (19) (in which it is considered that the values of $a^{(0)}_{n_1n_2}$ are given) is known, we get

$$\psi_{k_1k_2}(x_1x_2) \equiv \psi_{k_1k_2}(x_1, x_2, a^{(0}_{n_1n_2}).$$

Equation (21) after this is an algebraic system of linear equations for the values of $a^{(0)}_{n_1n_2}$. If the values of $a^{(0)}_{n_1n_2}$ are found from this system, it is possible to calculate the values of $a_{l_1l_2}$ from equation (20), and thus all the matrix elements $(n_1n_2|t|k_1k_2)$ are known.

The system of equations (21) is an infinite algebraic system. It may be solved if we confine ourselves to a finite number of coefficients $a^{(0)}_{n_1n_2}$. This means that the Pauli principle will not be considered exactly, but the effect of this limitation may be investigated. In Eden's work, the system examined was limited by taking into consideration only the values of $a^{(0)}_{n_1n_2}$ corresponding to points situated below the dash line (see Fig. 1).

In our calculations, the limitation was modified and was applied only in the region in which the Pauli principle operated. In Fig. 1, the continuous lines indicate an example of the limitation we employed. Since the potential $v(x_1x_2)$ depends only on the relative coordinates $x = x_1 - x_2$, the wave function $\psi_{k_1k_2}(x_1x_2)$ may be expanded according to functions of the center of gravity of the two particles situated in the field of the harmonic oscillator.

$$\psi_{k_1k_2}(x_1x_2) = \sum_{N=0}^{\infty} \varphi_N(X)\,\psi_{Nk_1k_2}(x),$$

where

$$X = \frac{x_1 + x_2}{2}.$$

We thus obtain from the original equation (19) the equation

$$v(x)\,\psi_{Nk_1k_2}(x) = (E_{k_1} + E_{k_2} - E_N - H_{\text{рел}})\,\psi_{Nk_1k_2}(x) + \sum_{n_1n_2}^{(0)} a^{(0)}_{n_1n_2} I_{Nn_1n_2}(x), \tag{22}$$

where

$$I_{Nn_1n_2}(x) = \int \varphi_N(X)\,\varphi_{n_1}\left(X + \frac{1}{2}x\right)\varphi_{n_2}\left(X - \frac{1}{2}x\right)dX.$$

Equation (22) is a single-particle problem and for simple potentials may be solved analytically. It is possible to show that the system of equations (21) and (22) has a solution for

$$N = 0, 1, \ldots N_{\max}, a^{(0)}_{00}, \ldots a^{(0)}_{0l_{\max}}, \ldots a^{(0)}_{l_{\max}l_{\max}}$$

and for

$$v(x) = +V_0, \quad x \leqslant x_c$$

$$v(x) = 0, \quad x > x_c.$$

It may also be shown that this solution has a limit for $N_{max} \to \infty$, $l_{max} \to \infty$. It is impossible to show analytically that the system of equations (21) and (22) also has a solution for

$$\lim v(x) = +\infty, \quad x \leqslant x_c$$

$$v(x) = 0, \quad x > x_c.$$

but this may be verified numerically. The solution of equations (21) and (22) for gradually increasing and large V_0's may be compared with the solution for a hard core, and it may be shown that this solution continuously becomes the solution with a hard core.

For potentials with a hard core, equation (22) is very simple. It is sufficient to find its solution in the external region ($x > x_c$) when the left-hand side is zero. [Boundary conditions:

$$\psi_{Nk_1k_2}(x_c) = 0, \quad \lim_{x\to\infty} \psi_{Nk_1k_2}(x) = 0].$$

In the inner region $(x \leqslant x_c)$ identically $\psi_{Nk_1k_2}(x)=0$.

When the proof has been carried out, the entire calculation may be performed as follows. First of all, we find the solution of the hard-core problem. Then, by means of equation (19) we determine the existing product $v(x)\psi_{Nk_1k_2}(x)$.

In equation (18), we substitute on the right-hand side the real potential in the outer region $(x > x_c)$ and the wave function $\psi_{Nk_1k_2}(x) \equiv \psi^{(0)}_{Nk_1k_2}(x)$, calculated with a potential containing only a hard core. From equation (18) we obtain the new wave-function $\psi_{Nk_1k_2}(x) \equiv \psi^{(1)}_{Nk_1k_2}(x)$; ($\psi^{(1)}_{Nk_1k_2}(x)$ is the wave function $\psi_{Nk_1k_2}(x)$ after the first iteration). By means of this function, we determine new values of new $a^{(0)}_{n_1n_2}$ and products $v(x)\psi_{Nk_1k_2}(x)$ for $x \leqslant x_c$. We then repeat the previous calculation with the new data and obtain the wave function $\psi_{Nk_1k_2}(x) \equiv \psi^{(2)}_{Nk_1k_2}(x)$ after the second iteration. It is easy to show numerically that identical functions are obtained already after a few iterations.

The results of our calculations are interesting; they reveal above all a strong dependence on the way in which the Pauli principle is taken into account. Since the numbers N_{max} and l_{max} are dependent, the limitation for N is determined by the limitation of l. This represents an exclusion of the Pauli principle above a certain limit; the limitation of N provides a limit up to which the motion of the center of gravity of the two particles in the potential field of the harmonic oscillator is taken into account. Subsequently, it will be sufficient to examine only the dependence on N.

If we put $N_{max} = 4$, we get the Eden approximation. In our calculations $N_{max} = 4, 6, \ldots, 22$. It follows from them that for a problem with a potential containing only a hard core, the functions lim $v(x)\psi_{Nk_1k_2}(x)$ for N up to 10 have considerable significance. For $N \geq 12$, they are already so small that they may be ignored.

Values of $a^{(0)}_{n_1n_2}$ for $N_{max} = 4$ and for $N_{max} = 22$ differ by 70%. This means that the Eden approximation is very rough. The relatively good result which he obtained was possible only because the errors in the different approximations compensated each other. This shows that any approximation concerning the Pauli principle should be employed with extreme caution.

Our calculations show that the error in calculating the values of $a^{(0)}_{n_1 n_2}$ for $N_{max} = 10$ is less than 2%.

In the iteration of equations with a potential containing an attractive part, three or four iterations were found to be sufficient. On the fourth iteration, the matrix element t was determined with an error of 1%.

In our calculations, we also investigated the effect of the repulsive Coulomb potential on the value of the matrix elements t, and also the effect of the potential due to the motion of the center of gravity of the entire nucleus. It was found that both effects were negligibly small.

Finally, we studied the convergence of the series contained in the expression for the energy. For example, in the second order we have

$$\overline{\Delta E}^{(2)} = \sum_{k_1 k_2 k_3} \sum_{l_1} \frac{(k_3 k_2 \mid t(E_{k_2} + E_{k_3}) \mid k_3 l_1)(k_1 l_1 \mid t(E_{k_1} + E_{k_2}) \mid k_1 k_2)}{E_{k_2} - E_{l_1}} .$$

Summation over $k_1 k_2 k_3$ is the summation over the occupied states ($k_i = 0.1$). Summation over l_1 is carried out through all numbers from 2 to $+\infty$. It was found that the series converges rapidly because the contribution of terms with $l_1 = 10$ is of the order of 0.05%.

The entire calculation was performed with a simplified model corresponding to the s state (in other states, the situation is practically the same). We used a potential containing a hard core of radius $0.4f$ and an attractive part in the form

$$-\alpha \frac{e^{-\mu x}}{\mu x}, \quad \alpha \sim 400 \text{ MeV} \quad \mu \sim 1.3 \text{ F}^{-1}.$$

The calculation was performed in a computer with 20,000 operations per second, and we came to the conclusion that the calculation could be repeated also for real nuclei.

Figure 2 shows the functions $\psi_{Nk_1k_2}$ for $k_1 = k_2 = 0$, N = 0, 2, 4, 6 (the subscripts k_1, k_2 are omitted). The dotted line is the unperturbed oscillator function; the dash line is a function calculated with a potential containing only a hard core; the continuous lines represent functions calculated for potentials containing an attractive part in addition to a hard core. The variation

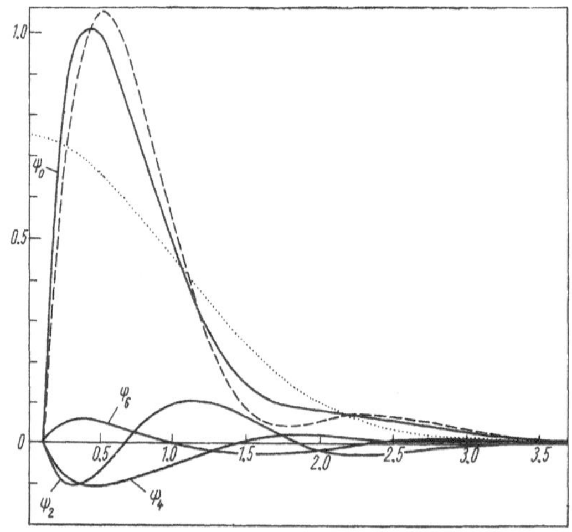

Fig. 2

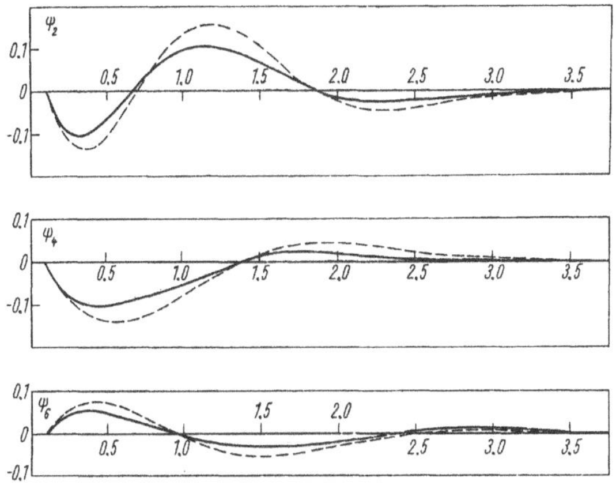

Fig. 3

of the wave function, caused by the attractive part of the potential, is shown in the same way in Fig. 3 also for the functions ψ_{200}, ψ_{400}, ψ_{600}.

These results show that by using computers with 20,000 operations per second, it is possible to calculate the basic physical constants of atomic nuclei, while taking exactly into account the Pauli principle and the motion of the center of gravity of the two particles.

Other interesting values which may be obtained directly from the theory are the so-called single-nucleon levels, density of the nucleons and radius of the nucleus. In the theory of the shell model, the levels on which the nucleons are situated correspond to the eigenvalues of the single-particle Hamiltonian. In the Brueckner theory, the total energy of the nucleus [see equation (10)] may be written in the form

$$E = E_0 + \Delta E = \sum_{k_1} [E_{k_1} + \sum_{k_2} (k_1 k_2 | t | k_1 k_2)]$$

and therefore the energies

$$E_{k_1} + \sum_{k_2} (k_1 k_2 | t | k_1 k_2) \equiv \mathcal{E}_{k_1}$$

may be regarded as single-particle energies. It follows from the perturbation theory that it is thus possible to write the energy only as far as second order terms.

In the third order are terms which already cannot be so written. Despite the fact that all terms of type (10) may be grouped together and the corresponding energies may be regarded as single-particle energies, the remaining terms represent an additional energy, determined by the structure of the nucleus. This energy will be small if the terms of third and higher orders are small.

The density operator $\rho(x)$ is equal to

$$\rho(x) = \sum_{n_1 n_2} \varphi^*_{n_1}(x)\, \varphi_{n_2}(x)\, \varphi^*_{n_1} \varphi_{n_2}.$$

Putting

$$\varphi^*_{n_1}(x)\, \varphi_{n_2}(x) = c_{n_1 n_2}(x),$$

the mean value of the density in the $|\psi\rangle$ state is

$$\bar{\rho}(x) = \frac{\langle\psi|\rho(x)|\psi\rangle}{\langle\psi|\psi\rangle} = \sum_{i=0}^{\infty} \bar{\rho}^{(i)}(x).$$

The zeroth approximation term has the form

$$\bar{\rho}^{(0)}(x) = \sum_{k} c_{kk}(x).$$

The first approximation term may be written in the form

$$\bar{\rho}^{(1)}(x) = \sum_{k_1k_2l_1}\left\{\frac{c_{k_2l_1}(x)\,(k_1l_1|\,t\,(E_{k_1}+E_{k_2})\,|k_1k_2)}{E_{k_2}-E_{l_1}} + \frac{(k_2k_1|\,t\,(E_{k_1}+E_{k_2})\,|k_2l_1)\,c_{l_1k_1}(x)}{E_{k_1}-E_{l_1}}\right\}.$$

The diagram

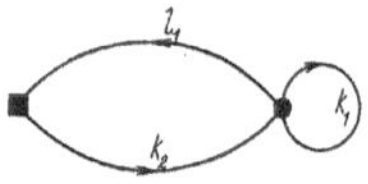

corresponds to the first term in this expression. Here ■ represents the quantity $c_{n_1n_2}(x)$, the diagram

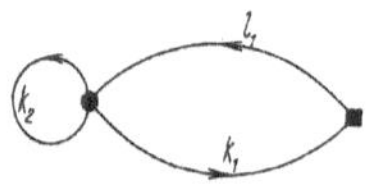

corresponds to the second term.

The term for the next approximation for the density is determined by 18 diagrams, the construction of which is obvious from the examples given. All that is necessary is to note that the diagrams of the next order include the above diagrams.

Up to second-order terms, the value of $\langle\psi/\psi\rangle$ is determined by the expression

$$\langle\psi|\psi\rangle = 1 + \sum_{k_1l_1} a^*_{k_1l_1} a_{n_1l_1} + 4 \sum_{k_1k_2l_1l_2} a^*_{k_1k_2l_1l_2} a_{k_1k_2l_1l_2},$$

$$a_{n_1l_1} = \sum_{k_2} \frac{(k_2l_1|\,t\,(E_{k_1}+E_{k_2})\,|k_2k_1)}{E_{l_1}-E_{k_1}}, \quad a_{k_1k_2l_1l_2} = \frac{(l_1l_2|\,t\,(E_{k_1}+E_{k_2})\,|k_1k_2)}{E_{l_1}+E_{l_2}-E_{k_1}-E_{k_2}}.$$

LITERATURE CITED

1. Gammel, J. L., and Thaler, R. M., Phys. Rev., 107:291, 1337 (1957).
2. Lassila, K. E., et al., Phys. Rev., 126:881 (1962).
3. Hamada, T., and Johnson, I. D., Nucl. Phys., 34:382 (1962).
4. Watson, K. M., Phys. Rev., 89:575 (1953).
5. Gell-Mann, M., and Low, F., Phys. Rev., 84:350 (1951).
6. Goldstone, J., Proc. Roy. Soc., A239:267 (1957).
7. Bethe, H. A., and Goldstone, J., Proc. Roy. Soc., A238:551 (1957).
8. Brueckner, K. A., Phys. Rev., 100:36 (1955); Bethe, H. A., Phys. Rev., 103:1353 (1956).
9. Rajaraman, R., Phys. Rev., 129:265 (1963).
10. Bethe, H. A., Phys. Rev., 138:B804 (1965).

11. Faddeev, L. D., Zh.Ékspérim. i Teor. Fiz., 39:1459 (1960).
12. Brueckner, K. A., Phys. Rev., 97:1344 (1955).
13. Brueckner, K. A., and Gammel, J. L., Phys. Rev., 109:1023 (1958).
14. Brueckner, K. A., and Masterson, K. S., Phys. Rev., 128:2267 (1962).
15. Razavy, M., Phys. Rev., 130:1091 (1963).
16. Brown, G. E., Nucl. Phys., 56:191 (1965).
17. Chun Wa Wong, Nucl. Phys., 56:213 (1964).
18. Brueckner, K. A., and Weitzner, H. Phys. Rev., 110:431 (1958); Brueckner, K. A., et al., Phys. Rev., 121:255 (1961); Masterson, K. S., Phys. Rev., 129:776 (1963).
19. Eden, R. J., et al., Proc. Roy. Soc., A248:266 (1958); 253:177 (1959); 253:186 (1959).
20. Blank, J., and Ulehla, I., Preprint E-2047, Dubna, 1965.

PAIR CORRELATIONS AND STRUCTURE OF DEFORMED NUCLEI

V. G. Solov'ev

Joint Institute for Nuclear Research

Introduction

Progress in the understanding of the structure of the nucleus began after the conception of the shell model of the nucleus by Mayer, Haxel, Jensen, and Suess. The next important step was made by Bohr and Mottelson, who proposed a generalized model of the nucleus, on the basis of which a large number of experimental facts were successfully explained. Both these models stimulated many experimental investigations. The superfluid model of the nucleus, or the model with pair and multipole–multipole forces, represents the further development of these models.

Before proceeding to give an account of the fundamental theorems of the superfluid model, we shall classify the experimental facts which cannot be explained, while remaining within the framework of the improved shell model and generalized model. Thus, some effects to which pair correlations lead, were earlier described phenomenologically: the term in the Weizsäcker formula allowing for the energy difference of the binding of odd–odd, even–even and odd–mass nuclei, the equality to zero of the spins of the ground states of even–even nuclei, etc.

We shall enumerate the experimental facts which it has not been possible to explain in the framework of the generalized and shell models (even by the introduction of the concept of two-body forces).

1. The gap in the spectra of excited states (associated with a variation in the internal structure) in even–even nuclei and the absence of such a gap in the spectra of odd and odd–odd nuclei. Figure 1 shows the spectra of excited nonrotational states of deformed nuclei. It will be seen from the figure that in odd and odd–odd nuclei, the energies of the first excited states are of the order of several tens of keV, while in even–even nuclei as a rule they exceed 1 MeV. A similar pattern is observed in all nuclei.

2. Moments of inertia of deformed nuclei. Measurement of the energies of the first rotational states provided information concerning the moments of inertia of a large number of deformed nuclei, some of which are given in Fig. 2. The moments of inertia $\mathcal{J}$, calculated on the basis of the independent-particle model for even–even nuclei, were found to be two or three times greater than the moments of inertia of neighboring even–even nuclei, this difference exceeding by many times the contribution from one additional nucleon.

3. Considerable difficulties in the explanation of the equilibrium form of nucleus. Thus, calculations on the basis of the independent-particle model lead to all the nuclei with nucleons

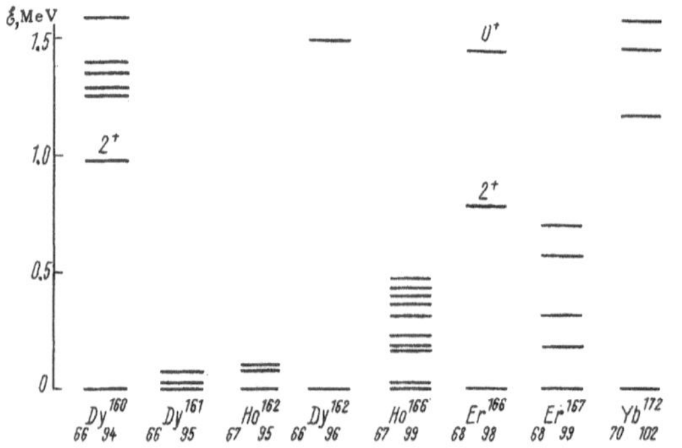

Fig. 1. Spectrum of excited nonrotational states of deformed nuclei, associated with a variation in the internal degrees of freedom (2^+ and 0^+ – collective γ- and β-vibrational states.

situated in unfilled shells, being deformed. However, the transition from the spherical to the ellipsoidal form occurs with the filling of approximately one quarter of the places in the last shell.

4. The density of single-particle equations in odd deformed nuclei, as determined from experimental data, was found to be approximately twice as large as the level density given by the Nilsson potential. It should be noted that the average density of single-particle levels in

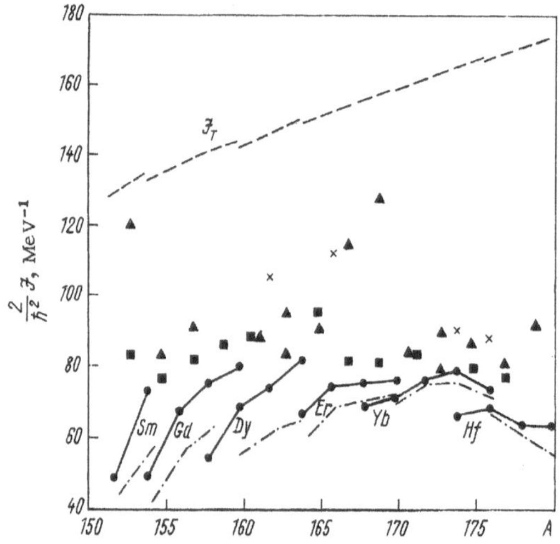

Fig. 2. Moments of inertia of deformed nuclei (according to the data of [12]). Experimental values of the moments of inertia: ●) Even – even nuclei; ■) odd-Z nuclei; ▲) odd-N nuclei; X) odd – odd nuclei. Calculations for even – even nuclei: – – – – according to the independent-particle model; – · – · – · according to the superfluid model.

the independent-particle model is connected to the density of nuclear matter and therefore a variation in potential of the average field may lead merely to a redistribution of the single-particle levels, but not to a variation of the average density.

5. Some difficulties exist in the α-decay theory. Thus, calculations according to the independent-particle model from the lifetimes relative to the α decay give high values for nuclei radii compared with data from nuclear reactions. In odd nuclei, it is not possible to explain the reduction in the probabilities of unfavorable α-transitions compared with favorable transitions, etc.

6. In the case of β decay, nonretarded transitions are observed which are solved according to single-particle matrix elements, but are strictly forbidden in the independent-particle model.

7. The description within the framework of the generalized model of collective vibrational states of odd–odd nuclei is very rough. It contains very many parameters; on the basis of this model it is impossible to calculate the energies of these states and other characteristics. In addition, the adiabatic condition, which requires $\omega_c \ll \omega_{in}$, i.e., that the frequencies of the collective vibration be much lower than the frequencies due to internal motion, is not satisfied in even–even deformed nuclei. Further, if the quadrupole and octupole excited states were to be described phenomenologically as oscillations of the nuclear surface, then these states would consequently be identical in all nuclei, and the magnitude of their energy could only vary slightly from one nucleus to another. Experimental data on the energies of the quadrupole and octupole states in deformed even–even nuclei do not satisfy this requirement. Thus, in the region of thorium, uranium and plutonium isotopes, the energies of the octupole states with $K\pi = 0^-$ are less than the energies of β- and γ-vibration states. In isotopes of dysprosium and erbium, the γ-vibration states are lower than the β-vibration states and so forth. The energies of the quadrupole and particularly the octupole states vary considerably from one nucleus to another.

The combination of these factors point to the necessity of finding a new approach to the description of the properties of the ground and excited states of atomic nuclei. Such a new approach is the microscopic approach of the superfluid model of the nucleus. The mathematical basis of this model is the method of approximate second quantization proposed by N. N. Bogolyubov in 1947 [1] and the methods of development used in the conception of the superconductivity theory [2].

N. N. Bogolyubov [3] was the first to point out the possibility of the superfluidity of nuclear matter, then Bohr et al. [4] postulated the question of the existence of superfluid states in atomic nuclei. The theory of pair correlations in atomic nuclei was conceived independently by S. T. Belyaev [15] and the present writer [16]. The superfluid model thus came into being. This model takes into account the residual interactions leading to pair correlations of superconducting type, and multipole–multipole interactions, whereby it is possible to describe those nuclei states which were formerly described phenomenologically as nuclear surface oscillations.

The present lectures deal with an explanation of the essence of the superfluid model of the nucleus and its application to the study of the structure of deformed nuclei.

1. Fundamental Postulates of the Superfluid Model of the Nucleus

A model is called superfluid which provides a microscopic description of the structure of the nucleus and which takes into account the residual interactions of the nucleons leading to pair correlations of a superconducting type, as well as to collective effects [7, 8]. This model is the further development of the independent-particle model. It should be noted that the model is not a consistently microscopic model, since phenomenological parameters are used in it.

The name "superfluid model" itself indicates the limited nature and approximate character of this method of studying the structure of the nucleus.

In the superfluid model, interaction between the nucleons in the nucleus is divided into three parts

$$H = H_{av} + H_{pair} + H_{coll}. \tag{1.1}$$

The average (or self-consistent) field of the nucleus H_{av} is separated. The remaining interactions between the nucleons are given by the term H_{pair} and H_{coll}, H_{pair} being an interaction leading to pair correlations of superconducting type, H_{coll} containing the multipole–multipole interaction responsible for the collective properties of the nucleus. As shown in [8], it is possible to separate strictly the average field of the nucleus and the interaction leading to pair correlations of superconducting type, from the most general form of interaction Hamiltonian.

The complete Hamiltonian also comprises the kinetic energy of rotation and the Coriolis interaction, describing the connection between the internal motion and rotation. Expression (1.1) should therefore be completed by two more terms

$$T_{rot} + \frac{C'}{2\mathcal{J}}(\mathbf{IJ}), \tag{1.2}$$

(where $\mathbf{J}$ is the moment of inertia; I is the total moment of the quantity of movement; $\mathbf{J}$ is the moment of internal motion), which will be disregarded.

Furthermore, the basic equations are found by means of the Bogolyubov variational principle, being the generalization of the well-known Hartree-Fock method. In solving them, use is made of approximate second quantization. It should be noted that these equations may also be obtained by the method of Green's functions.

In the superfluid model, the problem of studying the structure of the nucleus is formulated like a many-body problem. The Hamiltonian, describing the interaction of nucleons in the nucleus, is not connected directly with the potential of the interaction of free nucleons. In fact, after separation of the average field and taking into account the pair correlations of superconducting type, an examination is made of the interaction of quasiparticles, the properties of which cannot in any way be identified with the properties of free nucleons.

We shall consider the form of H_{pair} and H_{coll} interactions. Our examination relates to medium and heavy nuclei in which, as shown in [8], there are no pair correlations of a superconducting type between neutrons and protons, and therefore the neutron and proton systems must be considered apart for H_{pair}. We write H_{pair} in the following form

$$H_{pair} = -G_N \sum_{ss'} a^+_{s+} a^+_{s-} a_{s'-} a_{s'+} - G_Z \sum_{\nu\nu'} a^+_{\nu+} a^+_{\nu-} a_{\nu'-} a_{\nu'+}. \tag{1.3}$$

Here $a^+_{s\sigma}$, $a_{s\sigma}$ are the creation and absorption operators of a neutron. By means of the set of quantum numbers $(s\sigma)$ we describe the state of a neutron and $(\nu\sigma)$ the state of a proton. From the set of quantum numbers, we separate the number $\sigma = \pm 1$ so that states carrying the sign of σ become conjugate with respect to the time reflection operation; for example, σ may be the sign of the projection of a momentum on the axis of symmetry of the nucleus.

The Hamiltonian (1.3) contains the specific part of the residual interactions G(s+, s−; s'−, s'+). This is due to the fact that strong correlations between nucleons are realised only when they are in states of the same energy and the same quantum numbers (except σ). The nature of the residual short-range forces is such that it leads to a much stronger interaction in states with zero momentum than in other states. The binding energies in the last neutron in light nuclei (where there are neutron–proton correlations of superconducting type) show that in states of type (s+, s−), the correlations are strong, while in other states they are insignificant. In fact, in cases where an odd neutron and an odd proton are situated in states having the same quantum numbers (Na^{22}, Al^{26}, P^{30}, Cl^{34}, and K^{38}), the binding energy of the last neutron is of the order of 11-12 MeV, i.e., of the same order as in nuclei having an even number of neutrons (Al^{27}, P^{31}, Cl^{35}, etc.), where two outer neutrons are a pair. In cases where an odd neutron and an odd proton are in states with different quantum numbers (Na^{24}, Al^{28}, P^{32}, Cl^{36}, and K^{40}), the binding energy of the last neutron is of the order of 7-8 MeV, as in the nuclei Ne^{21}, Mg^{25}, Si^{31} and others, in which the last nucleon does not participate in pair correlations. Further, since the forces leading to pair correlations of superconducting type, are short-range forces, they may be represented approximately as $G \sim \delta(r_1 - r_2)$. This means that in momentum space, G is constant. In the independent particle model, therefore, we may assume approximately that G(s+, s–; s'–, s'+) is independent of s and s', i.e.,

$$G = \text{const.} \tag{1.4}$$

The multipole–multipole interaction is described by the Hamiltonian H_{coll}, which has the form

$$H_{coll} = -\frac{1}{2}\sum_{\lambda=2,3}\sum_{\bar{\mu}\geq 0}\{\varkappa_n^{(\lambda)} Q^{+}_{\lambda\bar{\mu}}(n)\, Q_{\lambda\bar{\mu}}(n) + \varkappa_p^{(\lambda)} Q^{+}_{\lambda\bar{\mu}}(p)\, Q_{\lambda\bar{\mu}}(p) + \varkappa_{np}^{(\lambda)}\left(Q^{+}_{\lambda\bar{\mu}}(n)\, Q_{\lambda\bar{\mu}}(p) + Q^{+}_{\lambda\bar{\mu}}(p)\, Q_{\lambda\bar{\mu}}(n)\right)\}. \tag{1.5}$$

This part of the Hamiltonian with simultaneously the neutron–neutron and proton–proton interactions also contains neutron–proton interactions.† Here $\varkappa_n^{(\lambda)}$, $\varkappa_p^{(\lambda)}$ and $\varkappa_{np}^{(\lambda)}$ are the constants of the multipole interaction. The maximum part is played by quadrupole–quadrupole ($\lambda = 2$) and octupole–octupole ($\lambda = 3$) interactions. The terms with $\lambda = 0$ describe the low-frequency vibrations at the average pole; the terms with $\lambda = 1$ describe the vibrations of the protons in relation to the neutrons in dipole–dipole interactions. They have no substantial effect on the properties of the ground and weakly excited states of atomic nuclei. Multipole–multipole interactions with $\lambda > 3$ exert a very slight influence on the energies of the corresponding states. Thus, H_{coll} comprises quadrupole–quadrupole and octupole–octupole interactions.

It should be noted that the method employed further is not connected with a concrete form of the functions $f(ss')$; only the factorized form of interaction Q+Q is essential for it.

The radial function $f(ss')$ appears in concrete numerical calculations and it is obtained as a heritage of the phenomenological method of examining the oscillations of the nuclear surface, and is closely bound up with stable quadrupole deformation.

2. Pair Correlations of Nucleons of Superconducting Type. Independent-Quasiparticle Model

We present the basic results of calculations with the Hamiltonian $H_{av} + H_{pair}$ in approximation of independent quasiparticles, i.e., in the model which takes into consideration the residual interactions of nucleons leading to pair correlations of superconducting type.

We shall examine the interactions between nucleons in the nucleus, which are described by the Hamiltonian

$$H_0 = H_{av} + H_{pair} = H_0(n) + H_0(p), \tag{2.1}$$

---

† $Q_{\lambda\bar{\mu}}(n) = \sum_{\substack{ss' \\ \sigma\sigma'}} f^{\lambda\bar{\mu}}_{\sigma\sigma'}(ss')\, a^{+}_{s\sigma} a_{s'\sigma}$, where $f^{\lambda 0} = r^{\lambda} Y_{\lambda 0}$, $f^{\lambda\bar{\mu}} = \frac{1}{\sqrt{2}} r^{\lambda}\left(Y_{\lambda\bar{\mu}} + (-1)^{\bar{\mu}} Y_{\lambda-\bar{\mu}}\right)$.

where

$$H_0(n) = \sum_{s\sigma} \{E_0(s) - \lambda_n\} a_{s\sigma}^+ a_{s\sigma} - G_N \sum_{ss'} a_{s+}^+ a_{s-}^+ a_{s'-} a_{s'+} \tag{2.2}$$

Here $E_0(s)$ are the single-particle (not fully renormalized) energies in the s states of the average field. The chemical potentials λ_n and λ_p are determined from the conditions of the conservation on the average of the number of neutrons and protons, i.e., from the equations

$$N = \sum_{s\sigma} < a_{s\sigma}^+ a_{s\sigma} >, \qquad Z = \sum_{\nu\sigma} < a_{\nu\sigma}^+ a_{\nu\sigma} >, \tag{2.3}$$

where <.....> signifies averaging according to the state considered. The operators $a_{s\sigma}^+$, $a_{s\sigma}$ satisfy the following commutation relations:

$$\left. \begin{array}{l} a_{s_1\sigma_1}^+ a_{s\sigma} + a_{s\sigma} a_{s_1\sigma_1}^+ = \delta_{ss_1} \delta_{\sigma\sigma_1}, \\ a_{s\sigma} a_{s'\sigma'} + a_{s'\sigma'} a_{s\sigma} = 0. \end{array} \right\} \tag{2.4}$$

It should be noted that N. N. Bogolyubov's method of canonical transformations may be used for solving the problem in its general form, where G(s+, s−; s'−, s'+) is not constant for all values of s and s'. However, in view of the fact that in nuclear theory, the approximation (1.4) appears to be fairly good, we shall use it from the very beginning.

We perform the linear canonical transformation of the Fermi amplitudes

$$a_{s\sigma} = u_s \alpha_{s-\sigma} + \sigma v_s \alpha_{s\sigma}^+, \tag{2.5}$$

i.e., we pass from particles to quasiparticles. If the transformation is not to disturb their commutational properties the following condition must be satisfied

$$\eta_s = u_s^2 + v_s^2 - 1 = 0. \tag{2.6}$$

We determine the ground state of the system from the condition

$$\alpha_{s\sigma} \Psi_0 = 0, \tag{2.7}$$

i.e., the ground state is a vacuum for quasiparticles. We find the average value of $H_0(n)$ from the condition Ψ_0.

$$<H_0(n)> = 2\sum_s \{E_0(s) - \lambda_n\} v_s^2 - G_N \sum_{ss'} u_s v_s u_{s'} v_{s'} - G_N \sum_s v_s^4 .$$

Since the term $G_N \sum_s v_s^4$ contributes to the self-consistent field, we perform renormalization

$$E(s) = E_0(s) - \frac{G_N}{2} v_s^2$$

and get

$$<H_0(n)> = 2\sum_s \{E(s) - \lambda_n\} v_s^2 - G_N \left(\sum_s u_s v_s\right)^2 . \tag{2.8}$$

We determine u_s and v_s from the condition of the minimum of (2.8), which we write

$$\delta\left\{<H_0(n)>_0 + \sum_s \mu_s \eta_s\right\} = 0, \tag{2.9}$$

where μ_s is the Lagrange multiplier. As a result, we get the following basic equation:

$$2\{E(s)-\lambda_n\}u_s v_s - G_N(u_s^2 - v_s^2)\sum_{s'} u_{s'} v_{s'} = 0. \tag{2.10}$$

Equation (2.10) has a trivial solution

$$u_s = 1-\theta_F(s), \quad v_s = \theta_F(s),$$

which corresponds to the independent-particle model. The function $\theta_F(s) = 1$, if $E(s) < \lambda_n$, and $\theta_F(s) = 0$, if $E(s) > \lambda_n$.

We introduce the correlation function

$$C_n = G_N \sum_s u_s v_s \tag{2.11}$$

and determine

$$u_s^2 = \frac{1}{2}\left\{1+\frac{E(s)-\lambda_n}{\varepsilon(s)}\right\}, \quad v_s^2 = \frac{1}{2}\left\{1-\frac{E(s)-\lambda_n}{\varepsilon(s)}\right\},$$

$$\varepsilon(s) = \sqrt{C_n^2 + \{E(s)-\lambda_n\}^2}, \quad u_s v_s = \frac{1}{2}\frac{C_n}{\varepsilon(s)}. \tag{2.12}$$

As a result, we get the following system of equations for the ground state of the system consisting of an even number of nucleons:

$$1 = \frac{G_N}{2}\sum_s \frac{1}{\sqrt{C_n^2 + [E(s)-\lambda_n]^2}}, \tag{2.13}$$

$$n = \sum_s \left\{1 - \frac{E(s)-\lambda_n}{\sqrt{C_n^2 + \{E(s)-\lambda_n\}^2}}\right\}. \tag{2.14}$$

The wave function of the ground state Ψ_0 is determined from equation (2.7), where

$$\alpha_{s\sigma} = u_s a_{s-\sigma} + \sigma v_s a_{s\sigma}^+, \tag{2.15}$$

in the form

$$\Psi_0 = \prod_s (u_s + v_s a_{s+}^+ a_{s-}^+)\Psi_0^{(0)}, \tag{2.16}$$

where $a_{s\sigma}\Psi_0^{(0)} = 0$. The ground state energy has the form

$$E_0 = \sum_s 2E(s)v_s^2 - \frac{C_n^2}{G_N}. \tag{2.17}$$

It should be mentioned that since pair interactions between nucleons are interactions of attraction, and since G_N is sufficiently large, the energy of the superfluid state, i.e., the state with $C \neq 0$, is appreciably less than the energy of the state corresponding to noninteracting particles ($C = 0$).

The lowest excited states of the system consisting of an even number of particles are the states with one disrupted pair, i.e., with two quasiparticles at the levels of the average field. The wave function of such two-quasiparticle states has the form:

$$\alpha_{s_1\sigma_1}^+ \alpha_{s_2\sigma_2}^+ \Psi_0. \tag{2.18}$$

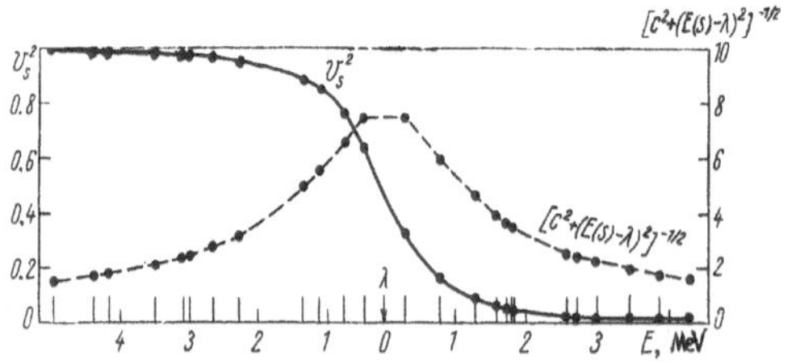

Fig. 3. Behavior of the functions v_s^2 (reckoned from the left-hand axis of ordinates) and $[C^2+(E(s)-\lambda)^2]^{1/2}$ (reckoned from the right-hand axis of ordinates) for a system consisting of N = 106.

The energy difference between the excited and ground states is

$$<\alpha_{s_2\sigma_2}\alpha_{s_1\sigma_1}H_0(n)\alpha^+_{s_1\sigma_1}\alpha^+_{s_2\sigma_2}> - <H_0(n)> = \\ = \varepsilon(s_1)+\varepsilon(s_2). \qquad (2.19)$$

Thus, the excited states of an even-number system are separated from the ground state by an energy gap greater than 2C.

The quantity v_s^2 represents the density of the number of particles on the s orbit. In the independent-particle model $v_s^2 = 1$ for $E(s) > \lambda_n$, i.e., all levels up to the level of the Fermi surface are filled, the remainder being empty. The pair correlations lead to a blurring of the value v_s^2, i.e., the interaction results in the particles passing part of the time on levels situated above the Fermi surface. Fig. 3 gives the values of v_s^2 for the case N = 106. The figure shows that the density of the number of particles differs appreciably from zero at (3-4) MeV above the energy of the Fermi surface. The wave functions of the ground states in the independent-particle model Ψ_{00} and in the superfluid model ($C_n \neq 0$) are shown diagrammatically in Fig. 4, in which the expressed levels of the average field and their filling with nucleons are shown twice.

It is interesting to note that the probabilities of a (dp) reaction are proportional to u_s^2; so are those of a (dt) reaction v_s^2 if the targets used are even – even nuclei (s denotes the quantum numbers of the excited state of an odd-A nucleus in the final state). Figure 5 shows the differential cross sections of the (dt) and (dp) reactions for the excited state with $IK\pi = 7/2, 5/2^-$ and the asymptotic quantum numbers [512] in odd-number isotopes of ytterbium. In light isotopes of ytterbium, the [512] state is a fractional state, furthermore the ground state, and in heavy isotopes, it is the excited hole state. These experimental values agree very well with similar curves with v_s^2 and u_s^2, if it is considered that an increase in the mass of the target corresponds to a decrease in the energy of the state.

We shall consider as example this model: n nucleons, located on Ω equidistant levels of the outer shell, interact only between themselves; the density of the levels ρ (ρ = const) is fairly high, so that in equations (2.13) and (2.14) summation may be replaced by integration. In that case, equations (2.13), (2.14) assume the form

$$1 = \frac{G_N}{2}\rho\int\frac{dE'}{\sqrt{C_n^2+(E'-\lambda_n)^2}}, \qquad (2.13')$$

$$n = \rho\int dE'\left\{1-\frac{E'-\lambda_n}{\sqrt{C_n^2+(E'-\lambda_n)^2}}\right\}. \qquad (2.14')$$

The solutions of these equations have the form

$$C_n = \frac{\sqrt{(2\Omega-n)\,n}}{\rho\left(e^{\frac{2}{G\rho}}-1\right)}e^{\frac{1}{G\rho}}, \qquad (2.20)$$

$$E_F-\lambda = \frac{\Omega-n}{\rho\left(e^{\frac{2}{G\rho}}-1\right)}, \qquad (2.21)$$

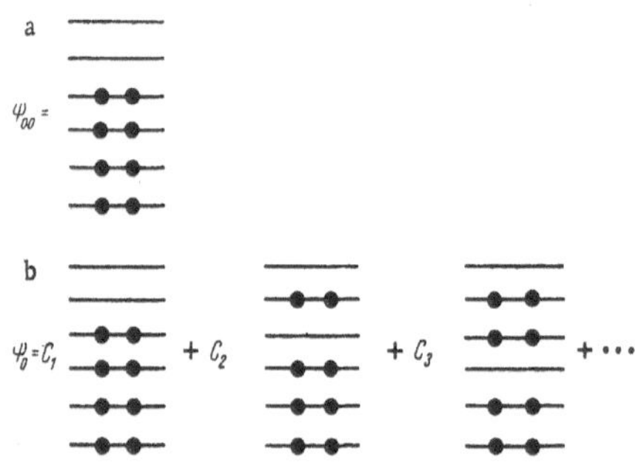

Fig. 4. Wave functions: a) In the independent-particle model; b) in the superfluid model (projected).

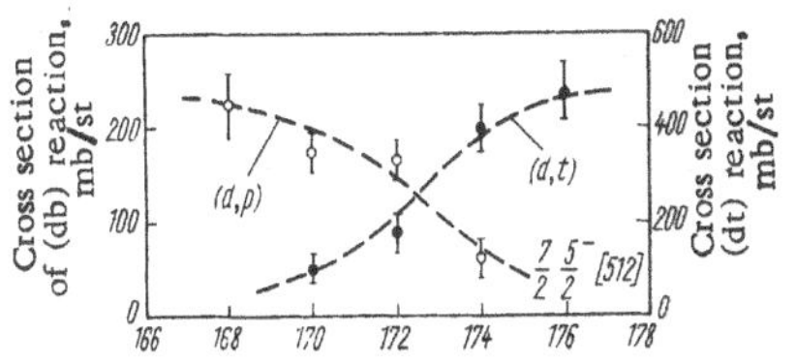

Fig. 5. Cross sections of (dp) and (dt) reactions in Yb isotopes for the $^7/_2$, $^5/_2{}^-$ [512] state.

where E_F is the energy of the Fermi surface. It will be seen from these formulas that for a closed shell, when $n=2\Omega$ or $n=0$, pair correlations of nucleons are absent. The chemical potential agrees with E_F when the shell is half filled.

We shall examine a system consisting of an odd number of neutrons. We determine the ground state of the system as $\alpha^+_{s'\sigma'}\ \Psi_0$, we find from this state the average value of $H_0(n)$ and make use of the variational principle. In this case, the equations for finding the correlation functions $C_n(s')$ and chemical potential $\lambda_n(s')$ have the form

$$\frac{2}{G_N} = \sum_{s\neq s'} \frac{1}{\sqrt{C_n(s')^2 + \{E(s) - \lambda_n(s')\}^2}}, \tag{2.22}$$

$$n = 1 + \sum_{s\neq s'} \left\{1 - \frac{E(s) - \lambda_n(s')}{\sqrt{C_n(s')^2 + \{E(s) - \lambda_n(s')\}^2}}\right\}, \tag{2.23}$$

if the quasiparticle is on the s' level of the average field. We write the wave function and energy of the system as follows

$$\alpha^+_{s'\sigma'}\Psi_0 = a^+_{s'-\sigma'} \prod_{s\neq s'} \left(u_s + v_s a^+_{s+} a^+_{s-}\right) \Psi_0^{(0)}, \tag{2.24}$$

$$\mathcal{E}(s') = E(s') + 2\sum_{s\neq s'} E(s)\, v_s^2 - \frac{C_n(s')^2}{G_N}. \tag{2.25}$$

The excited states of odd nuclei are not separated from the ground state by a gap, since

$$\mathcal{E}(s') - \mathcal{E}(s_0) \approx \varepsilon(s') - \varepsilon(s_0). \tag{2.26}$$

It is clear from this that pair correlations lead to a qualitative difference in the excitation spectra of even–even and odd nuclei. The results obtained are not dependent on a set of concrete quantum numbers and thereby not on the form of the independent model used; therefore, this method may be used in studying the properties of both spherical and deformed nuclei, including axially unsymmetrical nuclei.

Comparison of equations (2.24), (2.25) with equations (2.13), (2.14) shows that the equations for the system consisting of an odd number of neutrons differ from equations (2.13), (2.14) by the fact that in them the level on which the odd neutron is situated is deleted.

The influence of unpaired particles on the superfluid properties of the system in each state of the atomic nucleus has been given the name of blocking effect. The superfluid state of the atomic nucleus is the result of the interaction of nucleons with the Hamiltonian (2.1) and therefore if there is one nucleon on any doubly degenerate level of the average field, this level according to the Pauli principle, cannot be occupied by a pair. The blocking effect is that in calculating by means of the variation principle the superfluid properties of definite states, those levels of the average field, on which quasiparticles are situated, are disregarded. This results in a variation in the values of the correlation functions $C_n(s')$ and chemical potentials $\lambda_n(s')$, compared with the states in which there are no quasiparticles, or they are disposed on other levels of the average field. The blocking effect is quite substantial in the region of strongly deformed nuclei, where the density of the levels of the average field is not high.

If the blocking effect is taken into account, the wave functions, energies and fundamental equations for the two-quasiparticle excited states of the even system (for $s_1 \neq s_2$) have the following appearance [8]:

$$\Psi_0(s_1, s_2) = a^+_{s_1\sigma_1} a^+_{s_2\sigma_2} \prod_{s \neq s_1, s_2} \{u_s(s_1 s_2) + v_s(s_1 s_2) a^+_{s+} a^+_{s-}\} \Psi^0_0, \tag{2.27}$$

$$\frac{2}{G_N} = \sum_{s \neq s_1, s_2} \frac{1}{\sqrt{C_n(s_1 s_2)^2 + \{E(s) - \lambda_n(s_1 s_2)\}^2}}, \tag{2.28}$$

$$N = 2 + \sum_{s \neq s_1, s_2} \left\{1 - \frac{E(s) - \lambda_n(s_1 s_2)}{\sqrt{C_n(s_1 s_2)^2 + \{E(s) - \lambda_n(s_1 s_2)\}^2}}\right\}, \tag{2.29}$$

$$\mathcal{E}(s_1 s_2) = E(s_1) + E(s_2) + \sum_{s \neq s_1, s_2} 2E(s) v_s(s_1 s_2)^2 - \frac{C_n(s_1 s_2)^2}{G_N}. \tag{2.30}$$

The influence of the blocking effect on the correlation functions and on the energies of single and two-quasiparticle states is studied in [7, 8]. The part played by the blocking effect is shown in an exact solution of the model problem carried out in [9]. It should be mentioned that equations (2.27), (2.28) are sometimes found to be insufficiently exact, low values being obtained for the correlation functions. To improve the accuracy of the calculations in such cases, the method proposed by I. N. Mikhailov [10] should be used. In most nuclei, the accuracy of calculations in the independent-quasiparticle model is limited by the poor knowledge of the behavior of the levels of the average field and their fluctuations, but not by the accuracy of the mathematical method [9].

In the framework of the present model, we shall discuss calculations of the properties of strongly deformed nuclei in the regions $150 \leq A \leq 188$ and $228 \leq A \leq 254$. The levels of the Nilsson potential [11, 12] were taken as the levels of the average field. The constants of the pair interactions G_N for the neutron system and G_Z for the proton system were calculated on the basis of experimental data of the difference of nuclear masses by means of the formula

$$P_N = \frac{1}{4}\{3\mathcal{E}(Z, N-1) + \mathcal{E}(Z, N+1) - 3\mathcal{E}(Z, N) - \mathcal{E}(Z, N-2)\}. \tag{2.31}$$

From a comparison of the calculated values of the pairing energies with the experimental data in both ranges of deformed nuclei, $150 \leq A \leq 188$ and $228 \leq A \leq 254$, the following values were obtained for the constants of the pair interaction [8, 13, 14]:

$$G_N = \frac{26 \text{ to } 27}{A} \text{ MeV}$$
$$G_Z = \frac{28 \text{ to } 29}{A} \text{ MeV} \tag{2.32}$$

The results of the analysis are shown in Figs. 6 and 7, from which it will be seen that satisfactory agreement has been obtained between the calculated and experimental values of the pairing energies. It follows firstly that the approximation (2.4) $G = \text{const}$ is fairly good, and secondly that G_N and G_Z vary from one nucleus to another according to the law A^{-1}.

By taking into consideration pair correlations of superconducting type, it was possible to describe quantitatively those phenomena which could not be fitted in the framework of the independent-particle model (see introduction). Not only has the phenomenon of the gap in the spectra of even-even nuclei been explained, and the absence of such a gap in odd and odd-odd nuclei, but a quantitative description has been given of the energies of two-quasiparticle excited states.

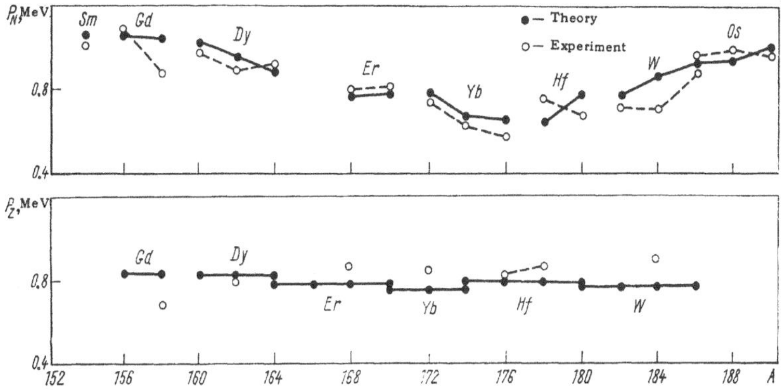

Fig. 6. Neutron and proton pairing energies in the range $154 \leq A \leq 188$.

As already mentioned above, it was shown in [15] that the density of single-particle equations in odd deformed nuclei, as found from experimental data, turned out to be approximately twice the density of the levels given by the Nilsson potential. In [16], the one-particle levels of many odd nuclei are calculated, the calculated values of the excitation energies agreeing with the experimental data somewhat better than values obtained by the Nilsson schemes. It was shown that in both the rare-earth and transuranium regions, the density of calculated low-energy single-particle levels agreed with the experimental data and was approximately twice the density of the levels obtained by the Nilsson scheme.

The moment of inertia for the ground state of an even–even nucleus in the independent-quasiparticle model has the following form [5]:†

$$\mathcal{J}_x = 2 \sum_{\substack{\mu\mu' \\ \sigma\sigma'}} \frac{|\langle\mu\sigma|j_x|\mu'\sigma'\rangle|^2}{\varepsilon(\mu)+\varepsilon(\mu')} (u_\mu v_{\mu'} - v_\mu u_{\mu'})^2, \tag{2.33}$$

where summation over μ, μ' means that it is performed over levels of the neutron and proton systems, j_x is the component of total angular moment. When taking the blocking effect into account, $\varepsilon(\mu)+\varepsilon(\mu')$ should be replaced by $\mathscr{E}(\mu,\mu')-\mathscr{E}_0$. In the independent-particle model, the expression for the moment has the form

$$\mathcal{J}_x = 2 \sum_i \frac{|\langle i|j_x|0\rangle|^2}{\mathscr{E}_i^0 - \mathscr{E}_0^0}, \tag{2.34}$$

where I_i^0, $\mathscr{E}_0^0$ are the energies of excited and ground states in the independent-particle model.

The influence of pair correlations on the moments of inertia appears in two effects. In the increase in the denominator from $E(\mu')-E(\mu)$ (difference in the energies between single-particle levels) to $\varepsilon(\mu)+\varepsilon(\mu')$, and in the introduction of a factor $(u_\mu v_{\mu'} - v_\mu u_{\mu'})^2 < 1$. Both these effects reduce the moments of inertia approximately equally.

The results of the calculations of the moments of inertia of the ground states of even–even pairs, cited in [17], are shown in Fig. 2. Compared with the rigid-body value $\mathcal{J}_T$, agreement with the experimental values has been much improved, but the theoretical values are 5-20% lower than the experimental values. The calculated values of the moments of inertia of single-quasiparticle states of odd nuclei and two-quasiparticle states of even–even nuclei agree with the experimental data [18] in the limits 10-30%. The increase in the moments of inertia of

† $\hbar = C = 1$.

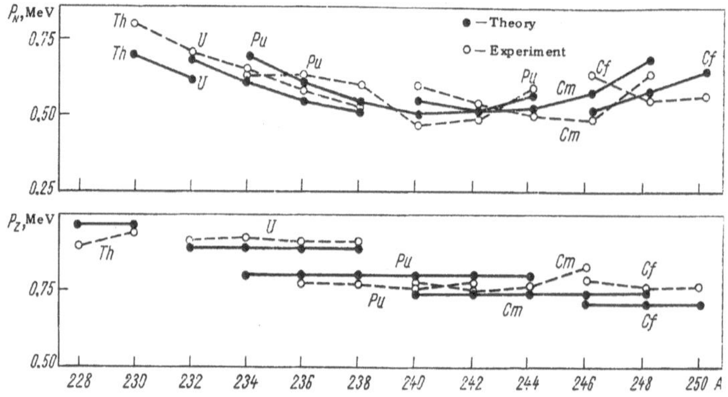

Fig. 7. Neutron and proton pairing energies in the range $228 \leq A \leq 250$.

odd nuclei compared with neighboring even–even nuclei is the principal consequence of the blocking effect (due to reduction in the correlation functions).

The pair correlations exert a greater influence on establishing the equilibrium form of the nucleus [5]. Thus, in the absence of pair correlations, all nuclei with nucleons in unfilled shells would be deformed. According to Bohr and Mottelson, there is rivalry between quadrupole–quadrupole forces, tending to deform the nucleus, and two-body forces (of superconducting type), which help to establish the spherical form of nucleus. Due to this rivalry the deformation of nuclei is sensitive to the behavior of the levels of the average field. Bes and Szymanski [19] calculated the equilibrium deformations of nuclei in the rare-earth range, using the energies and wave functions of the Nilsson potential and taking pair correlations into account. In most cases, they obtained fairly good agreement between theory and experimental results.

Pair correlations of nucleons of superconducting type exert a strong influence on the properties of the ground and excited states of atomic nuclei. They therefore play a great part in α, β, and γ transitions and in stripping and capture reactions.

Investigations [20, 21] have shown that pair correlations have a strong influence on the absolute probabilities of α decay and particularly on the forbiddenness factor F. Thus, the matrix element of α decay of an even–even nucleus between ground states has the form

$$M = \sum_{s,\,\nu} W(\nu,\ \nu\,|\,s,\ s)\,\xi_\nu\,\xi_s, \tag{2.35}$$

where summation of s(ν) is performed over neutron (proton) levels of the average field. The function W(ν, ν|s, s) describes the probability of the formation of an α particle from nucleons in the ν, ν, s, s states, and ξ_s and ξ_ν have the form

$$\xi_s = u_s(N-2)\,v_s(N). \tag{2.36}$$

Pair correlations result in the α particle being formed from pairs of nucleons situated on many levels of the average field, both above and below the level of the Fermi surface. Figure 8 shows the values of ξ_s for α decay of an even–even nucleus having 146 neutrons, and for favorable α decay of an odd nucleus having 147 neutrons. It will be seen from Fig. 8 that a considerable contribution is made in the formation of an α particle by states fairly remote from the Fermi surface. In Fig. 8, the continuous line indicates the values of ξ_s for favorable α-decay of a nucleus having 147 neutrons. This variation in the values of ξ_s is connected with the blocking effect and as shown in Fig. 8 plays an important part.

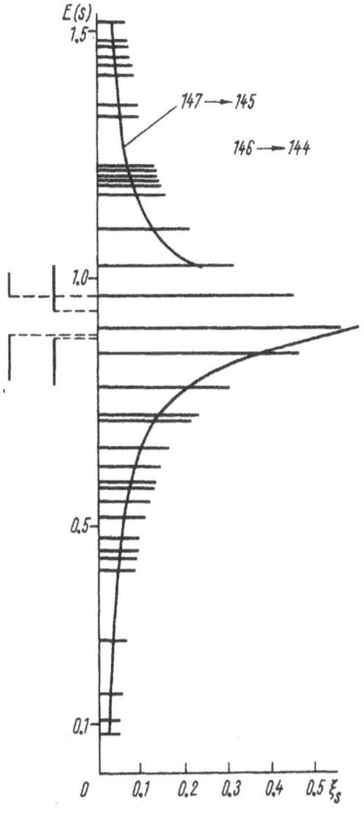

Fig. 8. The values of $\xi_s = v_s(N)u_s(N-2)$ for the α decay of an even–even nucleus with N = 146, and for favorable decay of an odd nucleus with N = 147. The energies of the levels of the average field have been plotted on the axis of ordinates in the units $\hbar\omega_0^0$, the lengths of the horizontal lines are equal to the values of ξ_s for each orbit. To the left of the axis of ordinates, the dash lines show the position of λ for the parent and daughter nuclei, the short lines for the 146 → 144 decay, and longer lines for the 147 → 145 decay. The vertical lines drawn from their ends are equal to the length of C_n, the top ones for the parent nuclei, the bottom ones for the daughter nuclei. The continuous curve shows the values of ξ_s for the 147 → 145 decay.

In α decay in two-quasiparticle states of even–even nuclei and in unfavorable α-decays of odd-A nuclei, the position varies sharply [8]. Thus, if the quasiparticles are neutron particles, the α particle is formed from proton pairs situated at many levels of the average field, and from neutrons situated only on those orbits on which the number of quasiparticles varied in the α-decay process. In this case, ξ_s becomes less than unity, which results in an increase in the forbiddenness factor F.

We shall examine the influence of pair correlations on the probabilities of β transitions [7, 8]. The matrix element describing β decay has the form

$$M \sim \langle \nu | \Gamma | s \rangle R_N^{1/2} R_Z^{1/2}, \tag{2.37}$$

where $\langle \nu|\Gamma|s \rangle$ is the single-particle matrix element. The quantities R_Z and R_N describe the rearrangement of the nucleus in β decay, the former describing the rearrangement of the proton system and the latter that of the neutron system. If in β decay, the number of paired particles is the same in the initial and final states, then

$$R_Z = u_{\nu'}^2, \tag{2.38}$$

and if in the decay process the number of paired nucleons varies by two, then

$$R_Z = v_{\nu'}^2, \tag{2.39}$$

where ν' relates to the level on which the quasiparticle disappeared or appeared. The functions $u_{\nu'}^2$, and $v_{\nu'}^2$ relate to systems with the least number of quasiparticles. For example in β decay between the single-quasiparticle states ${}^{102}_{72+1}Ta^{175} \rightarrow {}^{102+1}_{72}Hf^{175}$, the number of paired particles does not vary and therefore $R_Z = u_\nu^2$, $R_N = u_s^2$, while in β decay ${}^{94}_{66+1}Ho^{161} \rightarrow {}^{92+1}_{68}Er^{161}$, the number of paired neutrons and protons varies, and therefore $R_Z = v_\nu^2$, $R_N = v_s^2$. The values of R_N and R_Z for nuclei in the region $152 \leq A \leq 185$ are given in [7], and in the region $228 \leq A \leq 254$ they are given in [13].

The pair correlations of superconducting type exert a particularly strong influence on β transitions belonging to the second group according to the classification given in [7]. The group includes those β transitions which are strictly forbidden in the superfluid model of the nucleus.

However, the rates of these β decays are retarded somewhat in transitions to high, excited states of the nuclei. An analysis of the experimental data showed that there were more than 20 firmly established β-transitions in strongly deformed nuclei belonging to the second group. The relative values of log ft, calculated from the superfluid model of the nucleus, are in satisfactory agreement with corresponding experimental data.

Such a type of correction comes within the cross section of direct nuclear reactions, for example the cross sections of dp and dt reactions proportional to R_N. Pair correlations exercise a considerable influence on the probabilities of electromagnetic transitions (see [7, 8, 22]).

3. Study of Collective Nonrotational States by the Method of Approximate Second Quantization

Much theoretical and experimental work has been devoted to the study of the collective excited states of even–even nuclei. A fairly large amount of experimental material has been obtained on the energies of the quadrupole and octupole states in strongly deformed even–even nuclei, and on the derived probabilities of electromagnetic transitions. As previously mentioned, in the superfluid model of the nucleus, multipole–multipole interactions are taken into account, whereby it is possible to describe the collective nonrotational states of atomic nuclei.

Originally, the method of microscopic description of collective states was applied to spherical nuclei; the results of this work are given in [23-24] and elsewhere. In the region of strongly deformed nuclei, the investigations carried out up to 1963 [25] were restricted merely to presenting the basic equations and to a study of the question concerning the exclusion of the spurious state. In 1963-1964, however, detailed calculations were made of the energies of collective nonrotational states and the reduced probabilities of transitions in even–even deformed nuclei in the regions $150 \leq A \leq 186$ and $228 \leq A \leq 254$ [26-32].

We transform the interaction Hamiltonian (2.1). We carry out the canonical transformation of the Fermi-amplitudes (2.5) and introduce the operators

$$A(\mu\mu') = \frac{1}{\sqrt{2}} \sum_{\sigma} \sigma \alpha_{\mu'\sigma} \alpha_{\mu-\sigma}, \qquad B(\mu\mu') = \sum_{\sigma} \alpha^{+}_{\mu\sigma} \alpha_{\mu'\sigma}. \tag{3.1}$$

In order not to make the calculations cumbersome, we shall merely imply the operators

$$\overline{A}(\mu\mu') = \frac{1}{\sqrt{2}} \sum_{\sigma} \alpha_{\mu'\sigma} \alpha_{\mu-\sigma}, \qquad \overline{B}(\mu\mu') = \sum_{\sigma} \sigma \alpha^{+}_{\mu-\sigma} \alpha_{\mu'\sigma}. \tag{3.1}$$

A full explanation of this method is given in [33]. We denote by μ a quantum number characterizing the levels of the average field of the neutron and proton systems. We introduce the photon operators

$$Q_i = \frac{1}{2} \sum_{\mu\mu'} \{\psi^{i}_{\mu\mu'} A(\mu\mu') - \varphi^{i}_{\mu\mu'} A(\mu\mu')^{+}\}. \tag{3.2}$$

For each case ($\lambda\mu$), the number of states ($\mu\mu'$) is equal to the number of i states, so that $\psi^{i}_{\mu\mu'}$ and $\varphi^{i}_{\mu\mu'}$ are square matrices. The operators $A(\mu\mu')$ may be expressed approximately by the phonon operators Q_i and Q_i^+ as follows.

$$A(\mu\mu') = \sum_i \{\psi^i_{\mu\mu'} Q_i + \varphi^i_{\mu\mu'} Q_i^+\}. \tag{3.3}$$

After transformation, we write the Hamiltonian (2.1) in the form

$$H = H_0(n) + H_0(p) + H_{coll}, \tag{3.4}$$

$$H_0(n) = H_0'(n) + H_0''(n) + H_0'''(n), \tag{3.5}$$

$$H_0'(n) = \sum_s 2\{E(s) - \lambda_n\} v_s^2 - G_N \left(\sum_s u_s v_s\right)^2 + \sum_s \varepsilon(s) B(ss), \tag{3.6}$$

$H_0''(n)$ contains terms used for the exclusion of the spurious state [33]. $H_0'''(n)$ comprises the remaining terms. $H_0(p)$ has the same form as $H_0(n)$. Further, in the case of

$$\varkappa^{(\lambda)}_{np} = \varkappa^{(\lambda)}_{n} = \varkappa^{(\lambda)}_{p} \equiv \varkappa^{(\lambda)}$$

$$H_{coll} = -\frac{1}{2} \sum_{\lambda, \bar{\mu} \geqslant 0} \left(H_1^{\lambda\bar{\mu}} + H_2^{\lambda\bar{\mu}} + H_3^{\lambda\bar{\mu}}\right), \tag{3.7}$$

$$H_1^{\lambda\bar{\mu}} = \varkappa^{(\lambda)} \sum_{ii'} \sum_{\mu\mu'} \sum_{\mu_2\mu_2'} u_{\mu\mu'} u_{\mu_2\mu_2'} f(\mu\mu') \left(\psi^i_{\mu\mu'} + \varphi^i_{\mu\mu'}\right) f(\mu_2\mu_2') \left(\psi^{i'}_{\mu_2\mu_2'} + \varphi^{i'}_{\mu_2\mu_2'}\right) Q_i^+ Q_{i'}, \tag{3.8}$$

$$H_2^{\lambda\bar{\mu}} = \frac{1}{\sqrt{2}} \varkappa^{(\lambda)} \sum_i \sum_{\mu\mu'} \sum_{\mu_2\mu_2'} u_{\mu\mu'} v_{\mu_2\mu_2'} f(\mu\mu') \left(\psi^i_{\mu\mu'} + \varphi^i_{\mu\mu'}\right) f(\mu_2\mu_2') \{(Q_i^+ + Q_i) B(\mu_2\mu_2') + B(\mu_2\mu_2')(Q_i^+ + Q_i)\}, \tag{3.9}$$

where

$$u_{\mu\mu'} = u_\mu v_{\mu'} + u_{\mu'} v_\mu,$$

$$v_{\mu\mu'} = u_\mu u_{\mu'} - v_\mu v_{\mu'}.$$

The remaining terms are included in $H_3^{\lambda\bar{\mu}}$, the superscripts $\lambda\bar{\mu}$ in $f^{\lambda\bar{\mu}}(\mu\mu')$ being omitted for simplicity.

We shall explain the approximate second quantization method. The commutation conditions between the operators $A(\mu\mu')$, $A(\mu\mu')^+$ have a complex form and contain terms proportional to $B(\mu\mu')$. The average value of the operators $B(\mu\mu')$ over the quasiparticle vacuum is equal to zero. It is known that the number of quasiparticles, occurring as the result of taking into account the interaction between them, is small in the ground state of an even-even nucleus. Therefore, an approximation in which in the commutation relations for the operators $A(\mu\mu')$, the terms containing the operators $B(\mu\mu')$ are ignored, appears to be sufficiently good. In this approximation, the operators $A(\mu\mu')$ are regarded as boson operators. We thus arrive at the method of approximate second quantization or the quasiboson approximation or the random phase approximation.

In the framework of the method of approximate second quantization, the commutation relations have the following form:

$$\left[A(\mu\mu'), A(\mu_2\mu_2')^+\right] = \delta_{\mu\mu_2} \delta_{\mu_1\mu_2'} + \delta_{\mu\mu_2'} \delta_{\mu'\mu_2}. \tag{3.10}$$

The wave function of the ground state of an even–even nucleus is determined as a nonphonon function, i.e.,

$$Q_i \Psi = 0. \tag{3.11}$$

The excited collective states are determined as one-phonon states with the wave function

$$Q_i^+ \Psi, \tag{3.12}$$

$$\left[Q_i, Q_{i'}^+\right] = \delta_{ii'}, \tag{3.13}$$

$$[Q_i, Q_{i'}] = \left[Q_i^+, Q_{i'}^+\right] = 0, \tag{3.13}$$

which ensures that orthonormalizing conditions of the wave functions in the ground and excited states are satisfied. It follows from the compatability of the commutation relations (3.10) and (3.13) that the functions $\psi^i_{\mu\mu'}$, $\varphi^i_{\mu\mu'}$ are connected together by the relationship

$$\sum_{\mu\mu'} \left(\psi^i_{\mu\mu'} \psi^{i'}_{\mu\mu'} - \varphi^i_{\mu\mu'} \varphi^{i'}_{\mu\mu'}\right) = 2\delta_{ii'}. \tag{3.14}$$

In the method of approximate second quantization, only those terms of the interaction Hamiltonian which give a coherent effect are taken into consideration. The Hamiltonian is selected in the form

$$H_c = H_0'(n) + H_0''(n) + H_0'(p) + H_0''(p) - \frac{1}{2} \sum_{\lambda, \bar{\mu} > 0} \left(H_1^{\lambda\bar{\mu}} + H_2^{\lambda\bar{\mu}}\right). \tag{3.15}$$

In this approximation, the mean value of the operators $B(\mu\mu')$ over the ground state of the even–even nucleus is assumed to be zero

$$\langle B(\mu\mu') \rangle = 0, \tag{3.16}$$

although the ground state is not a quasiparticle vacuum.

In some investigations, the collective excited states are considered in an approximation, in which the wave function of the ground state of the even–even nucleus is a nonquasiparticle function. This method has been called the Tamm-Dancoff approximation; in this case, the phonon operator has the form

$$\mathfrak{Q}_i = \frac{1}{2} \sum_{\mu\mu'} \psi^i_{\mu\mu'} A(\mu\mu'). \tag{3.17}$$

The interactions between the quasiparticles are considered only in the excited states. It should be noted that the formulas of this method are obtained from the formulas of the method of approximate second quantization by equating to zero the functions $\varphi_{\mu\mu'}$.

The fundamental equations of the method of approximate second quantization are obtained by means of the variational principle. We shall examine the case for those $\lambda\bar{\mu}$, where the diagonal matrix elements $f(\mu\mu)$ are absent. The general case is considered in [30]. We introduce the new notations $g^i_{\mu\mu'} = \psi^i_{\mu\mu'} + \varphi^i_{\mu\mu'}$, $w^i_{\mu\mu'} = \psi^i_{\mu\mu'} - \varphi^i_{\mu\mu'}$. We find the mean value of H_C from the single-phonon excited state $Q_i^+ \Psi$:

$$\langle Q_i H_c Q_i^+ \rangle = \sum_s 2\{E(s) - \lambda_n\} v_s^2 - \frac{C_n^2}{G_N} + \frac{1}{4} \sum_{ss'} \left(\varepsilon(s) + \varepsilon(s')\right)\left(g_{ss'}^{i2} + w_{ss'}^{i2}\right) + \sum_\nu 2\{E(\nu) - \lambda_p\} v_\nu^2 -$$

$$-\frac{C_p^2}{G_Z}+\frac{1}{4}\sum_{\nu\nu'}(\varepsilon(\nu)+\varepsilon(\nu'))\left(g^{i2}_{\nu\nu'}+w^{i2}_{\nu\nu'}\right)-\frac{\varkappa^{(\lambda)}}{2}\left(\sum_{\mu\mu'}u_{\mu\mu'}\,g^{i}_{\mu\mu'}\,f(\mu\mu')\right)^2. \tag{3.18}$$

The variational principle is formulated as follows:

$$\delta\left\{<Q_iH_cQ_i^+>-\frac{\omega_i}{2}\left(\sum_{\mu\mu'}g^{i}_{\mu\mu'}\,w^{i}_{\mu\mu'}-2\right)\right\}=0. \tag{3.19}$$

We here introduce the Lagrange multiplier ω_i to take into account compliance with condition (3.14). In studying the collective excited states having a definite value of $K\pi$, only the term of H_C with given $\lambda\bar{\mu}$ is taken into account, where $K=\bar{\mu}$, $\pi=(-1)^{\lambda}$.

By using the variational principle, we obtain the following equations for the determination of $g^{i}_{\mu\mu'}$ and $w^{i}_{\mu\mu'}$:

$$\left(\varepsilon(\mu)+\varepsilon(\mu')-\frac{\omega_i^2}{\varepsilon(\mu)+\varepsilon(\mu')}\right)g^{i}_{\mu\mu'}=\varkappa^{(\lambda)}f(\mu\mu')\,u_{\mu\mu'}D_i, \tag{3.20}$$

$$w_{\mu\mu'}=\frac{\omega_i}{\varepsilon(\mu)+\varepsilon(\mu')}\,g^{i}_{\mu\mu'}, \tag{3.21}$$

where

$$D_i=2\sum_{\mu\mu'}f(\mu\mu')\,u_{\mu\mu'}g^{i}_{\mu\mu'}. \tag{3.22}$$

From equation (3.20), we find

$$g^{i}_{\mu\mu'}=\varkappa^{(\lambda)}D_i\frac{f(\mu\mu')\,u_{\mu\mu'}}{\varepsilon(\mu)+\varepsilon(\mu')-\dfrac{\omega_i^2}{\varepsilon(\mu)+\varepsilon(\mu')}}$$

and substituting in equation (3.22), we obtain the following secular equation for finding ω_i:

$$1=2\varkappa^{(\lambda)}\sum_{\mu\mu'}\frac{f(\mu\mu')^2u^2_{\mu\mu'}\left(\varepsilon(\mu)+\varepsilon(\mu')\right)}{\left(\varepsilon(\mu)+\varepsilon(\mu')\right)^2-\omega_i^2}\equiv\mathcal{T}(\omega). \tag{3.23}$$

To find $g^{i}_{\mu\mu'}$ we use the condition (3.14) and as result we get

$$g^{i}_{\mu\mu'}=\frac{\sqrt{2}}{\sqrt{Y^i}}\cdot\frac{f(\mu\mu')\,u_{\mu\mu'}}{\varepsilon(\mu)+\varepsilon(\mu')-\dfrac{\omega_i^2}{\varepsilon(\mu)+\varepsilon(\mu')}}, \tag{3.24}$$

where

$$Y^i=\sum_{\mu\mu'}\frac{f(\mu\mu')^2u^2_{\mu\mu'}\,\omega_i\left(\varepsilon(\mu)+\varepsilon(\mu')\right)}{\left[\left(\varepsilon(\mu)+\varepsilon(\mu')\right)^2-\omega_i^2\right]^2}. \tag{3.25}$$

We further find $\psi^{i}_{\mu\mu'}$ and $\varphi^{i}_{\mu\mu'}$ in the form

$$\Psi^{i}_{\mu\mu'}=\frac{1}{\sqrt{2}}\cdot\frac{1}{\sqrt{Y^i}}\cdot\frac{f(\mu\mu')\,u_{\mu\mu'}}{\varepsilon(\mu)+\varepsilon(\mu')-\omega_i}, \tag{3.26}$$

$$\varphi^{i}_{\mu\mu'} = \frac{1}{\sqrt{2}} \cdot \frac{1}{\sqrt{Y^i}} \cdot \frac{f(\mu\mu') u_{\mu\mu'}}{\varepsilon(\mu) + \varepsilon(\mu') + \omega_i}. \tag{3.26'}$$

The energy of the excited states with the wave functions $Q_i^+\Psi$ is

$$< Q_i H_c Q_i^+ > - < H_c > = \omega_i, \tag{3.27}$$

where i = 1, 2, 3,... and so forth, in the order of increasing excitation energy. Having solved the secular equation (3.23), we find the energies $\omega_1, \omega_2, ..., \omega_i, ...$ and the wave functions of the corresponding states.

We shall discuss the properties of the solutions of the secular equation (3.27). For this purpose, in Fig. 9 we plot the values of F(ω) as functions of ω for states with $K\pi$ = 2+ for Pu^{240} and Cf^{250}. The points of intersection of the straight line $1/\varkappa$ with the F(ω) curve are the first, second etc. roots of the secular equation. Up to the first pole of equation (3.23) there can be only one root of this equation. The second root of (3.23) is situated between the values of the first and second poles, and so forth. The number $u^2_{\mu\mu'}$ determines the preferred value of a particle–hole transition compared with particle–particle and hole–hole transitions.

In this treatment, collective nonrotational excited states are considered at the same time as two-quasiparticle excitations, both the former and the latter being associated with a change in the internal structure of the nucleus. The wave functions of the collective excited states are the superposition of two-quasiparticle states. In this treatment there is no adiabatic condition for the collective states. In [33] it is shown how it is possible from equation (3.23) to pass to an adiabatic limit and obtain expressions for the mass parameter B and the coefficient of rigidity C.

In [33], the results of calculations of the first two states of the quadrupole (λ = 2, $\bar{\mu}$ = 0.2) and octupole (λ = 3, μ = 0, 1, 2, 3) types are summed in the region $150 \leq A \leq 186$ and $228 \leq A \leq 254$. In the calculations, the wave eigenfunctions† and energy of the Nilsson potential with parameters close to the data of [17] were used. The constants of the multipole–multipole interaction were written in the form

$$\varkappa^{(\lambda)} = k^{(\lambda)} A^{-4/3} \hbar\overset{0}{\omega}_0, \quad k^{(2)} = 10, \quad k^{(3)} = 0.8 - 1.0, \tag{3.28}$$

i.e., the constant of the octupole–octupole interaction was found to be approximately $1/10$ of the constant of the quadrupole–quadrupole interaction [26]. By taking the blocking effect into account, the accuracy of the calculations was appreciably improved [30].

4. Structure of Excited States of Deformed Even – Even Nuclei

In the superfluid model, the ground state of the even–even nucleus is nonquasiparticle, if the interaction between quasiparticles is not taken into account. In studying the collective effects, the ground state is regarded as nonphonon, in this case it contains small admixtures of four-quasiparticle, eight-quasiparticle, and so forth, states.

† The notations, the same as in [11], are based on asymptotic quantum numbers: N is the total number of oscillatory quanta; n_z is the number of oscillatory quanta along the axis perpendicular to the axis of symmetry of the nucleus; Λ is the component of the angular moment of a particle on the axis of symmetry of the nucleus; Σ is the projection of the spin of the particle on that axis; $K = \Lambda \pm \Sigma$; π is the parity. The state is written as $K\pi[Nn_z\Lambda]$ or more briefly $Nn_z\Lambda\uparrow$, if $K = \Lambda + \Sigma$ and $Nn_z\Lambda\downarrow$ if $K = \Lambda - \Sigma$; $\hbar\overset{0}{\omega}_0 = 41A^{-1/3}$ MeV.

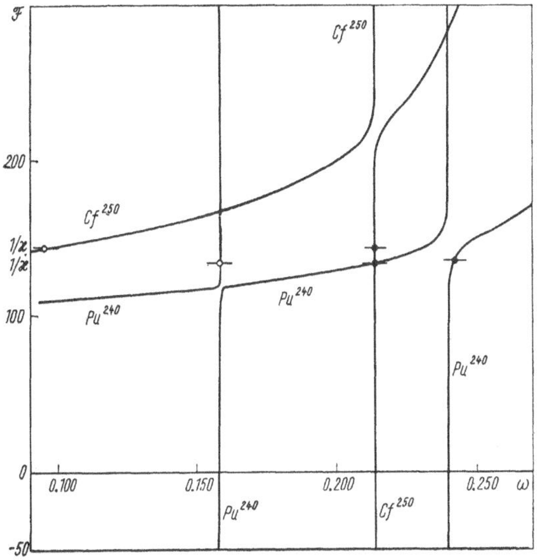

Fig. 9. Behavior of the function F(ω) for states with $K\pi = 2^+$ for Pu^{240} and Cf^{250} (intersections with the straight lines $1/\varkappa$ are denoted by dots).

The excited states of the even–even nucleus, for which H_{coll} plays no decisive part, are the two-quasiparticle, four-quasiparticle, and so forth. H_{coll} is significant for quadrupole states with $K\pi = 0^+$, 2^+ and octupole states with $K\pi = 0^-$, 1^-, and 2^-. States with other values of $K\pi$ may be regarded as quasiparticle for relatively low energies.

In [13, 14, 34, 35] the energies of two-particle levels in even–even nuclei are calculated for the regions $152 \leq A \leq 186$ and $228 \leq A \leq 254$, and compared with the corresponding experimental data. Some of these results are shown in Table 1. Comparison of the calculated values of the energies of the levels of even–even nuclei with the experimental data showed that most of the calculated low two-quasiparticle levels, which ought to be rapidly filled by corresponding β decays, were found experimentally. The problem is to find experimentally all the calculated levels (or to prove that some of the levels are missing). The agreement of the calculations, performed on the basis of the superfluid model of the nucleus without taking H_{coll} into account, with the experimental data with regard to the energies of the excited states and the probabilities of β transitions confirms the two-quasiparticle structure of some excited states of strongly deformed even–even nuclei. It should be emphasized that currently there are no experimental data which would be in conflict with the conclusions following from the superfluid model of the nucleus. Furthermore, new experimental data on two-quasiparticle states are in good agreement with calculations made in 1960-1961. The energies of all the two-quasiparticle states up to 2 MeV are given in [13, 14, 34, 35] for practically all deformed even–even nuclei, and the characteristics of these states are discussed in detail in [8, 36].

Collective nonrotational states of deformed even–even nuclei were investigated in [26-32], and all the information has been systematically arranged in a review [33]. The energies of collective states were calculated and compared with corresponding experimental data. Some of the results are given in Figs. 10 and 11. Thus, a natural explanation has been provided for the fall in the energies of the γ-vibrational states below the energies of the β-vibrational states in isotopes of dysprosium and erbium, the fall in the octupole $K\pi = 0^-$ states in isotopes of thorium, uranium and plutonium below the β- and γ-vibrational states and so forth. The calculated

TABLE 1. Energies of Two-Quasiparticle States

Nucleus	$K\pi$	Configuration	Energy, MeV Calc.	Energy, MeV Exp.	Nucleus	$K\pi$	Configuration	Energy, MeV Calc.	Energy, MeV Exp.
Cm^{246}	1−	*nn* 624↓ 734↑	1.1	1.080	Hf^{176}	1−	*pp* 404↓ 514↑	1.2	1.159
	1−	*pp* 523↓ 633↑	1.3	1.351		3+	*nn* 514↓ 521↓	1.6	1.647
Cm^{244}	6+	*nn* 622↑ 624↓	0.92	1.042	Yb^{174}	6+	*nn* 514↓ 512↑	1.6	1.530
U^{234}	6−	*nn* 633↓ 743↑	1.2	1.417		5−	*pp* 514↑ 411↓	1.8	1.900
	1−	*nn* 633↓ 743↑	1.2	1.438	Yb^{172}	3+	*nn* 521↓ 512↑	1.1	1;174
	3+	*nn* 633↓ 631↓	1.4	1.495		3+	*pp* 411↓ 404↓	1.4	1.644
	5−	*nn* 631↑ 743↑	1.5	1.688	Er^{168}	4−	*nn* 633↑ 521↓	1.1	1.095
Pt^{184}	8−	*nn* 514↓ 624↑	1.7	1.838	Er^{166}	6−	*nn* 523↓ 633↑	1.6	1.785
Os^{182}	8−	*nn* 514↓ 624↑	1.7	1.833		1−	*nn* 523↓ 633↑	1.6	1.826
W^{184}	7−	*pp* 514↑ 402↑	1.3	1.503	Dy^{164}	6−	*nn* 523↓ 633↑	1.7	1.680
	2−	*pp* 514↑ 402↑	1.3	1.150	Dy^{162}	5−	*nn* 642↑ 523↓	1.3	1.485
W^{182}	2−	*pp* 514↑ 402↑	1.3	1.290	Dy^{160}	4+	*nn* 521↑ 523↓	1.6	1.694
	4−	*nn* 624↑ 510↑	1.5	1.554	Dy^{158}	4+	*nn* 521↑ 523↓	1.7	1.672
	6−	*nn* 624↑ 512↓	1.9	1.830	Gd^{156}	4+	*pp* 413↓ 411↑	1.4	1.511
W^{180}	8−	*nn* 514↓ 624↑	1.5	1.531		3−	*nn* 651↑ 521↑	1.9	1.935
Hf^{178}	8−	*pp* 514↑ 404↓	1.0	1.148					
	8−	*nn* 514↓ 624↑	1.5	1.480					

energies of the quadrupole states with $K\pi = 0^+$ and 2^+ and the octupole states with $K\pi = 0^-$, 1^-, and 2^- are in fairly good agreement with the corresponding experimental data in both regions of deformed nuclei.

The study of the reduced probabilities of electromagnetic transitions from and to collective nonrotational states in even–even nuclei has been the subject of much theoretical work and experimental research. This is due to the fact that data on the reduced probabilities of electromagnetic transitions provide the most direct information on the structure of a given excited state. Thus, the criterion of the collective character of a definite state is the increase in the reduced probability of an electromagnetic transition compared with the value corresponding to a single-particle transition.

The results of calculations of the reduced probabilities of E2 and E3 transitions [31] are in sufficiently good agreement with the experimental data. Thus, B(E3) values in the region of thorium and uranium isotopes are 15-20 times greater than single-particle values, in agreement with experimental data.

In the microscopic treatment of the superfluid model of the nucleus, collective nonrotational excited states are considered simultaneously with two-quasiparticle excitations, both the

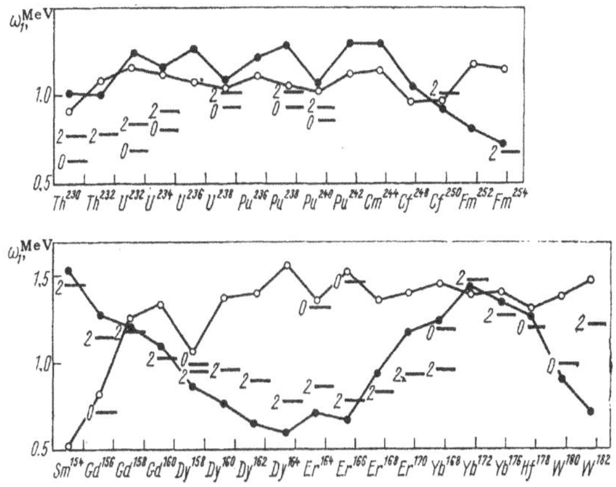

Fig. 10. Energies of the first excited states $K\pi = 0^+$ and 2^+ for $\varkappa^{(2)} = 10A^{-4/3}\hbar\mathring{\omega}_0$, $\varkappa^{(2)}_{np} = \varkappa^{(2)}$: 0) Energies of states with $K\pi = 0^+$, experimental; O) energies of states with $K\pi = 0^+$, theoretical; 2) energies of states with $K\pi = 2^+$, experimental; ●) energies of states with $K\pi = 2^+$, theoretical.

former and the latter being associated with a variation in the internal structure of the nucleus. In this treatment, the wave functions of the collective states are a superposition of different kinds of two-quasiparticle states. Investigation has shown [30] that most of the low states with $K\pi = 0^+$ and $K\pi = 2^+$ have clearly pronounced collective properties, and a large number of two-quasiparticle states contribute to their wave functions. Table 2 shows the contribution of two-quasiparticle states to collective states with $K\pi = 2^+$ in Dy^{158}, Er^{166}, and Yb^{176}. For Dy^{158} the table shows the contribution of two-quasiparticle states in the second $K\pi = 2^+$ state, which was also found to be collectivized to a considerable degree. It should be noted that the structure of the $K\pi = 0^+$ state is mathematically more complex and physically less clear than other collective nonrotational states.

The structure of octupolar states was examined in [26]. It was shown that the first $K\pi = 0^-$ states are mainly collective, their wave functions contain the contribution of many two-quasiparticle states, and the energies are lower by (0.5-1.0) MeV relative to the first poles of the secular equations. States with $K\pi = 1^-$ and 2^- are much less strongly collectivized than $K\pi = 0^-$ states. Lower $K\pi = 1^-$, 2^- states are collective only in individual cases (isotopes of thorium and uranium), in most cases their properties resemble two-quasiparticle properties. Thus, the admixture of residual states with the two-quasiparticle state corresponding to the first pole of the secular equation is 2-20%. The energies of the $K\pi = 1^-$ and 2^- states in most cases are lower by (0.1-0.3) MeV relative to the first pole. States with $K\pi = 3^-$ are practically two-quasiparticle, since the admixture of the residual states does not exceed 1%.

Expressions are given in [33] for the probabilities of α, β, and γ transitions to collective nonrotational states and the corresponding experimental data are summarized. It should be noted that all these states in deformed even nuclei must satisfy the following conditions:

a) for electromagnetic transitions

$$\frac{B(E2)}{B(E2)_{s.p}} > 1 \tag{4.1}$$

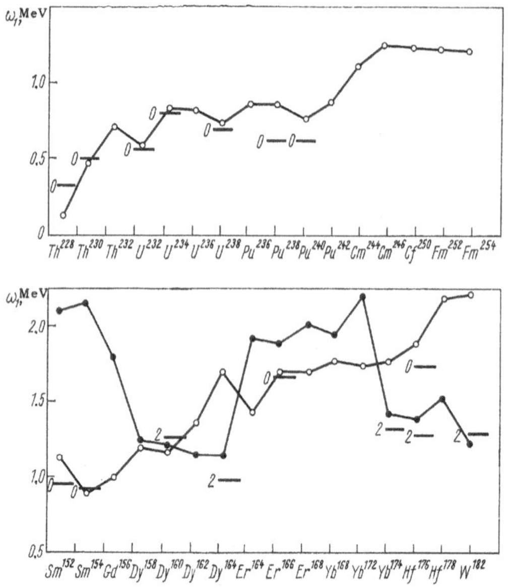

Fig. 11. Energies of first excited octupole states with $K\pi = 0^-$ and 2^- for $\varkappa^{(3)} \approx 1 A^{-4/3} \hbar \overset{0}{\omega}_0$ for $\varkappa_{np}^{(3)} = \varkappa^{(3)}$: 0) Energies of states with $K\pi = 0^-$ experimental; O) energies of states with $K\pi = 0^-$ theoretical; 2) energies of states with $K\pi = 2^-$ experimental; ●) energies of states with $K\pi = 2^-$ theoretical.

TABLE 2. Contribution of Two-Quasiparticle States to Collective States, with $K\pi = 2^+$ for $\varkappa^{(2)} = 10\ A^{-4/3}\ \hbar\overset{0}{\omega}_0 (\varkappa_{np}^{(2)} = \varkappa^{(2)})$

Configuration of the two-quasiparticle states	$f(\mu\mu')$	Dy^{158}		Er^{166}	Yb^{176}
		$i=1$	$i=2$	$i=1$	$i=1$
Neutron states					
523↓ —521↓	—1.28	1.7	0.01	22.2	0.2
512↑ —521↓	+0.014	10^{-3}	10^{-5}	0.03	0.02
512↑ —510↑	—1.82	0.07	10^{-4}	0.4	36.5
514↓ —512↓	—1.50	0.02	10^{-4}	0.1	34.0
505↑ —503↑	—1.70	1.0	0.01	0.8	0.6
642↑ —660↑	—0.77	7.3	1.4	0.5	0.01
633↑ —651↑	—0.68	1.5	0.33	2.5	0.03
532↓ +530↑	—1.06	2.0	0.04	0.1	10^{-3}
521↑ +521↓	—1.45	4.1	0.06	13.3	0.3
512↓ +510↑	—1.82	0.02	10^{-4}	0.06	5.0
651↑ +660↑	+0.85	9.5	6.0	0.15	0.01
Proton states					
413↓ —411↓	+1.28	25.6	40.7	18.1	3.9
402↑ —411↓	—0.11	0.01	10^{-4}	0.03	0.2
402↑ —400↑	+1.86	0.2	10^{-3}	0.5	1.4
413↑ —411↑	+1.28	1.1	0.01	0.2	0.08
404↑ —402↑	+1.52	1.8	0.01	1.2	2.7
523↑ —541↑	—0.50	0.5	0.01	0.1	0.04
411↑ +411↓	+1.47	40.0	51.7	35.2	9.0

or

$$\frac{B(E3)}{B(E3)_{s.\,p}} > 1; \tag{4.1'}$$

b) for β decay

$$\log (ft)_{coll} > \log (ft)_{s.\,p}; \tag{4.2}$$

c) for α decay

$$1 < F_{coll} < F_{s.\,p}. \tag{4.3}$$

Thus, a single description of two-quasiparticle and collective nonrotational states of deformed even nuclei has been obtained in the framework of the superfluid model of the nucleus. It has been shown that the average field of the nucleus is responsible for the variation in collective properties on transition from one nucleus to another.

The ground and low excited states of odd–odd nuclei are two-quasiparticle states, one quasiparticle being a proton particle and the other a neutron particle. Among the higher excited states, there ought to be four quasiparticle and collective nonrotational states, and it is an important problem to discover these states experimentally.

5. Structure of the Excited States of Deformed Odd Nuclei

The independent quasiparticle model is characterized not only by its sequence and self-consistency but also by the simplicity of the deductions regarding the atomic nucleus [36]. It follows from this model that the ground state and some excited states of an odd A nucleus have a single-quasiparticle structure, higher excited states, a three-quasiparticle structure and so forth. According to this model, the behavior of single-quasiparticle levels of strongly deformed odd nuclei is determined primarily by the average field of the nucleus. The success of the Nilsson potential [11] is due to this, both in explaining the behavior of the levels of odd-A nuclei as well as in complying with the rules of selection of β and γ transitions based on asymptotic quantum numbers. It should be noted that the single-quasiparticle structure of the low-lying states of deformed odd nuclei has been confirmed by experimental data on β decay according to (dp), (dt), and other reactions.

In odd nuclei, there ought to be observed two types of three-quasiparticle states. The first (3n) and (3p), when all three quasiparticles are either neutron or proton particles, the second (2n, p) and (2p, n), when two quasiparticles are neutron particles and one is a proton particle or, on the contrary, when two quasiparticles are proton particles and one is a neutron particle. Attention was drawn as long ago as 1961 [36] to the necessity of detecting such levels experimentally. The most direct experimental evidence of the three-quasiparticle structure of excited states is firstly the very large values of K, larger than the maximum K for single-particle states, such as for example, with $K\pi = {}^{23}/_{2}{}^{-}$ in Lu^{177}, demonstrated in [37], and secondly the very fast β-decays classified as au to three-quasiparticle states, when there are no auβ decays to single-particle states, for example [38] the state with $K\pi = {}^{3}/_{2}{}^{+}$ having an energy of 1.427 MeV in Er^{165}. Three-quasiparticle states, which can be found most simply experimentally, are described in [36]. There are signs of the existence of three-quasiparticle states in the range of isotopes of holmium, erbium, thulium, luthenium, hafnium, and others.

We shall consider the collective nonrotational states in deformed odd nuclei. In the odd nucleus, there is one quasiparticle in addition to the quasiparticles and phonons of the even–even nucleus. The problem amounts to taking into consideration the interaction of the quasiparticles with the phonons describing the collective states of odd–odd nuclei. We write the Hamiltonian H_c' (3.15) taking into account the secular equation (3.23) in the form

$$H_c' = \sum_{\mu'} \varepsilon(\mu') B(\mu'\mu') - \frac{1}{2} \sum_{\lambda,\,\mu>0} H_1^{\lambda\mu} - \frac{1}{4} \sum_{\lambda\mu>0} \times$$

$$\times \sum_i \frac{1}{\sqrt{Y^i(\lambda\mu)}} \sum_{\mu_2\mu_2'} f^{\lambda\mu}(\mu_2\mu_2') v_{\mu_2\mu_2'} \{(Q_i^+(\lambda\mu) + Q_i(\lambda\mu)) B(\mu\mu') + B(\mu\mu')(Q_i^+(\lambda\mu) + Q_i(\lambda\mu))\}. \tag{5.1}$$

The wave function, for example for the odd-Z nucleus, describing the states with given $K\pi$ is written in the form [39]

$$\Psi(K\pi) = \Omega(K\pi)^+ \Psi_0, \tag{5.2}$$

$$\Omega(K\pi) = \frac{1}{\sqrt{2}} C_\rho \left\{ \sum_\sigma \alpha_{\rho\sigma}^+ + \sum_{\lambda\bar{\mu}i} \sum_{\nu\sigma} D_{\rho\nu\sigma}^{\lambda\bar{\mu}i} \alpha_{\nu\sigma}^+ Q_i(\lambda\bar{\mu})^+ \right\}, \tag{5.3}$$

where ρ are the levels of the average field with given value of $K\pi$. The interactions of the quasiparticles with phonons having different values of λ, $\bar{\mu}$, and i are taken into account here. The normalizing condition of the wave function (5.3) may be written

$$C_\rho^2 \left\{ 1 + \frac{1}{2} \sum_{\lambda\bar{\mu}i} \sum_{\nu\sigma} \left(D_{\rho\nu\sigma}^{\lambda\bar{\mu}i} \right)^2 \right\} = 1. \tag{5.4}$$

We find the mean value of H_c' according to $\Psi(K\pi)$:

$$< \Omega(K\pi) H_c' \Omega(K\pi)^+ > = C_\rho^2 \left\{ \varepsilon(\rho) + \frac{1}{2} \sum_{\lambda\bar{\mu}i\nu\sigma} \left(\varepsilon(\nu) + \omega_i^{\lambda\bar{\mu}} \right) \left(D_{\rho\nu\sigma}^{\lambda\bar{\mu}i} \right)^2 - \frac{1}{2} \sum_{\lambda\bar{\mu}i} \frac{1}{\sqrt{Y^i(\lambda\bar{\mu})}} \sum_{\nu\sigma} D_{\rho\nu\sigma}^{\lambda\bar{\mu}i} v_{\rho\nu} f^{\lambda\bar{\mu}}(\rho\nu) \right\}$$

and determine C_ρ and $D_{\rho\nu\sigma}^{\lambda\bar{\mu}i}$, making use of the variation principle in the form

$$\delta \left\{ < \Omega(K\pi) H_c' \Omega(K\pi)^+ > - \eta_j \left[C_\rho^2 \left(1 + \frac{1}{2} \sum_{\lambda\bar{\mu}i} \sum_{\nu\sigma} \left(D_{\nu\sigma}^{\lambda\bar{\mu}i} \right)^2 \right) - 1 \right] \right\} = 0, \tag{5.5}$$

where η_j is the Lagrange multiplier. After calculations, we obtain the secular equation in the form [39]

$$0 = \frac{1}{4} \sum_{\lambda\bar{\mu}i} \sum_\nu \frac{v_{\rho\nu}^2}{Y^i(\lambda\bar{\mu})} \frac{(f^{\lambda\bar{\mu}}(\rho\nu))^2}{\varepsilon(\nu) + \omega_i^{\lambda\bar{\mu}} - \eta_j} - (\varepsilon(\rho) - \eta_j) \equiv P(\eta_j). \tag{5.6}$$

It should be noted that terms with $\lambda > 3$, $i > 2$ provide a very small contribution, since the value of $Y^i(\lambda\bar{\mu})^{-1} \to 0$, when the state approaches the two-quasiparticle state. The behavior of $P(\eta)$ for the state with $K\pi = 7/2^+$ in Tb^{157} is shown in Fig. 12, where the values of $P(\eta) = 0$ correspond to the roots of (5.6). The functions C_ρ and $D_{\rho\nu\sigma}^{\lambda\bar{\mu}i}$ have the form

$$C_\rho^{-2} = 1 + \frac{1}{4} \sum_{\lambda\bar{\mu}i} \sum_\nu \frac{v_{\rho\nu}^2}{Y^i(\lambda\bar{\mu})} \cdot \frac{(f^{\lambda\bar{\mu}}(\rho\nu))^2}{\left(\varepsilon(\nu) + \omega_i^{\lambda\bar{\mu}} - \eta_j\right)^2}, \tag{5.7}$$

$$D_{\rho\nu\sigma}^{\lambda\bar{\mu}i} = \frac{1}{2} \frac{v_{\rho\nu}}{\sqrt{Y^i(\lambda\bar{\mu})}} \cdot \frac{f^{\lambda\bar{\mu}}(\rho\nu)}{\varepsilon(\nu) + \omega_i^{\lambda\bar{\mu}} - \eta_j}. \tag{5.8}$$

The secular equation (5.6) may be refined as follows: 1) By taking into account the fact that there are several single-particle states with given $K\pi$, i.e., in expression (5.3) by summing over P; the secular equation for this case is given in [39]; 2) $\varepsilon(\mu)$ is calculated by using the values of $C_p(\mu)$ and $\lambda_p(\mu)$ for the odd nucleus, but $\varepsilon(\mu)$ should be replaced by $E(\mu)-E_0$ [see equation (2.25) and (2.17)]; 3) by taking into account the influence of the blocking effect on the phonons, i.e., by calculating the values of $\omega_i^{\lambda\bar{\mu}}$ and $Y^i(\lambda\bar{\mu})$, starting from the fact that the wave function of the ground state is $\alpha^+_{\nu_0\sigma}\Psi_0$, and of the excited state is $Q_i(\lambda\bar{\mu})^+\alpha^+_{\nu_0\sigma}\Psi_0$.

The secular equation (5.6) has no free parameter and the entire arbitrariness consists in how many values of λ, $\bar{\mu}$ and roots i have to be taken into consideration. The first and second roots of equation (5.6) were obtained for nuclei in the region $153 \leq A \leq 185$ and the role of terms with different $\lambda, \bar{\mu}$, and i was studied. As result of the investigation, it was shown that the drop in energies relative to the single-quasiparticle values $\varepsilon(p)$ and relative to the first poles $\varepsilon(\nu)+\omega_i^{\lambda\bar{\mu}}$ was determined primarily by the terms of (5.6) with $\lambda = 2$, $\bar{\mu} = 2$, $i=1$ and $\lambda = 3$, $\bar{\mu} = 0$, $i=1$, the octupole phonons being often no less important than the quadrupole phonons. For certain states of some nuclei, an important part is played by the terms of equation (5.6) with $\lambda = 2$, $\bar{\mu} = 2$, $i = 2$ and $\lambda = 2$, $\bar{\mu} = 0$, $i = 1$. In by far the most cases, the total contribution to equation (5.6) of states with $\lambda = 2$, $\bar{\mu} = 0$, $i = 2$; $\lambda = 3$, $\bar{\mu}=0$, $i=2$; $\lambda=3$, $\bar{\mu}=1$ and $\bar{\mu}=2$, and $i=1$ and $i=2$ is not large. It may be concluded on the basis of this investigation that approximations which take into account (although also more accurately than in the present case) only phonons with $\lambda = 2$, $\bar{\mu} = 2$, $i = 1$ and do not take into account the remaining phonons (primarily with $\lambda = 3$, $\bar{\mu} = 0$, $i = 1$) are poor for the majority of deformed odd nuclei. There is no need to take terms with $\lambda > 3$ and $i > 3$ into account.

An analysis of the secular equation (5.6) shows that if η_1 is very close to $\varepsilon(\rho)$, the state will be practically single-quasiparticle If η_1 differs appreciably from $\varepsilon(\rho)$ and from the first pole $\varepsilon(\nu)+\omega_1^{\lambda\bar{\mu}}$, the structure of such a state is quite complex, since simultaneously with single-quasiparticle states, states with different quasiparticles and phonons contribute to the wave function. If, however, η_1 approaches the pole, i.e.,

$$\Psi(K\pi)\Big|_{\varepsilon(\nu)+\omega_1^{\lambda\bar{\mu}}-\eta_1=0} = \sum_{\sigma}\frac{1}{\sqrt{2}}\alpha^+_{\nu\sigma}Q_1(\lambda\bar{\mu})^+\Psi_0, \tag{5.9}$$

such a state may be called γ vibrational if $\lambda = 2$, $\bar{\mu} = 2$, or octupole if $\lambda = 3$, $\bar{\mu} = 0$, and so forth. At this limit, the probability $B(E/\lambda)$ of a transition tends to the value which it has in the corresponding odd–odd nucleus.

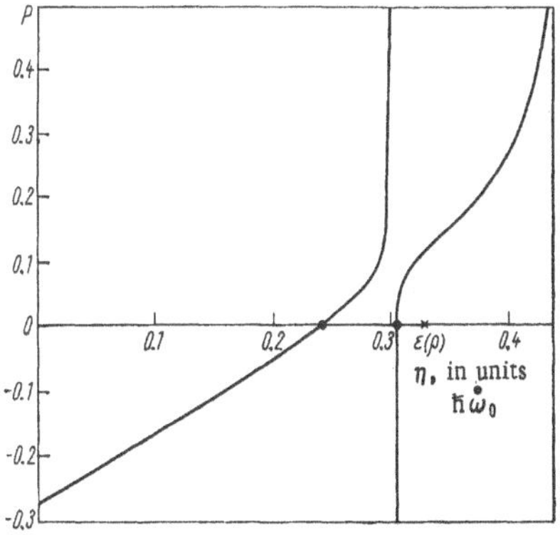

Fig. 12. Behavior of the function $P(\eta)$ for the state with $K\pi = 7/2^+$ in Tb^{157} (roots for $P(\eta) = 0$).

It should be noted that, as will be seen from equation (5.4), the value of C_ρ^2 determines the contribution of the single-particle state with given ρ to the $K\pi$ state considered. If C_ρ^2 is close to unity, the state is single-particle, if C_ρ^2 is rather less than unity, the state has a complex structure.

We shall examine the structure of nonrotational states of odd nuclei. We consider the case where

$$\varepsilon(\rho) \ll \varepsilon(\nu)+\omega_1^{\lambda\bar{\mu}}, \tag{5.10}$$

i.e., when the quasiparticle energy is much less than the value of the first pole of secular equation (5.6). In this case, states with $K = {}^{11}/_2$, ${}^9/_2$, and sometimes ${}^7/_2$ are close to single-particle states, and the contribution of the state ρ to the wave function is 95-99%. For example, the contribution to the $K\pi = {}^7/_2$ state of the single-particle 523↑ state in Tb^{161} is 99%, in Tb^{159} 98% and in Tb^{155} 97%. Another example: In isotopes of luthenium and tantalum, the $K\pi = {}^9/_2^-$ state is very close to the single-particle state and its energy is only very slightly less than $\varepsilon(\rho)$. At the same time in these nuclei in the $K\pi = {}^7/_2^+$ state (close to 404↓) admixtures play an important part and its energy is appreciably lower than $\varepsilon(\rho)$. In isotopes of luthenium and tantalum, therefore, the ${}^7/_2^+$ [404] state is the ground state and the ${}^9/_2^-$ [514] state is in every case the excited state. Similarly, the ${}^{11}/_2^-$ [505] state cannot be the ground state in odd N nuclei in the region $N = 91-97$.

Thus, the interaction of quasiparticles with phonons has a slight influence on states with $K = {}^{11}/_2$, ${}^9/_2$, close to $\varepsilon(\rho)$, and a stronger influence on states with lower values of K. As the result, states with $K = {}^{11}/_2$, ${}^9/_2$ are not ground states in deformed odd nuclei if in the system of levels of the average field there are levels with lower values of K very close to these states.

From conditions (5.10), states with $K = {}^1/_2$, ${}^3/_2$, ${}^5/_2$ are comparatively close to single-quasiparticle states and the contribution of the ρ state is 80-95%, i.e., admixtures of quasiparticle plus phonon states play a greater part and their energies are lower in relation to $\varepsilon(\rho)$ than for states with $K = {}^{11}/_2$, ${}^9/_2$. For example, the contribution to the ground state with $K\pi = {}^3/_2^+$ of the single-quasiparticle state 411↑ in Tb^{161} is 93%, in Tb^{159} 89%, and in Tb^{157} and Tb^{155} 90% each.

The investigations made have shown that the interaction of quasiparticles with phonons results in an unequal drop relative to $\varepsilon(\rho)$ of the different states close to the one-particle state. In some nuclei, therefore, the sequence of the excited states may differ from the sequence of the excited states in the Nilsson system. Further, the interaction of quasiparticles with phonons leads to a variation in the differences between the energies of states close to single-particle states in different nuclei which have the same number of odd neutrons or odd protons.

Let us consider the case

$$\min\left\{\varepsilon(\nu) + \omega_1^{\lambda\bar{\mu}}\right\} \ll \varepsilon(\rho), \tag{5.11}$$

i.e., when the energy corresponding to the first pole is much less than the single-quasiparticle state. In this case, in states with $K = {}^{11}/_2$, ${}^9/_2$, the predominant part is played by the term of equation (5.6) corresponding to the first pole, and these states have a quasiparticle pole phonon structure. The contribution of ρ to them is 0.1-3.0%. Only in this case can one speak of γ-vibrational octupole and so forth states in deformed odd nuclei. Thus, states with $K\pi = {}^9/_2^+$ with an energy of 0.97 MeV in Re^{185}; where the contribution of ρ is 1.4%, the state with $K\pi = {}^{11}/_2^-$ with the energy 0.687 MeV in Ho^{165} where the contribution of ρ is 0.1% and others are γ-vibrational states.

Under the condition of (5.11), states with $K = {}^1/_2$, ${}^3/_2$ have a rather more complex nature; in equation (5.6), some terms with different λ and $\bar{\mu}$ play a greater part and the contribution of ρ is 1-10%. States with $K = {}^5/_2$, ${}^7/_2$ occupy an intermediate position.

We shall consider the intermediate case

$$\varepsilon(\rho) = (0.5 - 2.0)\ \min\left\{\varepsilon(\nu) + \omega_1^{\lambda\bar{\mu}}\right\}, \tag{5.12}$$

where the maximum drop occurs in the roots of equation (5.6) in relation to both $\varepsilon(\rho)$ and to the first pole of the secular equation. In this case, the interaction of quasiparticles with phonons leads to the occurrence of deformed odd nuclei of collective nonrotational states of complex structure. The contribution of ρ is 30-80%.

The energies of states with K= $^1/_2$, $^3/_2$ are low in relation to $\varepsilon(\rho)$ and those of the first pole are stronger than the energies of states with K = $^9/_2$, $^{11}/_2$. In the secular equation many terms with different values of λ, $\bar{\mu}$, i, and ν play an important part. For example, for the state with $K\pi = {}^1/_2{}^+$ in Eu^{153} with an energy of 0.635 MeV, calculations give an energy of 0.6 MeV, contribution of $\rho = 411$ (66%), the drop in relation to $\varepsilon(\rho)$ is 0.63 MeV, in relation to the first pole, 0.95 MeV.

The energies of the collective nonrotational states satisfy the following rule:

$$\mathcal{E}_{coll}(K_0 - 2) < \mathcal{E}_{coll}(K_0 + 2), \tag{5.13}$$

i.e., the energies of the collective states with $K = K_0 - 2$ are less than the energies of the collective states with $K' = K_0 + 2$ if the first pole of equation (5.6) corresponds to $\lambda = 2$, $\mu = 2$, where K_0 relates to the ground state of the odd nucleus. Compliance with this rule is associated with two circumstances. Firstly, for small values of K in equation (5.6), there are many more terms in summing over ν than in the case of larger values of K. Secondly, in the system of single-particle levels of the average field, there are few states with $K = {}^{11}/_9$, $^9/_2$ compared with $K = {}^1/_2$, $^3/_2$ states, and therefore, for large values of K, the cases [condition (5.12)], where the energy of the first nonrotational state is very low are much less frequent than for small values of K. The results of the calculations of the splitting $\mathcal{E}_{coll}(K_0+2) - \mathcal{E}_{coll}(K_0-2)$ are in satisfactory agreement with experimental data.

It should be noted that the interaction of quasiparticles with phonons has a strong influence on the decoupling factors a for states with $K = {}^1/_2$. Available experimental data [40] on the energies of collective levels, reduced probabilities B(E2) and decoupling factors *a* in deformed odd nuclei satisfy the deductions enumerated above.

Conclusion

It should be pointed out that the superfluid model is rather crude. In particular, the introduction of multipole–multipole interactions, as well as the methods of solving the many-body problem in this case are open to criticism. However, the following conclusion may be drawn from a comparison of calculated results with experimental data: Agreement is remarkably good. The principal properties of the quadrupole and octupole states are correctly described. Such results as the correct description of the two-particle states, the drop in the energies of γ-vibrational states below the β-vibrational states in isotopes of dysprosium and erbium, as well as the drop of the octupole $K\pi = 0^-$-states below the β- and γ-vibrational states in isotopes of thorium, uranium, plutonium and others cannot be doubted. All these facts indicate that it has been possible, in the framework of the superfluid model, to describe correctly those residual interactions between nucleons which are of maximum value in atomic nuclei for comparatively small excitation energies.

It should be noted that the average field of a nucleus plays a decisive part in the determination of the properties of nonrotational excited states in deformed nuclei. In all nuclei, pair and multipole–multipole forces are operative, and these forces are identical in neighboring nuclei, and their constants decrease monotonically with increase in A. Lowering and consequently collectivization of a state with given $K\pi$ in one nucleus, its raising and its approximation to the two-particle state in another nucleus are determined by the average field. Thus, the average field controls the pair and multipole–multipole forces and determines the specific characteristics of each nucleus.

It should be pointed out that the single description of one-quasiparticle, two-quasiparticle and collective nonrotational states in deformed nuclei has been obtained in the framework of the superfluid model.

It may be concluded from the good agreement of theory with experiment that the average field is quite correctly described by the Nilsson potential.

In conclusion, it should be noted that the investigation of complex nuclei on the basis of the superfluid model of the nucleus, has not solved every problem. These investigations have merely shown that the microscopic approach to the description of the properties of nuclei is very promising. Very important, therefore, are various attempts to escape from the method of approximate second quantization and also to develop new methods. It should, however, be borne in mind that the accuracy of calculating the different characteristics of nuclei on the basis of the superfluid model is limited mainly by the crudeness of the description of the average field. In addition to the development of more modern methods of solving the many-body nuclear problem, therefore, it is essential to describe the average field of the nucleus more exactly.

LITERATURE CITED

1. Bogolyubov, N. N., Lectures on Quantum Statistics, Kiev, Izd. "Radyan'ska shkola," 1947.
2. Bogolyubov, N. N., Zh. Éksperim. i Teor. Fiz., 34:58, 73 (1958); Bogolyubov, N. N., et al. A New Method in Superconductivity Theory, Moscow, Izd. Akad. Nauk SSSR, 1958.
3. Bogolyubov, N. N., Dokl. Akad. Nauk SSSR, 119:52 (1958).
4. Bohr, A., Mottelson, B., and Pines, D., Phys. Rev., 110:936 (1958).
5. Belyaev, S. T., Mat. Fys. Medd. Dan. Vid. Seelgk., Vol. 31, No. 11 (1959).
6. Solov'ev, V. G., Zh. Éksperim. i Teor. Fiz., 35:823 (1958); 36:1869 (1959); Nucl. Phys., 9:655 (1958/59).
7. Solov'ev, V. G., Dokl. Akad. Nauk SSSR, 133:325 (1960); Mat. Fys. Skr. Dan. Vid. Selsk., Vol. 1, No. 11 (1961).
8. Solov'ev, V. G., Influence of Pair Correlations of the Superconducting Type on the Properties of Atomic Nuclei, Gosatomizdat, Moscow, 1963; Selected Topics in Nucl. Theory, Intern. Atomic Energy Agency, 1963.
9. Volkov, M. K., Pavlikovski, A., Rybarska, V., Solov'ev, V. G., Izv. Akad. Nauk SSSR, Ser. Fiz., 27:878 (1963).
10. Mikhailov, I. N., Zh. Éksperim. i Teor. Fiz., 45:1102 (1963); Bang, E., Mikhailov, I. N., Izv. Akad. Nauk SSSR, Ser. Fiz., 29:113 (1965).
11. Nilsson, S. G., Mat. Fys. Medd. Dan. Vid. Selsk., Vol. 29, No. 16 (1955). Mottelson, B., and Nilsson, S., Mat. Fys. Skr. Dan. Vid. Selsk., Vol. 1, No. 8, 1959.
12. Nathan, O., and Nilsson, S. G., Alpha, beta and gamma rays spectroscopy, Suegbann, C., North Holland Publishing Company, Amsterdam, 1964.
13. Veresh, T., et al., Izv. Akad. Nauk SSSR, Ser. Fiz., 26:1045 (1962).
14. Pyatov, I. N., Solov'ev, V. G., Izv. Akad. Nauk SSSR, Ser. Fiz., 26:11, 1617 (1964).
15. Bakke, F. H., Nucl. Phys., 9:670 (1958/59).
16. Solov'ev, V. B., Zh. Éksperim. i Teor. Fiz., 40:1654 (1961), Liu Yüan, Pyatov, I. N., et al., Zh. Éksperim. i Teor. Fiz., 40:1503 (1961).
17. Nilsson, S. G., and Prior, O , Mat. Fys. Medd. Vid. Selsk., Vol. 32, No. 16 (1960).
18. Liu Yüan, Izv. Akad. Nauk SSSR, Ser. Fiz., 28:18 (1964); Safarov, R., Kh. Izv. Akad. Nauk UsbSSR, Ser. Fiz.-Mat., No. 1, p. 86 (1965).
19. Bes, D. R., and Szymanski, Z., Nucl. Phys., 28:42 (1961).
20. Solov'ev, V. G., Dokl. Akad. Nauk SSSR, 144:1281 (1962); Phys. Lett., 1:202 (1962).
21. Mang. H. I., and Rasmussen, J. O., Mat. Fys. Skr. Dan. Vid. Selsk., Vol. 2, No. 3 (1962).
22. Gadetskii, O. G., and Pyatov, N. I., Preprint Joint Inst. Nuclear Res., P-1907 (1964).
23. Belyaev, S. T., Selected Topics in Nucl. Theory, Intern. Atomic Energy Agency, 1963, p. 291.
24. Baranger, M., Phys. Rev., 120:957 (1960); Kisslinger, L. S., and Sorensen, R. A., Mat. Fys. Dan. Vid. Selsk., Vol. 32, No. 9 (1960); Rev. Med. Phys., 35:853 (1963).

25. Zaretskii, D. F., and Urin, M. G., Zh.Éksperim. i Teor. Fiz., 41:898 (1961); 42:304 (1962); 43:1021 (1962); Bes, D., and Szymanski, Z., Nuovo Cimento, 26:787 (1962).
26. Solov'ev (Soloviev), V. G., and Vogel, P., Phys. Lett., 6:126 (1963); Proc. Congr. Intern. de Physique Nucléaire, 11:594 (1964); Solov'ev (Soloviev), V. G., et al., Izv. AN SSSR, Ser. Fiz., 28:1599 (1964).
27. Marshalek, E. R., and Rasmussen, I. O., Nucl. Phys., 43:438 (1963).
28. Liu Yüan et al., Zh. Éksperim i Teor. Fiz., 42:252 (1964).
29. Bes, D. Nucl. Phys., 49:544 (1963); Bes, D., et al., Nucl. Phys., 65:1 (1965).
30. Solov'ev (Soloviev), V. G., Nucl. Phys., 69:1 (1965).
31. Vogel, P., Yad. Fiz., 1:752 (1965).
32. Zheleznova, K. M., et al., Preprint Joint Inst. Nuclear Res., D-2157 (1965).
33. Solov'ev, V. G., Preprint Joint Inst. Nuclear Res., P-1973 (1965); Atomic Energy Review, 3:117 (1965).
34. Gallagher, C. J., and Solov'ev (Soloviev), V. G., Mat. Fys. Skr. Dan. Vid. Selsk., Vol. 2, No. 2 (1962).
35. Solov'ev (Soloviev), V. G., and Siklos, T., Nucl. Phys., 59:145 (1964).
36. Solov'ev, V. G., Zh. Éksperim. i Teor. Fiz., 43:145 (1964).
37. Jorgensen, M., et al., Phys. Lett., 1:321 (1962).
38. Gromov, K. Ya., et al., Izv. Akad. Nauk SSSR, Ser. Fiz., 47:195 (1963); Preibisz, Z., et al., Phys. Lett., 14:206 (1965).
39. Solov'ev (Soloviev), V. G., Phys. Lett., 16:308 (1965).
40. Gnatovich, V., and Gromov, K. Ya., Preprint Joint-Inst. Nuclear Res., P-2086, 1965.

THE QUASIPARTICLE METHOD IN NUCLEAR THEORY

A. B. Migdal

USSR

Introduction

Almost all problems of theoretical physics for some time past have been associated with the many-body problem, i.e., a problem in which it is necessary to take into account the interaction of many particles. Such a problem of physics was encountered in quantum electrodynamics, where it was possible to apply the perturbation theory, i.e., to utilize the fact that γ quanta interact weakly with electrons. This means that the charge of the electrons is small in dimensionless units. The interaction of electrons with quanta is characterized by the small dimensionless quantity

$$\alpha = e^2/\hbar c = \frac{1}{137}.$$

In other problems connected with elementary particles, the interaction cannot be regarded as weak. This so-called nucleon charge g, determining the interaction of a neutron or proton with π mesons, is already not small in dimensionless units

$$g^2 e/\hbar c = 12 - 14$$

and the perturbation theory cannot be used. It was necessary to develop methods of calculation which did not presuppose the weakness of the interaction between the particles. The dispersion theory of elementary particles was created; in this theory, weakness of the interaction is not presupposed, but instead use is made of the fact that some values are subject to little charge in the energy region with which we are concerned and may be replaced by constants.

After introducing into the theory constants which are taken from experiment (charges and masses of the particles, determining their interaction with each other), we obtain strict relationships between the various observed quantities.

Several years ago, the method developed in the theory of elementary particles, began to be applied very intensively to the study of systems of strongly interacting particles such as the electrons of a metal, liquid helium or the atomic nucleus.

If a system of strongly interacting particles contains, as in the case of the nucleus, 100 or 200 particles, it is clear that there is no hope of finding an exact solution to the problem. Even the problem of three interacting particles cannot be solved in a general form.

Were the system reminiscent of a gas, where the particles rarely interact, it would be possible to develop an approximate method of treatment which would take into account only paired collisions of the particles. In the nucleus, however, the distance between the particles is of the same order as the radius of action of the forces and several particles interact at once simultaneously. In addition, however, the energy of this interaction is comparable with the kinetic energy, so that the interaction cannot be regarded as weak.

Other approximate methods must be developed. The idea of these methods is the same as in the dispersion theory of elementary particles where the constants introduced are the masses of the particles and constants characterizing their interaction. It is necessary to introduce constants characterizing the motion of neutrons and protons in the nucleus, as well as constants describing the interaction of nucleons with each other when they are situated in the nucleus. This interaction differs greatly from the interaction of two nucleons in a vacuum.

After the introduction of such constants, all nuclear phenomena associated with low energies (with energies of less than 40 MeV), may be calculated exactly, i.e., it is possible by finding the constants of some phenomena, to explain quantitatively all the other experimental facts of low-energy nuclear physics.

In a finite system, for characterizing single-particle excitations, it is necessary to intoduce, in addition to the effective mass of the quasiparticles, also the parameters of the effective potential well in which the quasiparticles are moving. For systems with short-range forces of radius r_0, such parameters include the depth and width of the well and the width of the layer ($\sim r_0$) on which the density passes from its value inside the system to zero.

Expressions may be obtained for all these quantities in the form of series of the perturbation theory, whereby an approximate estimate may be made even in cases where the interaction is not small. In the first order of interaction between the particles, the effective potential is transformed into the so-called Hartree-Fock self-consistent field.

In the nucleus, where interaction is not small, the parameters of the potential well must be found from a comparison of the theoretical and experimental single-particle energies.

In nuclear physics, the so-called shell model is widely used and gives good results. In this model, the energy levels of the nucleons are determined as if a gas of noninteracting Fermi particles "poured" into the potential well. The theoretically good results obtained in the shell model are quite inexplicable, since there is a strong interaction between the particles and it is not possible to ignore cases where several particles interact simultaneously.

The method of Green's functions explains this fact. The part played by repeated collisions means that single-particle states in the well – quasiparticle states – ought to be calculated with a somewhat altered mass. (Comparison with experiment shows that the variation in mass does not exceed 10-20%.)

The problem of determining the reactions of the system on the external field (transition intensities and frequencies, magnetic and quadropole moments, etc.), amounts to the problem of the way in which the gas of interacting quasiparticles, situated in the potential well, behaves in the external field. It is sufficient here to take into account only the pair collisions of the particles. Repeated collisions of the particles are taken accurately into account by the theory, but only result in a variation in the interaction between the quasiparticles, and in the occurrence of a "charge" for the interaction of the quasiparticles with the external field. In most cases, this charge can be determined from general considerations (from the law of conservation of the charge, energy, momentum, etc.).

All these results have a very simple and clear explanation. Let a rather weak field act on the system so that the variation in energy of each particle in this field is small compared with its kinetic energy. The state of the system then corresponds to the occurrence of several quasiparticles and several quasiholes. The number of quasiparticles occurring forms a small proportion of the particles in the system. If the average distance between the particles is of the order of the radius of action of the forces, the average distance between the quasiparticles is much greater than the radius of the interaction forces and consequently the quasiparticles form a gas, i.e., it is possible to ignore cases where three or more quasiparticles collide simultaneously. The interaction between the quasiparticles is of the same order as the interaction between the particles, but differs substantially from it. As we shall see, in some cases, attraction may be replaced by repulsion, due to the influence of the residual nucleons of the medium which, unlike the quasiparticles, are present in large numbers and are situated alongside the two quasiparticles considered.

Thus, although the interaction between quasiparticles is not small, an enormous simplification of the problem is obtained, since it is sufficient to consider the collision of only two quasiparticles at a time.

The charge of the quasiparticle in relation to the external field is described by the interaction with the field of the cluster of particles forming the quasiparticle and results in the difference of the effective mass from the mass of a particle. Let us assume that an electric field, acting only on protons, is applied to the nucleus. Since the charge is retained in the interaction of the proton with the other particles, the entire cluster forming the proton quasiparticle has the same charge as the proton. At the same time the effective charge of a quasiparticle is equal to the charge of a particle. For other external fields, for example a magnetic field, the interaction of a quasiparticle with the field differs from the corresponding quantity for the particle.

A neutron moving in a vacuum interacts only with a magnetic field because of its internal magnetic moment, whereas a neutron quasiparticle in its movement also entrains the protons in movement, as the result of which an electric current is set up and the interaction with the magnetic field is varied. In the case of neutron quasiparticles orbital magnetism is produced, i.e., magnetism due to their motion in an orbit. In the absence of interaction, orbital magnetism is present only in the case of the protons.

Thus, for many phenomena, the nucleus may be regarded as a gas composed of two types of interacting quasiparticles situated in a potential well. The effective interaction between the neutron and proton quasiparticles is characterized by several constants, which are the same for all nuclei and all types of transitions with an accuracy which is the same as the constancy of the average density of nuclear matter.

Graphical Method and Green's Function

Feinman's graphical method is used for obtaining the various relationships in the theory of elementary particles. This method is also used in other many-body problems, in particular in the approach to nuclear theory, which will be described in what follows.

Various physical processes occurring with particles may be represented in the form of diagrams. The motion of a light quantum is represented by a dotted line

- - - - - - - - - - - - - - - - - - - -

the motion of particles by a straight line

————————————————

The following diagram

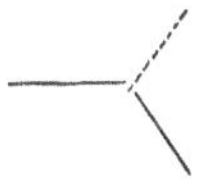

means that a charged particle, let us assume an electron, has emitted a light quantum. The solid line with the inflection means that after the emission of the light quantum, the electron assumed another momentum.

Let there be two noninteracting particles

On their interaction we have

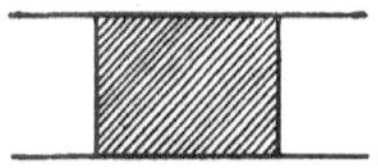

If their interaction takes place by means of light quanta (this means that the interaction is a Coulomb interaction) they are connected together by a dotted line

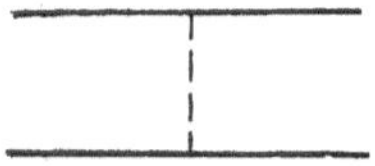

If there are two nucleons and the interaction occurs with a π-meson transfer, the particle lines are connected by a wavy line

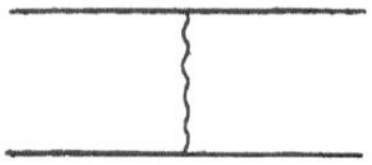

This means that two nucleons have reacted with each other. If they interacted twice we have

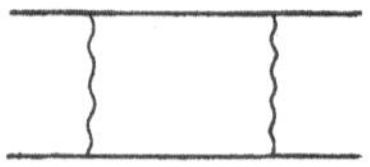

The following diagram

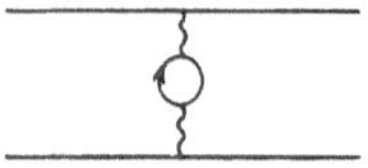

represents a more complex process – a nucleon has emitted a π meson which has then disintegrated into a nucleon and an antinucleon. These two particles are transformed into a π meson again, which is absorbed by the second nucleon.

Even more complex processes occurring with particles may be similarly represented.

For these diagrams to have not only an illustrative meaning but also a quantitative meaning, we shall understand by each diagram the amplitude of the transition from one state in an initial moment of time to another state in a final moment of time. The square of the amplitude of the transition determines the probability of finding the final state at the final moment of time. Thus, for example, the diagram shown above of the emission of a quantum denotes the amplitude of the transition of a charged particle with momentum p to the state with a quantum of momentum q and a particle with momentum p'.

In accordance with the superposition principle, the total amplitude of the transition, or as it is usually called, Green's function, represents the sum of all the possible physically different amplitudes of the transition. To illustrate the graphical method, we obtain the relationship connecting the scattering amplitude of two particles with the interaction potential.

According to the superposition principle, the scattering amplitude may be represented graphically by the sum of the diagrams

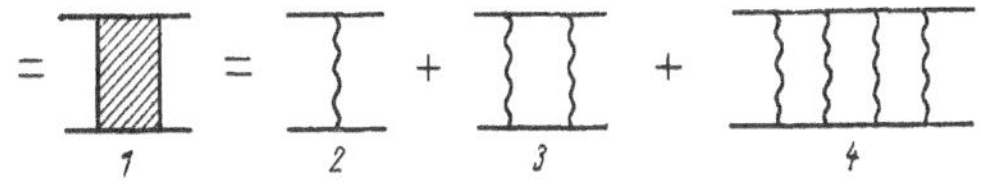

The first diagram represents the interaction between the particles, the second corresponds to the twofold interaction of the particles. The amplitude of the transition between the acts of interaction corresponds to the movement of two noninteracting particles.

We associate the first diagram with the potential U of the interaction between the particles

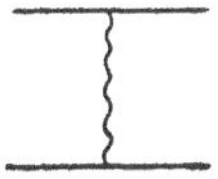 $= U$

and the line with Green's function, i.e., the amplitude of the transition of a free particle G. We then write the second diagram arbitrarily

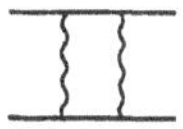 $= UGGU,$

since the transition amplitude of two free particles is equal to the product of Green's function of each of the particles. For the scattering amplitude, we obtain the series

$$\Gamma = U + UGGU + UGGUGGU + \ldots$$

The expression in the second and subsequent terms on the right of UGG again forms a sum giving Γ. For Γ we get the equation

$$\Gamma = U + UGG\Gamma.$$

The function G in this equation can easily be found. If the Ψ function of a particle in the initial moment is the superposition of different eigenfunctions, the problem of finding G is reduced to the problem of the spreading of the wave packet. If, however, the particle is situated at the initial moment in a state having a definite energy, the transition amplitude is determined quite simply. Evidently, the expression for Γ represents a symbolic record of the equation, known from quantum mechanics, for the scattering amplitude

$$\Gamma(p_1 p_2) = U(p_1 - p_2) + \int U(p_1 - p') \frac{\Gamma(p' p_2)}{\varepsilon_{p_1} - \varepsilon_{p'} + i\gamma} \frac{d^3p'}{(2\pi)^3}.$$

By comparing with equations for Γ, it is easy to establish exact agreement between the graphical and analytical expressions.

Similarly, Green's function of a particle in the external field $\tilde{G}$ may be associated with Green's function of a free particle G. Green's function in the field $\tilde{G}$ is represented by the sum of the partial transition amplitudes

Where the points with the dotted line represent the event of the action of the external field

$$V = \quad \text{—}\!\!\bullet\!\!\text{—}$$

By collecting all the diagrams in $\widetilde{G}$ to the right of V, we again obtain $\widetilde{G}$. Thus

$$\widetilde{G} = G + GVG + GVGVG + \ldots = G + GV\widetilde{G}.$$

Comparing the correction to G in the first order of the perturbation theory according to V

$$G^{(1)} = GVG$$

with the known quantum mechanics expression, it is easy to ascertain in what sense the multiplication in the symbolic formula for $\widetilde{G}$ is to be understood.

Thus, the idea of the graphical method is to establish by means of simple examples agreement between the elements of the diagrams and the analytical expressions, after which any diagrams consisting of these elements may be decoded. It is so much simpler to obtain the relationships by the graphical method than by analytical deductions that it is easy to investigate complex problems which appear insoluble if the approach is analytical. In what follows, we shall establish analytical agreement for all the diagrams describing processes occurring in a system of interacting particles, whereby the quantitative relationships can easily be obtained.

Nuclei in an External Field

1. Effective Field

We shall determine the variation of Green's function of a quasiparticle in an external field. For simplicity, we shall confine ourselves to a first approximation of the field perturbation, but shall consider exactly the interaction between the particles. We write down some diagrams appearing in Green's function $\widetilde{G}$ of a quasiparticle in the field

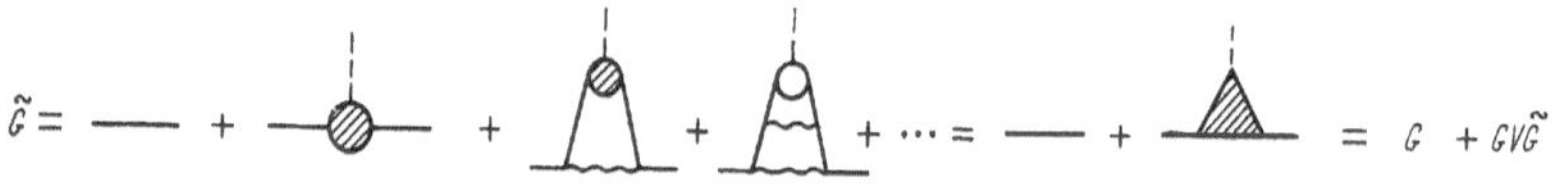

$$\widetilde{G} = \quad \text{———} \; + \; \text{—}\!\!\oslash\!\!\text{—} \; + \; \triangle \; + \; \triangle \; + \; \cdots = \quad \text{———} \; + \; \blacktriangle \; = \; G \; + \; GV\widetilde{G} \qquad (1)$$

where the circle denotes direct interaction of quasiparticles with the external field V^0

$$\text{—}\!\!\oslash\!\!\text{—} \; = \; e_q V^0$$

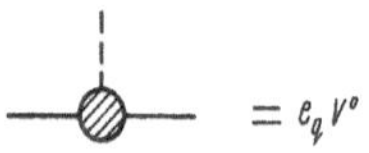

e_q being the charge of a quasiparticle. As we shall see, for certain types of fields $e_q \neq 1$, which denotes the difference of the external field acting on the quasiparticles from the external field applied to the particles. For noninteracting particles, we have

$$G_0 = \quad \text{———} \; + \; \text{—}\!\!\bullet\!\!\text{—} \; = \; G_0 + G_0 V^0 G_0$$

Thus, the shaded triangle in formula (1) replaces a point in this diagram and represents the effective field acting on the quasiparticle.

We obtain the equation for the field V. Among the diagrams included in V, there is one which does not contain an interaction between quasiparticles $(e_q V^0)$. All the other diagrams have the following structure. On moving from the base to the apex of the triangle, all the diagrams commence with interaction, then come two lines of free movement and then the set of diagrams representing the effective field. We introduce the block F not containing parts connected by two lines. The effective field is then determined by the equation

$$V = e_q V^0 + FGGV \tag{2}$$

or graphically

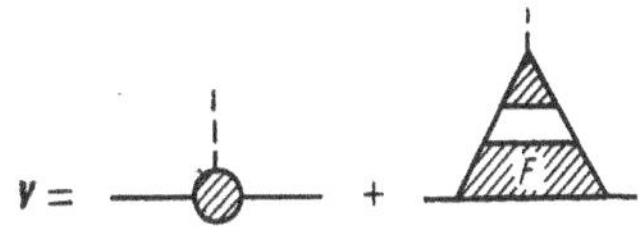

The first term in V describes the direct action of the external field on the quasiparticle. The second term gives the additional field resulting from the polarization of the medium, i.e., caused by the force action of the redistributed nucleons of the nucleus.

In the representation of the eigenfunctions Q_λ, the potential well, we obtain for the external field of frequency ω,

$$V_{\lambda_1\lambda_2} = e_q V^0_{\lambda_1\lambda_2} + \sum (\lambda_1\lambda_2 | F | \lambda'\lambda) A_{\lambda\lambda'} V_{\lambda\lambda'}, \tag{3}$$

where

$$A_{\lambda\lambda'} = \int G_\lambda (t) G_{\lambda'} (-t) e^{-i\omega t} dt. \tag{4}$$

In those cases where pair correlation is unimportant, calculation gives

$$A_{\lambda\lambda'} = \frac{n_\lambda - n_{\lambda'}}{\varepsilon_\lambda - \varepsilon_{\lambda'} - \omega}. \tag{4'}$$

In the case where pair correlation is important, or where there are close, competing levels, the expression for $A_{\lambda\lambda'}$ becomes complicated.

We have omitted the isotope indices in the expression for V. Equation (3) describes the effective field acting on protons (V^p) and neutrons (V^n).

We write equation (3) showing the isotope indices. Let the external field act on protons (V^{0p}) and neutrons (V^{0n}). Then

$$\begin{aligned} V^p &= e_q^{pp} V^{0p} + e_q^{pn} V^{0n} + F^{pp}\delta G^p + F^{pn}\delta G^n, \\ V^n &= e_q^{nn} V^{0n} + e_q^{np} V^{0p} + F^{nn}\delta G^n + F^{np}\delta G^p. \end{aligned} \tag{3'}$$

2. Effective Interaction of Quasiparticles in an Unfilled Shell

For processes connected with transitions inside the last unfilled shell, it is convenient to carry out a renormalization of equation (3) so that summation is performed only over the states of the last shell. We divide A into two components

$$A = A_1 + A_2.$$

$(A_2)_{\lambda\lambda'}$ differs from zero if both states are situated outside the last shell; $(A_1)_{\lambda\lambda'}$ differs from zero if one or both states lie in the last shell. Since in A_2 both states are situated fairly remotely from the Fermi surface, pair correlation in them, and generally the influence of Fermi states close to the surface, may always be ignored. A_2 is therefore determined by the simply expression (4').

We write the equation for V in symbolic form

$$V = e_q V^0 + FA_1V + FA_2V.$$

In the third term we replace V by the first part of this equation. By repeating this operation, we get

$$V = e_q \{1 + FA_2 + FA_2FA_2 + \ldots\} V^0 + \{F + FA_2F + FA_2FA_2F + \ldots\} A_1V.$$

We denote the expression in the square brackets at A_1V by F'

$$F' = F + FA_2F + FA_2FA_2F = F + FA_2F'. \tag{5}$$

In λ-representation, we have

$$(\lambda_1\lambda_2 | F' | \lambda_3\lambda_4) = (\lambda_1\lambda_2 | F | \lambda_3\lambda_4) + \sum{}'' (\lambda_1\lambda_2 | F | \lambda'\lambda) \frac{n_\lambda - n_{\lambda'}}{\varepsilon - \varepsilon_{\lambda'} + \omega} (\lambda\lambda' | F' | \lambda_3\lambda_4). \tag{5'}$$

The primes signify that neither λ nor λ' lie in the last shell. This is the equation for effective interaction of quasiparticles in the last shell.

We select the local interaction F included in this equation as follows:

$$F = \left(\frac{dn}{d\varepsilon_F}\right)^{-1} \{f_0 + f_0'\tau_1\tau_2 + (g_0 + g_0'\tau_1\tau_2)\sigma_1\sigma_2\}\delta(\mathbf{r}_1 - \mathbf{r}_2).$$

Here $f_0 f_0'$, $g_0 g_0'$ are constants to be found experimentally.

We now denote by means of V' the quantity

$$V' = e_q\{1 + FA_2 + FA_2FA_2 + \ldots\} V^0 = e_qV^0 + FA_2\{e_qV^0 + FA_2V^0 + \ldots\} = e_qV^0 + FA_2V'. \tag{6}$$

In this notation, the equation for V assumes the form

$$V = V' + F'A_1V. \tag{7}$$

In equation (7) summation is taken over states, one or both of which lie in the last shell. The value of A_1 in equation (7) should be determined while taking into account the distortions of Green's functions due to the influence of close levels.

Nuclear Moments

1. System of Calculating the Moments

The variation in the nucleus in a static external field, for example in the electrical magnetic field of atomic electrons, is determined by the different moments. The energy variation in a uniform magnetic field is determined by the dipole magnetic moment of the nucleus which it is customary to call simply the magnetic moment. If the heterogeneity of the magnetic field is substantial, octupole magnetic moments must be introduced. The interaction with the electric field of atomic electrons is determined by two moments, by the mean square of the electrical radius of the nucleus, which is included in the formula for the isotopic shift of atomic spectral lines, and by the quadrupole moment, which is found from hyperfine splitting. All these quantities are expressed as a mean, according to the ground state of the nucleus of the corresponding operator. The moment Q is equal to

$$Q = \left(\varphi_0 \sum_n Q_n \Phi_0\right)$$

or by means of operators of the second quantization we obtain

$$Q = (\Phi_0 \Sigma a_\lambda^+ a_{\lambda'} Q_{\lambda\lambda'} \Phi_0) = \sum_{\lambda\lambda'} \rho^a_{\lambda\lambda'} Q_{\lambda\lambda'} = Sp\rho^a Q,$$

where ρ^a is the matrix of the density of the particles.

We shall show that the variation of the averages with variation in the number of particles in the nucleus may be calculated by finding the variation of the quasiparticle density matrix by $\delta\rho$, or more simply by finding the effective field produced by the external field Q (i.e., by an addition of the form $H' = \Sigma a_\lambda^+ a_{\lambda'} Q_{\lambda\lambda'}$ to the Hamiltonian), and determining the variation in the filling numbers of quasiparticles occurring on transition from one nucleus to another. The variation in the moment Q is

$$\delta Q = Spe_q \delta\rho \hat{Q} = \sum_\lambda V_{\lambda\lambda}[Q]\,\delta n_\lambda,$$

where e_q is the charge of the quasiparticle with respect to the field Q. Since even–even nuclei have no magnetic moments, it is sufficient for calculating the magnetic moment of an even–odd or odd–even nucleus to find the variation in the density matrix on the addition of one particle to the odd–odd nucleus. The quadrupole moments in the region of spherical nuclei are zero for even–even nuclei; therefore, for calculating the quadrupole moment of a neighboring nucleus it is also sufficient to know the variation in the density matrix on the addition of one particle.

Thus, the system of calculating the static moments is as follows.

The effective field V[Q], corresponding to the field Q, is found. The field Q is equal to $r^2P_2(\cos\theta)$ in the case of quadrupole moments, to r^2 in the case of an isotope shift, or finally to the operator of the magnetic moment of one particle when the magnetic moment of the nucleus is found. The variation in the number of quasiparticles at the level δn_λ on the addition of one particle is found. After that, Q is calculated by means of formula (1).

In the simplest case, the addition of one particle to the doubly magic nucleus gives

$$\delta n_\lambda = \delta_{\lambda\lambda_0},$$

where λ_0 is the state in which an odd quasiparticle appears.

In this case

$$Q = V_{\lambda_0\lambda_0}[Q].$$

We expand the isotope indices in these relationships.

Let the operator Q act only on protons, as occurs in the case of quadrupole moments and isotope shift. Then

$$\delta Q = \delta Q^p = Q^p \delta\rho^p = \sum V^p_{\lambda\lambda}[Q^p]\,\delta n^p_x + V^n_{\lambda\lambda}[Q^p]\,\delta n^n_x.$$

The magnetic moments are also calculated similarly, both proton and neutron parts of the moment operator contributing.

2. Magnetic Moments

The operator of the dipole magnetic moment of a nucleon may be written in the form of the sum of two components

$$\mu = \mu^j + \mu^s$$

$$\mu^j = \frac{1+\tau_z}{2} j, \quad \mu^s = \left[\frac{1+\tau_z}{2}(\gamma_p - {}^1/_2) + \frac{1+\tau_z}{2}\gamma_n\right]\sigma_z,$$

where $(1+\tau_z)/2$ and $(1-\tau_z)/2$ are matrices corresponding to the proton and neutron states, and the total moment $j = l + {}^1\!/_2\sigma$, where l is the orbital moment; γ_p and γ_n are the proton and neutron gyromagnetic ratios. It is necessary to find the effective field $V[\mu]$, corresponding to the operator μ, and to calculate the magnetic moment from the formula

$$<\mu> = \sum_\lambda V_{\lambda\lambda}[\mu]\,\delta n_\lambda.$$

The magnetic moments of higher multipolarity are calculated according to a similar system. The operator of arbitrary multipolarity has the form

$$\mathfrak{Q}^L = \sum a_\lambda^+ a_{\lambda'} \mu_{\lambda\lambda'}^L,$$

$$\mu^L = \left[\frac{1+\tau_z}{2}\left[\left(\gamma_p - \frac{1}{L+1}\right)\sigma + \frac{2}{2L+1}j\right]' + \frac{1-\tau_z}{2}\gamma_n\sigma\right] \nabla\left(r^L Y_L^M\right).$$

For L = 1, we get the reduced operator of the dipole magnetic moment. L = 3 corresponds to the octupole moment.

In the same way as for the dipole magnetic moment, the orbital part of the octupole moment varies little as the result of the inclusion of interaction.

The equation for the spin part after separation of the angular variables is reduced to a system of equations for two functions of r^2. Tables of experimental and calculated values of the matrix elements of octupole magnetic moments are given in [1].

3. Quadrupole Moments and Isotope Shift

For calculating the quadrupole moments, it is essential to find the effective field produced by the external scalar field equal to $V^0 = r^2 P_2(\cos\theta)$. The isotope shift is determined by the field $V^0 = r^2$.

The variation in quadrupole moments and the value of $<r^2>$ were calculated in [9]. The equation for the effective field was solved on computers. The function $\hat{f}_0(r)$ was selected in the form

$$\hat{f}_0(r) = \hat{f}_0 + [\hat{f}_{0ex} - f_0]\,\frac{n(0) - n(r)}{n(0)},$$

where n(r) is the density of the particles in the nucleus.

Tables comparing the theoretical and experimental values of $\delta<r^2P_2>$ and $\delta<r^2>$ are given in [2].

Agreement with the experimental values is obtained in every case in the limits of 30-40% (with the exception of the light elements).

Unless the function $\hat{f}_0(r)$ is taken into account, the data cannot agree.

The following values were obtained for the constants $f_0, f_0', f_{0ex}, f_{0ex}'$:

$$f_0 = 1 \pm 0.2, \quad f_0' = 0.4 \pm 0.2, \quad f_{0ex} = -3.5 \pm 1,$$

$$f_{0ex}' = 0.4 \pm 0{,}2.$$

The discrepancies are obviously due to the roughness of the formula for $\hat{f}_0(r)$.

Electromagnetic and β-Decay Transitions

1. Dipole Excitations

We shall consider dipole transitions under the effect of γ quanta of the low frequency $\omega \ll 40$ MeV. It is then possible to ignore nonuniformity of the field over the nuclear radius ($k^2R^2 \ll 1$). For studying the excitation of the nucleus and not its motion as a whole, it is convenient to pass to the inertia center system, when in addition to the electric field acting on the protons, there is an inertia field, as the result of which the perturbation of the Hamiltonian of the system assumes the form (e=1)

$$H' = \mathbf{E}\left\{\frac{N}{A}\sum_p \mathbf{r}_i - \frac{Z}{A}\sum_n \mathbf{r}_i\right\}.$$

The effective field, occurring on the perturbation of H', satisfies the equation

$$V^p = \frac{N}{A} Ex + \bar{f}^{pp}A^pV^p + \bar{f}^{pn}A^nV^n,$$

$$V^n = -\frac{Z}{A} Ex + \bar{f}^{nn}A^nV^n + \bar{f}^{np}A^pV^p.$$

For the position of the maximum of the giant resonance curve we get

$$\omega_s^2 = \omega_0^2\left(1 + 2f_0'\right) = \omega_0^2 \frac{3\beta}{\varepsilon_F}.$$

We get an analogous relationship also for the width of the maximum [3]

$$\Gamma^2 = \Gamma_0^2 \frac{3\beta}{\varepsilon_F},$$

where Γ_0 is the width for a system of noninteracting particles in a well having a diffuse edge. These results are in satisfactory agreement with experiment.

2. Quadrupole Transitions

There is a large number of well-studied low-frequency quadrupole transitions. It is necessary in this case to use expressions which take pair correlation into account.

For studying quadrupole transitions, we must determine the effective field produced by the external field of frequency ω in the form

$$V^0 = r^2P_2(\cos\theta) \equiv Q.$$

The equation for the effective scalar field has the form

$$V^p(\mathbf{r}) = V^0(\mathbf{r}) + \left[f_0^{pp}(\mathbf{r})\,\delta n^p(\mathbf{r}) + f_0^{pn}(\mathbf{r})\,\delta n^n(\mathbf{r})\right]\left(\frac{dn}{d\varepsilon_F}\right)^{-1},$$

where $\delta n^p(\mathbf{r})$, $\delta n^n(\mathbf{r})$ are the variations in density of the protons and neutrons in the field.

As a rule, in studying quadrupole collective oscillations, what is referred to as the quadrupole–quadrupole interaction is used. This interaction has the form

$$(\lambda_1\lambda_2|\,F^Q\,|\lambda_3\lambda_4) = -\varkappa Q_{\lambda_1\lambda_2}Q_{\lambda_3\lambda_4}.$$

The only reason for such a choice of interaction is that equation (1) is then reduced to an algebraic equation.

The interaction F^Q leads to an effective field of the form

$$V = C(\omega) V^0,$$

i.e., equivalent to the assumption that the variation in density in the field is

$$\delta n(\mathbf{r}) = \alpha(r) P_2(\cos\theta) = C_1 r^2 P_2(\cos\theta).$$

Such an assumption distorts the radial function $\delta n(r)$ and consequently also the value of $V - V^0$. There is a considerable error in cases where high-moment states are important. The principal contribution to $V - V^0$ is then determined by the values of r close to the surface of the nucleus where the values of $f_0(r)$ and $\delta n(r)$ vary sharply with r.

The intensities of single-particle quadrupole transitions were calculated in [4] for some cases where pair correlation was not important. Satisfactory agreement with experimental data was obtained. The value of $f_0(r)$, appearing in the solution, was the same as in calculating the quadrupole moments and isotope shift. For pair correlation, in addition to single-particle quadrupole states, there is also a state with the energy $\sim\Delta$, usually referred to as the collective state. The intensity and frequency of these transitions, like those of single-particle quadrupole transitions, are calculated as follows. In the equation for the effective field and in the equations for d (d is the variation in amplitude of the pair correlation of Δ in the external field), the angular variables are eliminated by the usual method. The equations for the radial matrix elements V and d are solved by means of computers with the function $f_0(\mathbf{r})$.

For a qualitative analysis, we assume

$$V = C(\omega) Q \quad d = D(\omega) Q.$$

By multiplying the equations for V and d by Q and integrating, it is not difficult to find the values of $C(\omega)$ and $D(\omega)$. It may be concluded from this approach that considerable errors are introduced by ignoring the quantity d, as is often done in nuclear calculations, for $\omega \ll 2\Delta$, while for $V/2\Delta \simeq 1$, the result is quite distorted. This is not suprising since to ignore d leads to nonconservation of the number of particles.

3. β Decay

For calculating the intensities of β transitions, we must solve the equations for the effective field produced by the field of weak interaction.

For Fermi transitions, the difference between the effective and external fields is caused only by Coulomb corrections since if the Coulomb field is ignored, the perturbation

$$H' = \sum a_\lambda^+ a_{\lambda'} (\tau_\pm)_{\lambda\lambda'}$$

due to isotope invariance, does not cause polarization of the medium, just as it is not caused by the perturbation

$$H' = \sum a_\lambda^+ a_{\lambda'} (T_z)_{\lambda\lambda'} = T_z.$$

The equation for $V(\tau_+)$ shows that only second-order corrections are involved with respect to V_Q/ε_F, where V_Q is the Coulomb field. Therefore, corrections to the Fermi matrix element calculated on the assumption of strict isotope invariance are negligibly small.

The equations for the effective field in the case of Gamow-Teller transitions are very similar to the equations for the effective field corresponding to the spin part of the magnetic moment. In summation over λ and λ', one state relates to neutrons and the other to protons. Just as for the spin part of the magnetic moment, the principal part is played by terms of the sum with λ and λ', differing only by the sign of the projection of spin on the direction of the moment j.

For mirror nuclei, the Gamow-Teller matrix element is expressed strictly by the magnetic moment of the daughter or parent nucleus in the ground state. Substitution of the observed moments gives amplitudes of β transitions which agree with experiment within the limits of the experimental error.

The probabilities of β transitions taking pair correlation approximately into account are calculated in [5]

A table comparing the theoretical and experimental values of the probabilities of transition is given in [6].

In all cases, where there are no perturbations of the configurations, satisfactory agreement is obtained between the absolute probabilities of transitions and the experimental values. The spin interaction involved in the calculation is the same as for magnetic moments

$$g_0' = \frac{g_{nn} - g_{np}}{2} \approx 0.5.$$

4. l-Forbidden Transitions

Among the magnetic and allowed β transitions, there are transitions in which there is a variation of orbital moment by two units.

Such l-forbidden transitions are impossible in the single-particle model, since the matrix element σ or $\sigma\tau_+$ is equal to 0 for transitions with a variation in orbital moment.

The explanation of l-forbidden transitions is as follows. The probability of a transition includes the matrix element not of the external field but of the effective field. The effective field for the external field σ and for the field $\sigma\tau_+$ has the form

$$V_\alpha = V_1(r^2)\,\sigma_\alpha + V_2(r^2)\,\frac{r_\alpha r_\beta \sigma_\beta}{r^2}.$$

The second term in this expression has the matrix elements for states with values of the orbital moment differing by two. The intensities calculated in this way are in good agreement with the experimental probabilities of l-forbidden transitions [6].

5. μ Capture

In μ capture, in contrast to β decay and K capture, the momentum carried away by the neutrons is high and an energy of the order of 10-15 MeV is transferred to the nucleus. In the perturbation Hamiltonian, the flux of light particles cannot be assumed to be independent of r. In addition, the perturbation must be supplemented by an induced pseudoscalar interaction which makes a negligibly small contribution to the β decay, but provides an appreciable correction in μ capture. For the rest, the calculation of the probability of a μ capture is performed in the same way as the calculation of dipole or quadrupole transitions at high excitation energies. A summation is performed in [7] in quasiclassical approximation over the states in an equation for the effective field. A formula is obtained for the dependence of the time of μ capture on A and Z, which is in good agreement with experiments. More exact, numerical results were obtained in [8], in which the equation for the corresponding effective field was solved by means of a computer.

Conclusion

The general aim of the approach to the calculation of nuclei which has been described is to formulate in a strict language and to express by means of the universal constants of theory all the problems which have been solved semiquantitatively by means of simple models.

Satisfactory agreement is obtained between the various phenomena of nuclear physics. For this purpose, it is sufficient to introduce the interaction between quasiparticles characterized by the constants f_0, f_0', g_0, g_0' inside the nucleus, and analogous constants outside the nucleus. To obtain more reliable results, the interaction between the particles should be refined. The most vulnerable point of the theory is the law of transitions of the constants from values inside the nucleus to values outside it. The theory may be improved in two directions. New constants α' and R' must be introduced, characterizing the transition law, or an attempt must be made to obtain the transition law from theoretical considerations. It is possible that even with a slight decrease in density at the surface of the nucleus, the gas approximation is applicable, and therefore it is reasonable to attempt to find the transition law on the basis of the gas approximation. Furthermore, the values of the interaction outside the nucleus are fairly sensitive to the value of the energy of the interaction particles, since a pole remains in the scattering amplitude at low energies. This pole has little effect on the interaction F, since this interaction appears in the sum, where the energy along the channel of two particles is not fixed but varies over a fairly wide range. This phenomenon may be easily taken into account by assuming, as interaction amplitude outside the nucleus, the dependence of the pole form on the sum of the energies of the particles.

The value of the interaction constants outside the nucleus may be found theoretically by examining the scattering of nucleons close to the surface of the nucleus and introducing the known interaction of free nucleons. The number of constants of the theory is then appreciably reduced.

A more exact comparison of theory with experiment enables the following harmonics to be found for the expansion of F over the angle between incident pulses of particles.

Let us enumerate the problems which have not yet been solved.

We commence with nuclear reactions.

For examining reactions with the formation of an intermediate nucleus, it is not difficult to reformulate existing calculations in the language of interacting quasiparticles, utilizing, as above, the gas approximation for the quasiparticles.

More interesting results may be obtained from an examination of the direct nuclear reactions, studied in detail in [9] by the method of separating the polar parts of corresponding diagrams. In doing this, it is necessary, as in the approach described, to introduce constants characterizing the nonpolar parts of the diagrams.

In the simplest cases, these constants may be expressed by the introduction of constants of the interaction between quasiparticles.

First of all, elastic and inelastic nucleon scattering is described by diagrams of the scattering amplitude

$\Gamma =$ 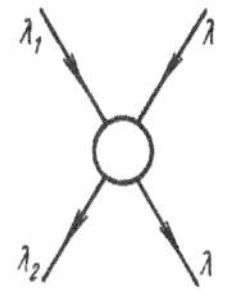

Here $\lambda_1\lambda_2$ are the initial and final states of the scattered nucleon; λ and λ' correspond to single-particle excitation.

Thus, this reaction is expressed by interaction constants appearing in F.

Furthermore, the block

$g =$ 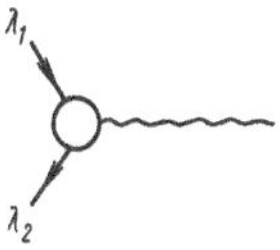

corresponding to scattering with the formation of collective excitation represents the residue of the scattering amplitude in the corresponding pole and is consequently expressed by F.

The reactions (γ, n) and (γ, p) do not require the introduction of fresh constants.

Reactions with the formation of deuterons or α particles for processes occurring close to the Fermi surface may be described by a supplementary constant.

For α particles, this constant appears in the refined theory of α decay. In fact, the irreducible block characterizing the formation of an α particle from quasiparticles situated close to the Fermi surface in coordinate representation is δ formed and is characterized by a constant.

Interacting phenomena are associated with discontinuous variations in the radii and shape of nuclei on the addition of particles.

On the addition of a small number of particles to a magic nucleus, the added particles for the time being do not produce any intersection of terms and there is a smooth redistribution of the density without variation in the radius of the nucleus. The density at the center of the nucleus exceeds the mean value corresponding to the formula $R = r_0A^{1/3}$. For a sufficiently large number of added particles, the radius or shape of the surface of the nucleus varies abruptly and the mean value of the density is restored. The phenomenon could be observed in experiments of the Hofstadter type.

The calculation of the interaction constants and the constants characterizing the potential well and resulting from the interaction of free nucleons is very difficult and is beyond the scope of the problems of the theory under discussion.

Currently existing approximate methods of solving this problem are based in some form or other on the assumption of the small magnitude of the interaction or the applicability of an approximation for which there is no theoretical justification.

However, if as the result of such calculations correct values were to be obtained for already known interaction constants, these calculations could be relied on after such a verification. A more detailed discussion of the method under consideration is given in [10].

LITERATURE CITED

1. Troitskii, M. A., and Khodel', V. A., Yad. Fiz., 1:205 (1965).

2. Bunatyan, G. G., and Mikulinskii, M. A., Yad. Fiz., 1:38 (1965).
3. Lushnikov, A. A., and Zaretsky, D. F., Nucl. Phys., 66:35 (1963).
4. Kamerdzhiev, S. P., Yad. Fiz., 2:415 (1965).
5. Gaponov, Yu. V., Yad. Fiz., 2:6:1002 (1965).
6. Kodel', V. A., Yad. Fiz., 2:24 (1965).
7. Novikov, V. N., and Urin, M. G., Yad. Fiz., 2:112 (1965).
8. Bunatyan, G. G., Yad. Fiz. 2:868 (1965).
9. Shapiro, I. S., Theory of Direct Nuclear Reactions, Moscow, Gosatomizdat, 1963.
10. Migdal, A. B., Theory of Finite Fermi Systems and the Properties of Atomic Nuclei, Moscow, "Nauka," 1965.

LIGHT NUCLEI CALCULATIONS BY THE SELF-CONSISTENT FIELD METHOD

D. M. Brink

England

Introduction

These lectures discuss calculations, based on the Hartree-Fock method, for finding the approximate solution of the quantum mechanical many-body problem. If we have a wave function in the form of the Slater determinant $\Phi = a\Pi\Phi_i(x_j)$ (i.e., the antisymmetrized product of single-particle wave functions), it may be used for calculating the mean value of the energy,

$$<E>_\Phi = <\Phi, H\,\Phi>,$$

where H is the Hamiltonian of the particle system. The Hartree-Fock method is equivalent to the following problem. The Slater wave function ought to give minimum $\langle E\rangle$. The practical determination of the minimum value of $\langle E\rangle$ is difficult, since the class of Slater wave functions is wide.

We shall consider three different approximations for the solution of the general problem. In each method the solution of the variational problem

$$\delta <E> = 0, \tag{1}$$

is determined, but with a limited set of Slater wave functions.

1. The first method is based on the physical idea that the majority of nuclei have ellipsoidal deformation. We construct the Slater wave function from single-particle functions, these being the solution of the problem with the potential of the deformed harmonic oscillator

$$H_{osc} = \frac{1}{2m} p^2 + \frac{1}{2} m \left(\omega_x^2 x^2 + \omega_y^2 y^2 + \omega_z^2 z^2\right). \tag{2}$$

We then calculate the matrix elements of the real Hamiltonian having this wave function. This average value of the energy $\langle E\rangle$ is a function of the parameters ω_x, ω_y, ω_z, determining the single-particle orbits. These parameters are varied to find the minimum of $\langle E\rangle$. Such a method was used recently by M. K. Volkov, Towles and Heywood, Brown and Brink.

2. The second method is based on the idea of the α clustering of light nuclei. The Slater wave functions, giving accurate α clustering, are constructed and the parameters of the wave functions are varied for producing the minimum value of $\langle E\rangle$.

3. The third method, the method of deformed orbits, has been used by Levinson et al., Dietrich et al., and Bassichis and Ripka.

In using the Hartree-Fock method, states are often obtained which are not eigenstates of conserved values, for example angular momentum and parity. We shall encounter such wave functions and find a method of projecting the deformed internal states on the eigenstates of angular momentum and parity.

Elaboration of the Wave Functions

For describing configurations with α clusters in a nucleus with 4N nucleons we introduce a set of N single-particle wave functions

$$\Phi_i(\mathbf{r}) = A\exp\left\{-\frac{(\mathbf{r}-\mathbf{R}_i)^2}{2b^2}\right\}, \qquad i=1, 2, \ldots, N, \qquad A=\left[b^3\pi^{3/2}\right], \tag{3}$$

dependent on N vectors $\mathbf{R}_i$, and parameters. The wave function Φ_i describes the motion of a nucleon in the 1s orbit of the harmonic oscillator with center at $\mathbf{R}_i$. The 4N-nucleon state may be constructed of these N functions, if it is required that each state should be occupied by two protons and two neutrons, and the corresponding 4N-particle normalized Slater wave function $\Phi(R_1, \ldots, R_N; b)$ is then constructed. The resultant wave function describes a state with maximum symmetry in the sense of the Wigner theory of supermultiplets (i.e., it has [4, 4, ..., 4]-symmetry and $T=0$ and $S=0$. Two neutrons and two protons have a tendency to cluster around the points R_i. This clustering, however, is broken down by the Pauli principle when the points $R_1, \ldots, R_N$ are close together. If there is a tendency to α clustering in light nuclei, the wave functions described here contain the possibility of describing such clustering. The basic set of wave functions is not an orthogonal set, but this does not give rise to any difficulties in the construction of the Slater wave functions.

Matrix Elements of Kinetic and Potential Energies

Let Φ and Ψ be two 4N-nucleon wave functions constructed from functions determined by equation (3), with one parameter b, but with different sets of parameters $R_1, \ldots, R_N$; S_1, S_N. We have to calculate the matrix elements of the standard operators.

a. Overlap matrix. We use the standard formula to obtain (Löwdin, 1955)

$$\langle\Phi, \Psi\rangle = \frac{(\det\langle\varphi_i, \psi_j\rangle)^4}{(\det\langle\varphi_i, \varphi_j\rangle \det\langle\psi_i, \psi_j\rangle)^2}.$$

In our case, the single-particle wave functions are simple and the overlap integrals may be calculated exactly

$$B_{ij} = \langle\varphi_i, \psi_j\rangle = \exp\left\{-\frac{(\mathbf{R}_i-\mathbf{S}_j)^2}{4b^2}\right\}, \tag{4}$$

$$\langle\varphi_i, \varphi_j\rangle = \exp\left\{-\frac{(\mathbf{R}_i-\mathbf{R}_j)^2}{4b^2}\right\}, \tag{5}$$

$$\langle\psi_i, \psi_j\rangle = \exp\left\{-\frac{(\mathbf{S}_i-\mathbf{S}_j)^2}{4b^2}\right\}. \tag{6}$$

b. Matrix elements of kinetic energy. We have a formula for the matrix elements of an arbitrary single-particle operator, independent of spin and isotopic spin. We use it to calculate the matrix elements of kinetic energy

$$<\Phi,\ T\ \Psi> = 4 <\Phi,\ \Psi> \sum_{i,j} <\varphi_i,\ t\ \psi_j> (B^{-1})_{ji}.$$

Here B^{-1} is the reciprocal matrix of B determined by equation (4). The single-particle matrix elements of kinetic energy in equation (7) may be calculated exactly

$$<\varphi_i,\ t\ \psi_j> = -\frac{\hbar^2}{2M} A^2 \int \exp\left\{-\frac{(\mathbf{r}-\mathbf{R}_i)^2}{2b^2}\right\} \nabla^2 \exp\left\{-\frac{(\mathbf{r}-\mathbf{S}_j)^2}{2b^2}\right\} d\mathbf{r} = \frac{\hbar^2}{Mb^2}\left[\frac{3}{4} - \frac{1}{8b^2}(\mathbf{R}_i - \mathbf{S}_j)^2\right]$$
$$\times \exp\left\{-\frac{(\mathbf{R}_i-\mathbf{S}_j)^2}{4b^2}\right\} = \hbar\omega B_{ij}\left[\frac{3}{4} - \frac{1}{8b^2}(\mathbf{R}_i - \mathbf{S}_j)^2\right]. \qquad (8)$$

c. Potential energy. Using Löwdin's method, we get an expression for the matrix elements of the central exchange potential in the general form

$$V = U(\mathbf{r})(W + BP_\sigma - HP_T - MP_\sigma P_T) <\Phi,\ V\ \Psi> = x_d V_d + x_l v_l, \qquad (9)$$

where $x_d = 8W + 4B - 4H - 2M$, $x_l = 8M + 4H - 4B - 2W$, and the direct and exchange interactions V_d and V_l have the form

$$V_d = <\Phi,\ \Psi> \sum_{ijkl} <\varphi_i\varphi_j | U | \psi_k\psi_l> (B^{-1})_{ki} (B^{-1})_{lj}, \qquad (10a)$$

$$V_l = <\Phi,\ \Psi> \sum_{ijkl} <\varphi_i\varphi_j | U | \psi_k\psi_l> (B^{-1})_{kj} (B^{-1})_{li}. \qquad (10b)$$

If $U(\mathbf{r})$ is the local potential, the orbital two-particle matrix elements may be calculated exactly

$$<\varphi_i\ \varphi_j | U | \psi_k\psi_l> = A^4 \int \exp\left\{-\frac{1}{2b^2}[(\mathbf{r}_1-\mathbf{R}_i)^2 + (\mathbf{r}_2-\mathbf{R}_j)^2 + (\mathbf{r}_1-\mathbf{S}_k)^2 + (\mathbf{r}_2-\mathbf{S}_l)^2]\right\} U(\mathbf{r}_1-\mathbf{r}_2)\, d\mathbf{r}_1 d\mathbf{r}_2.$$

This expression for the matrix elements may be simplified by using the identity

$$(\mathbf{r}_1-\mathbf{R}_i)^2 + (\mathbf{r}_1-\mathbf{S}_k)^2 = 2\left[\mathbf{r}_1 - \frac{1}{2}(\mathbf{R}_i + \mathbf{S}_k)\right]^2 + \frac{1}{2}(\mathbf{R}_i - \mathbf{S}_k)^2$$

and an analogous identity for $\mathbf{r}_2$, $\mathbf{R}_j$, and $\mathbf{S}_l$. We get

$$<\varphi_i\varphi_j | U | \psi_k\psi_l> = B_{ik}B_{jl}A^4 \int \exp\left\{-\frac{1}{b^2}\left[\mathbf{r}_1 - \frac{1}{2}(\mathbf{R}_i + \mathbf{S}_k)\right]^2 + \left[\mathbf{r}_2 - \frac{1}{2}(\mathbf{R}_j + \mathbf{S}_l)\right]^2\right\} U(\mathbf{r}_1-\mathbf{r}_2)\, d\mathbf{r}_1 d\mathbf{r}_2 = \qquad (11)$$
$$= B_{ik}B_{jl} I(\mathbf{R}_i + \mathbf{S}_k - \mathbf{R}_j - \mathbf{S}_l),$$

where

$$I(\rho) = \frac{A^2}{\sqrt{8}} \int \exp\left\{-\frac{1}{2b^2}\left(r - \frac{\rho}{2}\right)^2\right\} U(\mathbf{r})\, d\mathbf{r}. \qquad (12)$$

If $U(\mathbf{r})$ is the Gaussian potential or is equal to the sum of the Gaussian potentials, the integral $I(\rho)$ may be calculated exactly. If

$$U(r) = U_0 \exp\left(-\frac{r^2}{\beta^2}\right),$$

then

$$I(\rho) = V_0 U^3 \exp\left\{-\frac{\rho^2(1-U^2)}{8b^2}\right\},$$

where (13)

$$U^2 = \frac{1}{1+2\left(\frac{b}{\beta}\right)^2} .$$

If U(r) is the local potential in general form, $I(\rho)$ may be expressed by means of Talmi integrals:

$$I(\rho) = \exp\left(-\frac{\rho^2}{8b^2}\right)\sum_n \left(\frac{\rho^2}{8b^2}\right)\frac{T_n}{n!}, \tag{14}$$

where

$$I_n = \frac{2}{(2b^2)^{n+3/2}\,\Gamma\left(n+\frac{3}{2}\right)}\int e^{-\frac{r^2}{2b^2}} r^{2n+2}\, U(r)\, dr.$$

The series (14) for $I(\rho)$ converges rapidly for all ρ's, if the dimensions of the potential U(r) are much greater than the parameter b used in the calculations.

Examples of α Clusters of Wave Functions

1. Be^8. Let there be a system of two α-clusters with the orbital wave functions

$$\Phi_1(\mathbf{r}) = A\exp\left[-\frac{1}{2b^2}(\mathbf{r}-\mathbf{d})^2\right],$$

$$\Phi_2(\mathbf{r}) = A\exp\left[-\frac{1}{2b^2}(\mathbf{r}+\mathbf{d})^2\right].$$

The eight-nucleon Slater wave function constructed from these two functions represents two α-clusters, the centers of which are situated at the points **d** and −**d**, separated by the distance 2d. We select the z axis parallel to **d** and examine the behavior of the Slater wave function for $d \to 0$. At this limit, the two orbital wave functions become identical and the nonnormalized Slater wave function becomes zero. However, at the same time, the normalized factor becomes infinite and the normalized Slater wave function tends to a definite nonzero limit. In order to perceive this, we introduce orthogonal linear combinations of the functions Φ_1 and Φ_2

$$\chi_1 = \Phi_1 + \Phi_2; \quad \chi_2 = \frac{1}{d}(\Phi_1 - \Phi_2),$$

when

$$d \to 0, \quad \chi_1 \to 2A\exp\left(-\frac{r^2}{2b^2}\right),$$

$$\chi_2 \to \frac{4A\mathbf{r}\mathbf{d}}{d2b^2}\exp\left(-\frac{r^2}{2b^2}\right) = 2\,\frac{AZ}{b^2}\exp\left(-\frac{r^2}{2b^2}\right).$$

In other words, χ_1 tends to the spherical 1s state and χ_2 to the 1p state. The limit form of the cluster wave function for $d \to 0$ is the wave function of the shell model with oscillator potential, with four nucleons in the 1s shell and four nucleons in the 1p shell in the (001) orbital state. This is the deformed internal state of the type considered by Elliott in his SU(3) model. If we project it on the eigenstate of the angular momentum (with L equal to 0, 2, 4), we get the lower states in the shell description of Be^8 in LS coupling.

2. C^{12}. We assume that there are three α-clusters, the centers of which form an equilateral triangle with the side d in the xy plane, and we examine the behavior of the normalized Slater wave function for $d_i \to 0$. As in the preceding case, we obtain at the limit a shell oscillator wave function with the configuration $(000)^4$, $(100)^4$, $(010)^4$(n_x, n_y, n_z represent the oscillator state with the quantum numbers n_x, n_y, n_z). This is the deformed internal state and by projecting it on the eigenstate of angular momentum (with $L = 0, 2, 4$), we obtain the lower states of C^{12} in the shell model with LS coupling. Let us consider a system of three α-clusters situated on a line parallel to the x axis. The shell limit is the configuration $(000)^4$ $(100)^4$ $(200)^4$ with four nucleons in each of the 1s, 1p, 2s1d shells. In other words, we have four-particle, four-hole excitation from the ground state of the shell model.

If we expand the 3α-particle configuration, intermediate between an equilateral triangle and a linear configuration, according to states of the oscillator shell model, we obtain a superposition of 1-particle–1-hole, 2-particle–2-hole, 3-particle–3-hole, and 4-particle–4-hole states.

3. O^{16}. The tetrahedral configuration of four α-clusters becomes a configuration with two closed shells in the limit case of overlapping clusters. The plane configuration in the xy plane leads to a 4-particle–4-hole excited state in which the four nucleons are excited from the (001) state of the p shell to a state in the sd shell. The orbital state of these nucleons depends on the geometric configuration of the α clusters. If the α clusters are arranged at the corners of a square, the orbital configuration of the sd nucleons is equal to $(1/\sqrt{2})[(200) - (020)]$. For another arrangement of the α clusters, the orbital wave function of the sd nucleons in the limit of the overlapping clusters will be $[\alpha(200) + \beta(020)]$, where the coefficient α and β depend on the arrangement of the clusters. The most deformed 4p–4h wave function in the SU_3 classification belongs to the (84) representation, and the internal state has the shell configuration $(000)^4$ $(100)^4$ $(010)^4$ $(200)^4$, similar to the coplanar configuration of four α-clusters.

It is clear from the above-mentioned examples that the shell model is a limit case of the α-cluster model. The α-cluster model may be regarded as an expansion of the shell model, just as the particle–hole model is an expansion of the shell model. Calculations in the shell model include diagonalization of the nuclear Hamiltonian in the finite space of the shell states selected from the total set. The choice of several α-cluster states may give a better approximation to the correct wave functions of the system than many 1p–1h, 2p–2h, 3p–3h, and 4p–4h states.

Nucleon – Nucleon Interaction

Let us now consider the important question of how to use interaction in our calculations. All the α-cluster states described have a closed-shell structure, since each orbital state is filled with four nucleons. They have orbital symmetry in the sense of Wigner supermultiplets and in all of them $T = 0$ and $S = 0$; if all the α-cluster states have zero spin the matrix elements of the spin-orbital and tensor forces become zero. (Spin orbital and tensor forces are a spin-variable tensor of first and second rank, respectively, and consequently their matrix elements between states with zero spin disappear according to the Wigner-Eckhardt theorem.) Both these noncentral interactions make a contribution of second order only by means of an addition of states with $S = 1$ and $S = 2$, which have no α-cluster structure. In the calculations, we ignore these additions and consider only the central forces. In the preceding section, it was shown that in experiments for the matrix elements of the potential-energy operator in the cluster model, there would be only definite combinations of parameters in the general expression for the central potential

$$V(r) = U(r)(W + BP_\sigma - HP_\tau - MP_\sigma P_\tau), \tag{15}$$

TABLE 1. Values of the Parameters x_1 and x_d for Different Potentials

Types of interaction	W	M	B	H	x_d	x_l	Y_l	Y_0
Serber	0.5	0.5	0	0	3	3	1	0
Soper	0.40	0.33	0.17	0.10	2.82	1.56	0.73	0.14
Ferrel and Visscher	0.32	0.50	0	0.18	0.84	4.08	0.82	−0.32
Gillet (for O^{16})	0.35	0.35	−0.1	0.4	0.1	4.1	0.7	−0.4
Gillet (for C^{12})	0.40	0.4	−0.2	0.4	0.0	4.8	0.8	−0.5
Rosenfeld	−0.13	0.93	0.46	−0.26	0.0	4.8	0.8	−0.5
Mixture Majorana	1−M	M	0	0	8−10M	10M−2	1	1−2M
Elliott					0.0	6	1	−0.6

namely

$$x_d = 8W + 4B - 4H - 2M$$

and

$$x_l = 8M + 4H - 4B - 2W.$$

The two different exchange forces with the same x_d and x_l will produce the same effect in the cluster model. This is because the α-cluster wave functions have maximum orbital symmetry.

In light nuclei, central nucleon−nucleon interaction with large space exchange interaction tends to lower a state with maximum orbital symmetry, despite the fact that the spin-orbital forces tend to destroy this symmetry. For that reason, probably, the calculated energies of low states are very sensitive to the x_d and x_l combinations of exchange parameters, since these parameters may well determine the fit of calculated energies of weakly excited states with experimental values. In many calculations recently made by the self-consistent field method for the Sd shell, Rosenfeld exchange mixing has been used. Probably the results will be almost the same as with the use of a simple mixture of Wigner and Majorana exchange forces with suitable values of x_d and x_l.

Table 1 gives the values of the parameters x_l and x_d for different mixtures.

The classical condition for saturation with exchange forces without hard core in a nucleon−nucleon interaction has the form $x_d \leq 0$, consequently interactions (5) and (6) lead to saturation, interactions (3) and (4) are close to saturation, and interactions (1) and (2) do not lead to saturation. The last two columns of the table give the values of the quantities

$$Y_l = \frac{1}{6}(x_d + x_l) \text{ and } Y_0 = \frac{1}{10}(x_d - x_l), \tag{16}$$

characterizing the interaction force in states with even or odd values of the relative angular momentum of two nucleons. The majority of known phenomenological interaction potentials lead to strong repulsions in states with odd momentum; this is due to their ability to lead to saturation.

Experimental data on nucleon−nucleon scattering indicate the presence of a repulsive core and it is usually assumed that saturation is due to a hard core and also to the exchange character of the forces. The classical explanation of the saturation problem consists in the assertion that the forces are repulsive in states with odd momentum, and compensate the attraction in even states. For nuclear forces with a repulsive part, saturation is explained by a hard core, which acts most effectively in the relative S-state and with a weaker repulsion acting in odd states, owing to exchange effects.

Let us now consider nucleon–nucleon interaction in the relative P-state obtained from an analysis of nucleon–nucleon scattering. It is repulsive in $^1P_1$ and $^3P_1$ states and attractive in $^3P_0$ and $^3P_2$ states. We shall endeavor to calculate the matrix elements of the interaction Hamiltonian in these states, using the well-known methods for taking the influence of the hard core into account and averaging over the spin states corresponding to the wave functions of a nucleus with maximum space symmetry. The results obtained indicate the low value of the interaction in the P state; it completely disappears or proves to be slightly repulsive, depending on what the nucleon–nucleon potential was and how the hard core was taken into account. These results indicate that the most likely cause of saturation is the hard core and not the repulsion in states with odd momentum, and that in the calculations, the central part of the interaction potential may be approximated by a Serber repulsive core and exchange mixture or even by a repulsive core and attraction in a relative s-state (Moszkowski-Scott, Kallio-Kottwitz).

The phenomenological forces used in practice must satisfy definite requirements and above all ought to result in saturation. In the particle–hole model, due to the characteristics of the calculations, this is already not so important. This model is based on the Hartree-Fock theory, and although single-particle energies and residual interactions ought to be obtained from the same nuclear forces, this is never done in practice; the single-particle energies are taken from experiment and the residual interactions between particles or particles and holes are parametrized independently. The energies of excited states are calculated relative to the ground state, the energy of which is assumed to agree with that found experimentally. It is clear that such calculations are not self-consistent, and that good results may be expected only when the core is not strongly excited. The absence of self-consistency appears, for example, in spurious states, which do not occur with zero energies if the calculations are correct.

In the α-cluster theory, it is impossible to carry out distribution among cores and valence nucleons, all nucleons being regarded as equivalent. There is, therefore, no reason for introducing the concept of the average field of the nucleus, of single-particle energy levels. Only the total energy of the nuclear states can be calculated, and to obtain the excitation energies, it is necessary to subtract the calculated ground-state energy. In this case, properties such as internucleon forces are important as being capable of resulting in saturation.

Let us consider, for example, the nucleus O^{16}. If we wish to ascertain whether the spherical shape of the core in the ground state is stable to deformation in the tetrahedral configuration of four α clusters, we must calculate the total energy of several deformations and find the one at which a minimum is obtained. However, if these results are to have any meaning, it is also necessary to obtain a reasonable value of the binding energy of the O^{16} nucleus.

A. B. Volkov studied the internucleon interactions also used in the present calculations. This is a mixture of Wigner and Majorana forces.

$$V(r) = U(r)(W + MP_M), \tag{17}$$

where

$$U(r) = -60 \exp\left[-\left(\frac{r}{1.80}\right)^2\right] + 60 \left| \exp\left[-\left(\frac{r}{1.01}\right)^2\right]\right. .$$

The interaction was selected such that its radial dependence was similar to the long-range part of the potential obtained by the Moszkowski-Scott separation method. The potential has the following properties.

1. It results in saturation in light nuclei and gives a satisfactory value for the binding energy and equilibrium volume of the H^4 and O^{16} nuclei, if the wave functions of the harmonic oscillator are used (for this, it is essential that M assumes a value in the range 0.6-0.65).

TABLE 2. Binding Energy (MeV) and Deformation Value for C^{12} and O^{16}

Nucleus	Spherical	Equilibrium	$h\omega_x$ equilibrium	$h\omega_y = \omega_z$ equilibrium	Increase in binding energy due to deformation ΔE
C^{12}	64.6	69.1	18	18	4.5
O^{16} (Ground state)	117	117	16	16	0
O^{16} (2p−2h)	85.4	88.8	12	12	3.4
O^{16} (4p−4h)	67.6	79.0	10	10	11.4

2. It determines the order of magnitude of the two-quasiparticle energy parameters.

The interaction (17) also has its defects. It does not result in saturation in infinite nuclear matter and does not give sufficient binding in Be^8 and C^{12}. Perhaps these two defects are connected and one would like to find a better interaction.

Calculation of Equilibrium Deformation

In the first part of this section, we consider the ellipsoidal deformation of light nuclei.

As already mentioned in the introduction, for all the calculations wave functions of the deformed harmonic oscillator are used. In the present work, the interaction (17) is used. Towles and also Hayward, however, worked with standard effective interactions (Gillet, Soper, etc.) and varied the parameters under the condition of conserving the volume. The binding energy was calculated using the standard formula of the shell model

$$<E> = \sum_{\alpha} <\alpha|t|\alpha> + \frac{1}{2}\sum_{\alpha\beta}\{<\alpha\beta|v|\alpha\beta> - <\alpha\beta|v|\beta\alpha>\}, \tag{18}$$

where summation is taken over the occupied (orthogonal) single-particle states α, to which are given the numbers n_x, n_y, n_z, characterizing the number of quanta in the direction x, y, and z, respectively. For calculating the two-particle matrix elements, it is convenient to carry out a Talmi transformation to the respective coordinates and the coordinates of the mass center. The result may be represented in the form

$$<V> = \sum C(n_x n_y n_z) V(n_x n_y n_z), \tag{19}$$

where the coefficients $C(n_x n_y n_z)$ are numbers depending solely on the state of the nucleus, but not on the form of the potential and wave functions, and

$$V(n_x n_y n_z) = <n_x n_y n_z|v|n_x n_y n_z>,$$

where the average is taken between the states of the deformed harmonic oscillator.

The calculation will be shown in the example of the nucleus C^{12}, assuming the configuration $(000)^4$ $(010)^4$ $(001)^4$. In the language of the shell model, this means that the orbits 1s, $1p_y$, and $1p_z$ are each occupied by four nucleons. The total energy is then equal to

$$<E> = <T> + Y_l <V_{even}> + Y_0 <V_{odd}>, \tag{20}$$

where $<T> = 2.75\ h\omega_x + 4.75(h\omega_y + h\omega_z)$ (kinetic energy of the mass center deducted); Y_l and Y_0 are the interaction constants in states with even and odd momenta, respectively, determined in equation (16) by means of exchange parameters.

Thus

$$<V_{even}> = 27V(000) + 3V(020) + 3V(002) + 3V(011),\ <V_{odd}> = 15V(010) + 15V(001).$$

TABLE 3. Energy Characteristics of Different States of C^{12} and O^{16}

Configuration of C^{12} and O^{16}	Binding energy (d = 0), MeV	Equilibrium binding energy, MeV	ΔKE	ΔE	Equilibrium d, Fermi
C^{12} (triangular)	64.6	67.6	19	3	2.5
C^{12} (linear)	23.5	59.8	66	35.3	3.5
O^{16} (tetrahedral)	117.0	117.0	0	0	0
O^{16} (square)	64	85.0	77	21	3.0
O^{16} (square)	67	88.5	58	21	3.6

The calculation of <E> as a function of $\omega_x\omega_y\omega_z$ with the potential (17) (M = 0.65) gives the following results:

1) For the configuration with minimum energy $\omega_y = \omega_z$, i.e., as was expected, the x axis is the symmetry axis of the nucleus.

2) The binding energy increases by approximately 4.5 MeV on passing from the best spherical configuration ($\hbar\omega_x = \hbar\omega_y = \hbar\omega_z \approx 16$ MeV) to the best deformed configuration ($\hbar\omega_x \approx 18$ MeV, $\hbar\omega_y = \hbar\omega_z \approx 14$ MeV). The volume of the nucleus remains approximately the same.

Similar calculations were made for determining the excitation energy of the state of the O^{16} nucleus with four particles and four holes with the configuration $(000)^4$ $(100)^4$ $(010)^4$ $(200)^4$. It was assumed that the ground state had the configuration $(000)^4$ $(100)^4$ $(010)^4$ $(001)^4$, i.e., the four nucleons were excited from state (001) to state (200). The results of the calculations are given in Table 2.

Towles and Hayward carried out detailed calculations for the same 4p–4h states of the nucleus O^{16}, using a wide range of interactions. They found that the deformation energy was very sensitive to the choice of such a mixture of potentials.

It was found that forces causing considerable repulsion in states with odd momentum, for example Rosenfeld or Gillet forces, gave a higher deformation energy, and did not take into account the 4p–4h state, since apparently it corresponds to the 6.06 MeV 0^+ state.

In this section, examples are given of the binding energy and the energy of equilibrium deformations of α-cluster configurations, and a comparison is made with calculations, using wave functions of the deformed ellipsoid. The radius of the harmonic oscillator $b = (\hbar/M\omega)^{1/2}$, determining the volume of the α clusters, was always taken to be equal 1.61 fermi ($\hbar\omega = 16$ MeV). In each example, a definite geometric arrangement of the α clusters was selected and the binding energy was calculated as a function of the distance between them. When the distance between the α clusters became small (~5 fermi), the wave function approximated to the wave function of the shell model for the spherical harmonic oscillator with $\hbar\omega = 16$ MeV.

The results of the calculations are given in Table 3.

In the triangular configuration of the nucleus C^{12}, the α clusters are arranged at the corners of an equilateral triangle with side d. For d = 0, the wave function of the clusters agrees with the wave function of the ellipsoid in the limit of zero deformation. The increase in the binding energy due to deformation, is actually greater for the ellipsoidal wave function than for the α-cluster one. However, the cluster wave function is not an eigenvector of the parity vector and ought to be projected on the state with the necessary value of the latter, rather than to be compared. In any event, however the wave functions of the oblate ellipsoid and of the clusters considerably overlap.

Interest is afforded by the linear configuration of the nucleus C^{12} (three equidistant clusters arranged along a straight line), since the separation of the nucleus into three α clusters in this case results in a high energy gain, as the result of which the equilibrium energy is 7-8 MeV more than the energy of the ground state of the triangular configuration. Evidently, this state corresponds to the well-known 0^+-state in the 7 MeV region. In the shell limit, the excited 4p−4h-state corresponds to linear configuration.

In considering the tetrahedral configuration of the nucleus O^{16}, it is found that the state with least energy is obtained in the limit $d=0$, when the α clusters overlap considerably. In other words, for the interaction used, the spherical form of the ground state obtained in the shell model is stable. It should be remembered that this cluster configuration is not an eigenvector nor an angular momentum, nor parity, and therefore it must be projected, before asserting that the spherical state obtained in the shell has the least energy. We shall return to this question.

Coplanar configurations of four α-clusters lie fairly high above the ground state of O^{16}. The configuration having the diamond form evidently lies below the others. It is approximately 10 MeV lower than the ellipsoidal 4p−4h-configuration.

It is clear from Table 3 that the kinetic energy of deformation is very high in those cases where the α clusters are sufficiently far apart. This decrease in kinetic energy is due to the Pauli principle. If two α-clusters are separated by a distance which is large compared with 2b, all the nucleons in each cluster are in the s state. If clusters are highly overlapping then, due to the Pauli principle, four nucleons ought to pass to the 1p state, and the kinetic energy ought to be increased by $2\hbar\omega$.

If the nucleon−nucleon interaction used in the calculations has a large component of Majorana forces, the repulsion between the nucleons in a state with odd relative momentum will assist the separation of the α clusters. The decrease in kinetic energy on deformation also assists the separation of α clusters. The strong attraction acting between nucleons in states with even moment brings them closer together.

The component of the Majorana exchange forces was varied in the calculations for the tetrahedral cluster configuration of the nucleus O^{16}. If the binding constant for these forces increases to 0.75, the spherical form of the nucleus is no longer stable and the deformed tetrahedral α-cluster configuration becomes more favorable, even without projection in a state with a definite value of angular momentum and parity. This calculation illustrates the effect of repulsion between clusters due to the high value of the component of Majorana forces in the nucleon−nucleon interaction.

To conclude this section, we shall point to a possible failure in calculations using the Hartree-Fock method in the subspace of only single-particle states. Such calculations may reveal a local minimum, which is really a saddle point in a wider range of states. Thus, for example, the best coplanar configuration has the form of the diamond crystal lattice, but on removal of the limitation of coplanarity, the tetrahedral configuration is the most favorable.

Projection of Angular Momentum and Parity on Eigenstates

a. Parity

If $\Phi(\beta)$, the wave function of a cluster, depends on the vectors $\beta = R_T, \ldots, R_N$, the reflected wave function $P\Phi(\beta)$ is obtained by inversion of the coordinates of each nucleon in the state i

$$r_i \to -r_i \quad i = 1, \ldots, 4N. \tag{21}$$

Since the coordinates of a particle are always in the combination $(r_i - T_j)^2$, where R_j is the coordinate of the center of any cluster, the transformation (21) is equivalent to the inversion

$$R_j \to -R_j \quad j = 1, \ldots, N$$

of the coordinates of the centers of all the clusters, i.e.,

$$P\Phi(\beta) = \Phi(-\beta), \tag{22}$$

where $-\beta$ corresponds to $(-R_1, \ldots, -R_N)$. If the coordinates of the centers of the clusters are symmetrical with respect to inversion, $\beta = -\beta$, and the wave function $\Phi(\beta)$ is even.

When the cluster is not symmetrical with respect to inversion, we can separate the components of positive and negative parity Φ^+ and Φ^-

$$\left.\begin{aligned} \Phi^+ &= \frac{1}{2\alpha_+}(\Phi(\beta) + \Phi(-\beta)); \\ \Phi^- &= \frac{1}{2\alpha_-}(\Phi(\beta) - \Phi(-\beta)) \\ &\text{or} \\ \Phi(\beta) &= \alpha_+\Phi^+ + \alpha_-\Phi^-, \\ \Phi(-\beta) &= \alpha_+\Phi^+ - \alpha_-\Phi^-. \end{aligned}\right\} \tag{23}$$

If normalizing of Φ^+ and Φ^- is required, then

$$|\alpha_+|^2 + |\alpha_-|^2 = 1.$$

Actually, $|\alpha_+|^2$ and $|\alpha_-|^2$ are portions of states of positive and negative parity in $\Phi(\beta)$. We have also

$$|\alpha_+|^2 - |\alpha_-|^2 = \langle\Phi(\beta), \Phi(-\beta)\rangle. \tag{24}$$

Denoting $\langle\Phi^+, H\Phi^+\rangle$ and $\langle\Phi^-, H\Phi^-\rangle$ respectively by E^+ and E^-, we get

$$\langle\Phi(\beta), H\Phi(\beta)\rangle = |\alpha_+|^2 E^+ + |\alpha_-|^2 E^-, \tag{25}$$

if H does not violate parity. Table 4 gives some values of these quantities calculated for the tetrahedral configuration of O^{16} and interaction (17) with M = 0.65 for different values of the distance a between the clusters. The parameter of the cluster radius f = 1.61 fermi ($\hbar\omega = 16$ MeV), and the binding energy for the spherical shell model (i.e., limit at d → 0) is 117.6 MeV. The projections of the energy of the 0^+ and 3^- states are also given.

TABLE 4. Dependence of Some Characteristics of the Nucleus O^{16} on the Distance Between Clusters

	Distance between clusters, Fermi			
	0	1	2	3
$\lvert\alpha_+\rvert^2$	1	0.78	0.55	0.50
$\lvert\alpha_-\rvert^2$	0	0.22	0.45	0.50
$\langle\Phi(\beta), H\Phi(\beta)\rangle$, MeV	−117.6	−115.5	−109.6	−99.9
E^+, MeV	−117.6	−118.6	−112.4	−100.1
E^-, MeV	−104	−104.4	−105.9	−99.4
E_0^+, MeV	−117.6	−119.7	−123.1	−119.1
E_3^-, MeV	−104	−104.4	−108.0	−104.9

b. Projection of Angular Momenta

We separate the eigenvalues of the angular momenta using the projection operator

$$P^J_{KM} = \int d\Omega\, D^{+J}_{KM}(\Omega)\, R(\Omega), \tag{26}$$

$$\Phi^J_{KM}(\beta) = P^J_{KM}\,\Phi(\beta), \tag{27}$$

where $R(\Omega)$ is the unitary operator of rotation which rotates the vector of the wave function through the Euler angles Ω. Two successive rotations correspond to the combined rotation $\Omega_2\Omega_1$

$$R(\Omega_1)\, R(\Omega_2) = R(\Omega_2\Omega_1). \tag{28}$$

The unitary rotation matrix is determined as

$$\begin{aligned} D^J_{KM}(\Omega) &= \langle JK\,|\,R(\Omega)\,|\,JM\rangle, \\ D^{+J}_{KM}(\Omega) &= \left(D^J_{MK}(\Omega)\right)^* = D^J_{KM}(\Omega^{-1}). \end{aligned} \tag{29}$$

The operator relationship (28) gives the multiplication rule of the rotation matrices

$$\begin{aligned} D^J_{KM}(\Omega_2\Omega_1) &= D^J_{KZ}(\Omega_1)\, D^J_{ZM}(\Omega_2), \\ D^{+J}_{KM}(\Omega_2\Omega_1) &= D^{+J}_{KZ}(\Omega_2)\, D^{+J}_{ZM}(\Omega_1). \end{aligned} \tag{30}$$

These equations summarize some group properties of the rotation matrices.

Using these properties of the rotation matrices, it is possible to prove the following relationships for the projection operators (26):

$$R(\Omega)\, P^J_{KM} = \sum_{M'} P^J_{KM'}\, D^J_{M'M}(\Omega), \tag{31}$$

$$P^J_{KM}\, R(\Omega) = \sum_{K'} D^J_{KK'}(\Omega)\, P^J_{K'M}, \tag{32}$$

$$(P^J_{KM})^+\, P^{J'}_{K'M'} = \left(\frac{8\pi^2}{2J+1}\right) \delta_{JJ'}\,\delta_{MM'}\, P^J_{K'K}. \tag{33}$$

For example:

$$\begin{aligned} R(\Omega)\, P^J_{KM} &= \int d\Omega D^{+J}_{KM}(\Omega')\, R(\Omega'\Omega) = \int d\Omega'' D^{+J}_{KM}(\Omega''\Omega^{-1})\, R(\Omega'') = \\ &= \sum_{M'} P^J_{KM'}\, D^{+J}_{M'M}(\Omega^{-1}) \quad \text{from equation (30)} \\ &= \sum_{M'} P^J_{KM'}\, D^J_{M'M}(\Omega) \quad \text{from equation (29).} \end{aligned}$$

The proof of relationship (33) also requires the condition of orthogonality for the rotation matrices with respect to integration over the Euler angles.

For calculating the energies of cluster states, the matrix elements must be calculated with projected cluster wave functions. If we introduce the matrices

$$H^J_{KK'} = \langle \Phi^J_{Kn}\,|\,H\,|\,\Phi^J_{K'n}\rangle, \quad B^J_{KK'} = \langle \Phi^J_{KM},\, \Phi^J_{K'M}\rangle, \tag{34}$$

the average value of the energy in the (JKM) state is

$$E^J_K = H^J_{KK}/B^J_{KK}. \tag{35}$$

Φ^J_{KM} states are not orthogonal with respect to K and if there are several values for a given value of J, it may be necessary to replace the average values of the energy in expression (35) by the eigenvalues of the system of equations

$$\sum_K (H^J_{KK'} - EB^J_{KK'})\, G_{K'} = 0. \tag{36}$$

Writing the projection operator P^J_{KM} in explicit form in the matrix elements (35) we get

$$H^J_{KK'} = <\Phi|(P^J_{KM})^+ HP^J_{K'M}|\Phi> = <\Phi|HP_{K'K}|\Phi> \Big| \left(\frac{8\pi^2}{2J+1}\right) = \frac{8\pi^2}{2J+1}\int d\Omega D^{+J}_{K'K}(\Omega) <\Phi|HR(\Omega)|\Phi>. \tag{37}$$

A similar expression is also obtained for $B^J_{K'K}$. The projection method used enables the overlap matrix elements to be calculated numerically

$$<\Phi|HR(\Omega)|\Phi> \text{ and } <\Phi|R(\Omega)|\Phi> \tag{38}$$

for given values of Ω, and $H^J_{KK'}$ and $B^J_{KK'}$ to be estimated by means of numerical integration. The cluster states used are axially unsymmetrical and the internal states $\Phi(\beta)$ have no definite values of K. Consequently, integration in formula (37) should be over the three variables α, β, γ. To avoid this difficulty, we separate the linear combinations of internal states with an approximate axis and definite values of K. Integration in equation (37) is then reduced to the form:

$$H_{KK'} = \frac{8\pi^2}{2J+1}\int \sin\beta d\beta d^J_{KK'}(\beta)\, H_{KK'}(\beta), \tag{39}$$

where

$$H_{KK'}(\beta) = <\Phi_{K'}|HR(\beta)|\Phi_K>.$$

In practice, it is found that the overlap matrix elements are similar to Gaussian functions

$$H^J_{KK'}(\beta) \cong Ae^{-\alpha_i\beta^2} \text{ and } B^J_{KK'} \cong e^{-\alpha_i\beta^2}. \tag{40}$$

Even for C^{12} these Gaussian functions are numerically correct with an accuracy of up to 5%, and much more accurate for deformed states of O^{16}. The energy and overlap matrices are equally well described by formulas (40).

Numerical calculations performed for the tetrahedral configuration of O^{16} show that after projection of the angular momenta and parity, the tetrahedral form is unstable for Volkov forces with a Majorana exchange component $M > 0.6$.

Method of Deformed Orbits

This approximation for the solution of the Hartree-Fock problem was proposed by Levinson et al. [3] and Dietrich et al. [4]. Since a general discussion of the Hartree-Fock problem is outside our intentions, it is assumed that we have the solution Φ_0^N of the Hartree-Fock problem for N particles in the spherically symmetrical case, and that we know the single-particle energies ε_i and corresponding wave functions $|j_i m_i>$ in Φ_0^N. We have a test wave function for the system $A = N + n$ nucleons in the form

$$\Phi = b_1^+ \ldots b_n^+ |\Phi_0^N>, \tag{41}$$

where the single-particle creation operators b_i^+ are linear combinations of creation operators for a set of states n_α of the shell model (excluding the core state)

$$b_i^+ = \sum_\alpha C_\alpha^i \eta_{|\alpha}^+. \tag{42}$$

Substituting the wave function Φ in the variational equation

$$\delta <\Phi, H\Phi> = 0, \tag{43}$$

we then get a system of equations for the coefficients C_α^i

$$\begin{gathered}\sum_{\alpha'} <\alpha|h|\alpha'> C_{\alpha'}^i = EC_\alpha^i,\\ <\alpha|h|\alpha'> = \varepsilon_\alpha \delta_{\alpha\alpha'} + \sum_x <\alpha x|V_a|\alpha' x>,\\ \sum_x <\alpha x|V_a|\alpha' x> = \sum_{\beta\beta' x} C_\beta^x C_{\beta'}^x <\alpha\beta|V_a|\alpha'\beta'>.\end{gathered} \tag{44}$$

Summation over x is taken over states outside the core Φ_0^N. In these equations the subscript ***a*** at V denotes the antisymmetrized matrix element

$$<\alpha\beta|V_a|\alpha'\beta> = <\alpha\beta|V|\alpha'\beta'> - <\alpha\beta|V|\beta'\alpha'.$$

The first equation of system (44) determines the single-particle energies E_i and the expansion coefficients C_α^i, from which we get the creation operators b_i^+ according to equation (42) for the self-consistent wave function Φ in expression (41). The Hamiltonian determining C_α^i depends on C_α^i according to the second and third equations. The group of equations (44) forms a self-consistent system, which may be solved by iterations.

In the self-consistent calculations of ε_j, the interaction energies of the particles, situated above the filled shells, with the combination of particles of the core ought to be determined from the two-particle interaction V. This is difficult to do, since the usual effective forces are not saturated or they contain spin-orbital interaction. In practice, ε_j is replaced by experimental single-particle energies.

The principal simplification is in limiting the expansion of the deformed orbits, i.e., the expansion of b_i^+ according to η_α^+ only to a few α states. In the majority of calculations for nuclei except O^{16}, expansion was limited to a combination of 2s, 1d states. We know that the internal wave function Φ does not necessarily posses the symmetry of an exact function of the given Hamiltonian but sometimes there are imposed on the class of variable functions additional limitations requiring invariance of the internal wave function with respect to certain transformations, such as parity transformation, rotation about certain axes or invariance in time reversal.

The model described was extended by Ripka [6] so as to include particle-hole excitations.

In this lecture, I wanted to discuss the results of a paper by Bassichis, Giraud, and Ripka [6] in which they used the method of deformed orbits. They employed the formula

$$E_J = <\Phi|HP_{KK}^J|\Phi>/<\Phi|P_{KK}^J|\Phi>,$$

where Φ is the solution of the Hartree-Fock equation (44) and P_{KK}^J is the projection operator of angular momentum for calculating the energies of projected states of rotational bands of F^{19}, relating to 3p* and (4p* − 1h*) states of the Hartree-Fock problem, and bands of Ne^{20} of the corresponding 4p* state†; (there are also calculations for O^{16} and F^{18}). The internal wave

† Here p* and h* denote particle and hole, respectively.

TABLE 5. Comparison of Theoretical Data [6] for E_J(MeV) with Experimental Data

Element	State		Experiment	Theory
—		1/2+	—23.7	—24.7
—	3p*	3/2+	—22.2	—22.6
—		5/2+	—23.5	—24.0
F^{19}		9/2+	—20.9	—22.0
—		1/2−	—23.1	—23.7
—	4p*—1h*	3/2−	—22.3	—22.2
—		5/2−	—22.4	—22.2
—		9/2−	—19.8	—21.0
—		0+	—40.1	—42.1
—	4p*	2+	—38.5	—40.9
Ne^{20}		4+	—35.9	—38.2
—		6+	—(32.5)	—34.4

function has definite parity and is axially symmetrical. Central interaction in the Rosenfeld mixture was used, the interaction force being the parameter. Some results of Bassichis et al. [6] are given in Table 5.

Ripka and Bassichis also investigated the excited states of O^{16}. They found a solution for the deformed orbits of the (4p* − 4h*) and 2p* − 2h* states with the same interaction as that used for F^{19} and Ne^{20}. They found that the 0^+ projected state fell approximately to 5 MeV and the 4p* − 4h* state was much higher at 18 MeV.

These calculations of the energy of states give good results. There are distinct achievements in the fit of the energies of excited states of O^{16} of positive parity, and of states of F^{19} and Ne^{20} using only one parameter.

At the same time, the method of deformed orbits uses fairly considerable approximations. One of the most important approximations is that the deformed orbits are linear combinations of spherical 1p and 2s-1d states. We have seen that allowing for deformation of the core by means of α-cluster ellipsoidal wave functions gives a considerable energy gain.

In conclusion, I should like to thank the Joint Institute for Nuclear Research for the invitation to lecture at this school, and the local organizers and participants for their cordial hospitality. I should also like to thank Professor Wildermuth and V. G. Neudachin for discussions stimulating my interest in α-clustering in light nuclei.

LITERATURE CITED

1. Löwdin, P. O., Phys. Rev., 97:1490 (1955).
2. Volkov, A. B., Phys. Rev. Lett., 12:118 (1962).
3. Kelson, E., and Levinson, C. A., Phys. Rev., 134:B269 (1964).
4. Dietrich, K., Mang, H. J., and Pradal, J., EUCRL, report 11213, p. 40 (1964).
5. Bassichis, W. H., and Ripka, G., Phys. Rev. Lett., 15:320 (1965).
6. Bassichis, W. H., Ripka, G., and Giraud, B., to be published.

GROUP THEORY METHODS IN NUCLEAR PHYSICS

G. Flach

German Democratic Republic

Introduction

Modern theoretical physics makes use of group theory methods for solving various problems. It should be borne in mind that the principle of their application resides in the possibility of describing different kinds of symmetry of physical problems by means of the group theory. It is known that after the appearance of quantum mechanics and the relativity theory, the purely algebraic approach to symmetry problems made it possible to realize and describe more fully the properties of some physical systems.

The application of algebraic methods in describing physical systems leads to the geometrization of the system and thus to a better understanding of the physical situation. Algebraic methods have undoubtedly contributed much in both the theoretical and practical sense.

The application of algebraic methods in nuclear theory is limited to the use of the shell model. There is quite a strong reason for this. The shell model employs as a basic system functions which are found to be quite degenerate even in the case of considerable uncertainty in nucleon interaction. The practical success of these methods encourages a more detailed study of these questions. The problems of the shell model are the following:

1. The basic system of functions employed represents the product of single-particle functions in the selected average potential. The multiplicity of degeneracy of single-particle functions is $4(2l + 1)$ for the case without spin-orbital interaction, and $2(2j+1)$ in the opposite case. The multiplicity of degeneracy for the configuration l^k is equal to $[4(2l + 1)]^k$. Since in the construction of the wave functions and the participation of this one configuration alone, it is necessary to take into account the symmetry of nucleon interaction and the Pauli principle, the necessity arises of passing to a vectorially connected basic system with the quantum numbers L, S and T. In doing this, degeneracy is partly eliminated, but the residual degeneracy is still so great that apart from some simple cases, further solution of the problem without symmetry assumptions is practically impossible.

2. It is essential to consider different variants of the symmetry of the interaction and to use symmetry types in the classification of the states.

3. After selection of a definite type of classification, it is necessary to determine the scheme of construction of the corresponding wave functions. After all this, the various matrix elements may be calculated and the data of the model compared with reality, etc. We shall take the following case as example. Let the interaction between nucleons be central

$$\left.\begin{aligned} H_{int}^{(ij)} &= V_c(r_{ij})\{W + MP_{ij}^{(r)} + BP_{ij}^{(\sigma)} + HP_{ij}^{(\tau)}\},\\ H_{int} &= \sum_{i<j} H_{int}^{(ij)}. \end{aligned}\right\} \quad (1)$$

It is not difficult to show that such interaction commutes with operators of orbital, spin and isospin moments individually. This means that H_{int} commutes with all rotations in orbital, spin and isospin spaces, i.e., the symmetry group of this interaction is the direct product of three rotational groups in the corresponding spaces

$$G_{int} = O_3^{(r)} \times SU_2^{(\sigma)} \times SU_2^{(\tau)}. \quad (2)$$

(Instead of the rotational group in spin and isospin spaces, the corresponding homomorphic group SU_2 is used, since the basic spin and isospin functions are spinors with respect to the rotational group.) The symmetry group (2) suggests that the initial basic system ought to have the quantum numbers L, S, T (total orbital moment, spin, and isospin). We now assume that we may put in expression (1)

$$B = H = 0.$$

As before, G_{int} is the symmetry group, but now there is a new group, which contains the group G_{int} as subgroup. The new symmetry group will now have the form

$$G'_{int} = O_3^{(r)} \times SU_4. \quad (3)$$

Therefore, from the start, we classify the states of the problem according to the symmetry G'_{int}, and then by means of reduction according to the subgroup G_{int}, we obtain the splitting of the states in the given supermultiplet. After this, the corresponding wave functions must be constructed. The example shows the kind of problem which we shall consider in the following.

Some Information on the Group Theory

A group is a set of elements (any elements, for example a set of linear operators) having the following algebraic properties.

a. With each ordered pair of elements a and b from a set g there is uniquely associated a third element from g. The operation of association is called multiplication.

$$ab = c \quad a, b, c \in g.$$

b. The multiplication operation satisfies the associative law of multiplication

$$(ab)c = a(bc).$$

c. In the set g there is an element e with the property $ae = ea$ for any $a \in g$. The element e is called the unit element.

d. For any element $a \in g$, an element $a^{-1} \in g$ is found, such that

$$aa^{-1} = a^{-1}a = e.$$

The element a^{-1} is called the reciprocal element. We shall give some examples of sets which are groups. Let us take all the rotations of three-dimensional euclidean space which retain the norm of any vector. Each rotation is given by a definite set of Euler angles and is represented by a three-dimensional matrix. It is not difficult to show that a set of all these rotations forms a group having an infinite number of elements. This group is called the three-dimensional rotational group or the orthogonal group in three dimensions O_3.

This group is the prototype of a large class of so-called continuous groups. The number of elements of these groups is infinite, but each element is determined by the assignment of a certain number of parameters, i.e., it is a function of these parameters. If the parameters pass independently of each other through some continuous range of values, the functions (i.e., the elements) pass through the entire group. It is clear that the parameters of a three-dimensional rotational group are Euler angles.

Another large class consists of what are called finite groups. As the name already implies, the number of elements is here finite. The number of elements is called the order of the group. A permutation group may be cited as an example. We know that altogether there are k! permutations of k elements. These permutations may be multiplied together, resulting in further permutations, and so forth. These groups are called symmetrical groups. They play a large part particularly in nuclear theory, since they are connected with questions of antisymmetrization.

In what follows, we shall consider briefly the approach to some types of continuous groups. Let us assume that we have a continuous group, the elements of which are given by the parameters

$$R(a) = R(a_1, a_2, \ldots, a_r) \in g.$$

If the parameters pass through certain ranges of values, the functions R(a) pass through the entire group. The composition law of the group has the form

$$\left.\begin{aligned} &R(c) = R(a)R(b), \\ &c_k = \varphi_k(a_1, \ldots, a_r;\ b_1, \ldots, b_r), \end{aligned}\right\} \tag{4}$$

where k = 1, 2, ..., r.

For our type of continuous groups, the functions φ_k should be analytical functions of the parameters a and b. In such a case, it is said that the group is a Lie group with r parameters.

We shall consider the Lie group of r parameters in n-dimensional vector space:

$$x_i' = f_i(x_1, \ldots, x_n;\ a_1, \ldots, a_r), \tag{5}$$

then evidently

$$x_i'' = f_i(x_1', \ldots, x_n';\ b_1, \ldots, b_r)$$

or

$$x_i'' = f_i(x_1, \ldots, x_n;\ \varphi_1(ab), \ldots, \varphi_r(ab)).$$

We shall now consider what are called infinitesimal transformations

$$\left.\begin{aligned} &x_i' = f_i(x_1, \ldots, x_n;\ a_1, \ldots, a_r), \\ &x_i' + dx_i' = f_i(x_1', \ldots, x_n';\ \delta a_1, \ldots, \delta a_r) \end{aligned}\right\} \tag{6}$$

We have

$$dx_i' = \sum_{k=1}^{r}\left[\frac{\partial f_i(x_1', \ldots, x_n';\ a_1, \ldots, a_r)}{\partial a_k}\right]_{a=0}\delta a_k = \sum_{k=1}^{r} u_{ik}(x')\,\delta a(k).$$

The variations of a certain function by the action of the infinitesimal transformation (6) are now represented in the form

$$dF = \sum_{i=1}^{r} \frac{\partial F}{\partial x_i} dx_i = \sum_{i=1}^{n} \frac{\partial F}{\partial x_i} \sum_{k=1}^{r} u_{ik}(x)\,\delta a_k = \sum_{l=1}^{r} \delta a_l X_l F.$$

The operators

$$X_l = \sum_{i=1}^{n} u_{il}(x) \frac{\partial}{\partial x_i} \tag{7}$$

are called the infinitesimal operators of the Lie group. It is evident that the transformation

$$1 + \sum_{l=1}^{r} X_l \delta a_l$$

represents an infinitely small transformation close to unit element of the group, so that by selecting $F = x_i$ we get

$$x_i' = \left[1 + \sum_{l=1}^{r} X_l \delta a_l\right] x_i = x_i + \sum_{l=1}^{r} u_{il}(x)\,\delta a_l.$$

As example we shall now consider some Lie groups which are important when applied to questions of the shell model. Since the basic functions of the shell model are single-particle functions having definite angular momenta, we shall consider the $(2l+1)$-dimensional space of the functions of the angular momentum l. Instead of the coordinates x_i, it is now necessary to take the functions $\varphi_m^{(l)}$. We investigate unitary transformation in this space:

$$\varphi_{m'}^{(l)} = \sum_m u_{m'm}\, \varphi_m^{(l)},$$

$$U^+U = E.$$

All kinds of unitary transformations represent the unitary group U_{2l+1}. It is not difficult to see that for an infinitesimal transformation, we must have

$$(E + A^+)(E + A) = E,$$

$$A^+ + A = 0,$$

whence we obtain the infinitesimal operators

$$E_{mm'}^l = \varphi_m^{(l)} \frac{\partial}{\partial \varphi_{m'}^{(l)}}. \tag{8}$$

There are $(2l+1)^2$ infinitesimal operators in all. For the matrix elements, we get

$$\langle l'm' | E_{m_1 m_1}^l | l'm'' \rangle = \delta_{l'l}\, \delta_{m_1 m''} \delta_{m_1 m'}.$$

It is possible to pass to the new infinitesimal operators

$$U_q^k = \sum_{m,\,m'} \langle lmlm' | kq \rangle E_{mm'}^l, \qquad k = 0, 1 \ldots, 2l \quad q = -k, \ldots, +k. \tag{9}$$

Then

$$<lm_1|U_q^k|lm_2> = <lm_1lm_2|kq>$$

and

$$<l\|U^k\|l> = 1/\sqrt{2l+1}.$$

Infinitesimal operators in this representation are called unit tensor Racah operators.

We shall now consider a subgroup of the unitary group U_{2l+1}. Evidently, a group of all unitary unimodular transformations with det u = 1 will be a subgroup of this group. Its infinitesimal operators will be the same u_q^k's with the exception of k = 0.

The following subgroup will be the orthogonal group O_{2l+1}, its infinitesimal operators are u_q^k with odd k's. Finally, we have the subgroup O_3 with the infinitesimal operators u_q^1 (the operators of angular momentum in the spherical basis).

We shall now briefly examine the spaces of single-particle functions with half-integral spin j (j is a half integer). The corresponding unitary groups will now be U_{2j+1}. As before, the infinitesimal operators will again be the same u_q^k's. The combination of u_q^k's with odd k's again describes the subgroup. However, it is no longer an orthogonal group, but what is called a simplectic group Sp_{2j+1}. In the transformations of this group, there remains the invariant bilinear form

$$\sum_m (-1)^{j-m}\varphi_m^{(j)}\varphi_{-m}^{(j)} \quad j \text{ — half-integer.}$$

This form is skew symmetric, in contrast to the symmetric form

$$\sum_m (-1)^{l-m}\varphi_m^{(l)}\varphi_{-m}^{(l)} \quad l \text{ — integer,}$$

which remains invariant in orthogonal transformations. In conclusion, it should be noted that the commutators of the infinitesimal operators of a Lie group are always expressed by linear combinations of the same infinitesimal operators

$$[X_i, J_k] = \sum_{j=1}^{r} c_{ikj}X_j. \tag{10}$$

This means that the set of infinitesimal operators of each Lie group is closed with respect to commutation operations. A set of elements having such properties is called a Lie algebra.

We now pass to the conception which plays a fundamental part in the application of group theory to problems of physics, the conception of representation. Let us assume we are given a certain group G and a certain linear vector space of dimensionality N. If with each element $g \in G$ there is associated a linear operator in the space R_N so that

$$D(g_1)D(g_2) = D(g_1g_2) \quad g_1, g_2 \in G,$$

which means that the set of operators D(g), $g \in g$ forms a linear representation in the space R_N. Since R_N is finite-dimensional, it is always possible to give the representation of D(g) in the form of square matrices.

We say that the representation is reducible if by similarity transformation it is possible to reduce all the matrices to the form

$$D(g) = \begin{pmatrix} D_1(g) & 0 \\ 0 & D_2(g) \end{pmatrix}.$$

If this is impossible, the representation is called irreducible. The fundamental problem of the representation theory is as follows:

1) to find irreducible representations of a given group (if possible all of them);

2) to solve the problem of the expansion of reducible representation to irreducible representations. For this, there is an extensive mathematical apparatus. We shall confine ourselves to a discussion of the results for some groups of interest to us.

We commence with finite groups and take as example the symmetrical groups $\mathbf{S}_k$. The linear vector space is formed by the combination of the functions:

$$P\varphi(oc_1, \ldots, x_k), \quad p \in S_k.$$

If the initial function has no definite symmetry, we obtain altogether k! different functions, and the corresponding vector space has the dimensionality k! It is not difficult to induce in this space a linear representation of the $\mathbf{S}_k$ group. For this purpose, it is sufficient to act on the column

$$\begin{pmatrix} \vdots \\ P_i\varphi \\ \vdots \end{pmatrix}, \quad i = 1, 2, \ldots, k!$$

of any permutation $P \in S_k$. As result, we get a new column with another order of initial functions. The matrix converting the old order into the new one will be the representative of the permutation P.

It is not difficult to see that there is actually obtained a k! dimensional representation of the group $\mathbf{S}_k$. This representation is called regular; it is found that it contains all the irreducible representations of the group $\mathbf{S}_k$. We at once isolate two irreducible representations. For this we pass to a new basis in which the two functions clearly appear, namely

$$\left.\begin{aligned} \Phi_1 &= \frac{1}{\sqrt{k!}} \sum_p p\varphi, \\ \Phi_2 &= \frac{1}{\sqrt{k!}} \sum_p (-1)^p p\varphi, \end{aligned}\right\} \tag{11}$$

so that the new basis now has the appearance

$$\Psi = \begin{pmatrix} \Phi_1 \\ \Phi_2 \\ \vdots \\ \text{Remaining functions} \\ \vdots \end{pmatrix}$$

If we now perform any permutation in this column we get

$$P\Psi = \begin{pmatrix} 1 & 0 & 0 & \ldots & 0 \\ 0 & (-1)^p & 0 & \ldots & 0 \\ 0 & 0 & & & \\ \vdots & \vdots & & A & \\ 0 & \vdots & & & \\ & 0 & & & \end{pmatrix} \begin{pmatrix} \Phi_1 \\ \Phi_2 \\ \vdots \end{pmatrix}$$

We have now found two irreducible representations called the unitary and alternating representations; both have unit dimensionality. The basis functions of these representations have a definite symmetry (the first is symmetrical, the second is antisymmetrical). The basis functions of the other representations also have a definite symmetry although more complex than unitary representations. It is found that each irreducible representation may be characterized uniquely by the symmetry type of the corresponding basis functions. Each symmetry type is described by a Young scheme, a definite method of constructing the corresponding function being associated with each Young scheme. Different possible Young schemes are obtained by division of the number k into integral components

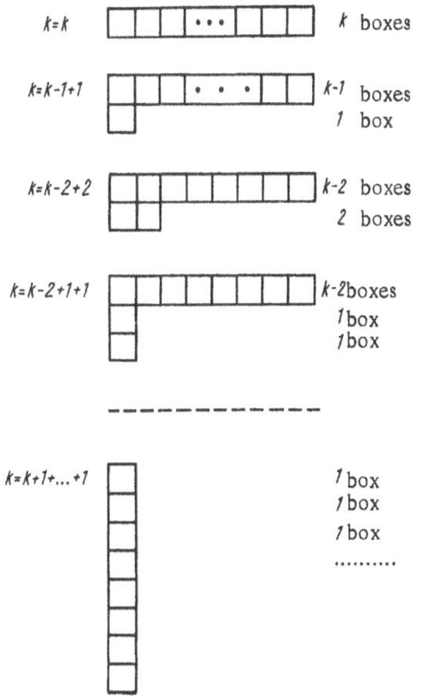

Thus, each irreducible representation of a group is given uniquely by the Young scheme

$$[\lambda] \equiv [\lambda_1, \lambda_2, \ldots, \lambda_k],$$
$$\lambda_1 \geqslant \lambda_2 \geqslant \ldots, \lambda_k; \sum_{i=1}^{k} \lambda_i = k. \tag{12}$$

The irreducible representations of the group $\mathbf{S}_k$ are exhausted by every possible Young scheme. For three particles, we have three possibilities:

$$[\lambda] = [3] \quad \text{symmetrical basis;} \qquad f_\lambda = 1,$$

$$\Phi_{[3]} = \frac{1}{\sqrt{3!}}\big(e + (12) + (13) + (23) + (123) + (132)\big)\, \varphi(x_1, x_2, x_3),$$

$$[\lambda] = [2, 1),$$

$$\Phi^{(1)}_{[2,1]} = \frac{1}{\sqrt{3}}\big(e + (12) - (13) - (132)\big)\varphi(x_1, x_2, x_3),$$

$$\Phi^{(2)}_{[2,1]} = \frac{1}{\sqrt{3}}\big((23) + (132) - (123) - (12)\big)\varphi(x_1, x_2, x_3),$$

$$\Psi^{(1)}_{[2,1]} = \frac{1}{\sqrt{3}}\big((23) + (123) - (132) - (13)\big)\varphi(x_1, x_2, x_3),$$

$$\Psi^{(2)}_{[2,1]} = \frac{1}{\sqrt{3}}\big(e + (13) - (12) - (123)\big)\varphi(x_1, x_2, x_3),$$

$$[\lambda] = [1^3] \quad \text{antisymmetrical basis;} \qquad f_\lambda = 1,$$

$$\Phi_{[1^3]}=\frac{1}{\sqrt{3!}}(e-(12)-(13)-(23)+(123)+(132))\ \varphi(x_1y_2y_3).$$

Similar results may be obtained for k > 3.

We now pass to the irreducible representations of the Lie group. Here, we can merely discuss some results without proof. We shall examine the space of function φ_m^1 for given l. Altogether, we have $2l+1$ basis functions. We now form the tensors of the k-th rank in this space. The basis tensors are evidently given by all possible products of the type

$$\varphi_{m_1}^{(l)}(1)\,\varphi_{m_2}^{(l)}(2)\ .\ .\ .\ \varphi_{m_k}^{(l)}(k),$$

where $m_1, m_2, \ldots, m_k$ pass independently of each other through the values from $-l$ to $+l$. Evidently, the dimensionality of this tensor space will be

$$(2l+1)^k.$$

If now in the space of the individual particles we perform the same unitary transformation, then in the tensor space the transformation takes place with a k-tuple direct product

$$[U(g)]^k \equiv U(g)\,xU(g)\,x\ .\ .\ .\ xU(g). \tag{13}$$

It then follows from the properties of the direct product that by means of the tensor space we have obtained the representation of the unitary group U_{2l+1}

$$g \in U_{2l+1} \Rightarrow [U_{2l+1}(g)]^k. \tag{14}$$

The dimensionality of this representation is $(2l+1)^k$. It is found to be reducible if $k>1$ and all the irreducible representations contained in it play an important part in the shell model. To understand better how this process of elimination of irreducible representations takes place, we shall examine the case of $k=2$. In this case, we have in all $(2l+1)^2$ basis tensors of the form

$$\varphi_{m_1}^{(l)}(1)\,\varphi_{m_2}^{(l)}(2).$$

We now form the tensors

$$\Phi_{m_1m_2}=\frac{1}{2}\{\varphi_{m_1}^{(l)}(1)\,\varphi_{m_2}^{(l)}(2)+\varphi_{m_1}^{(l)}(2)\,\varphi_{m_2}^{(l)}(1)\},$$

$$\Psi_{m_1m_2}=\frac{1}{2}\{\varphi_{m_1}^{(l)}(1)\,\varphi_{m_2}^{(l)}(2)-\varphi_{m_1}^{(l)}(2)\,\varphi_{m_2}^{(l)}(1)\},$$

where m_1 and m_2 pass through the entire range of possible values. The tensor Φ is symmetrical, the number of its independent components is $(l+1)(2l+1)$; Ψ is an antisymmetrical tensor, the number of its independent components is $l(2l+1)$. The Young scheme [2] may be ascribed to the tensor Φ and the scheme $[1^2]$ to the tensor Ψ. It is found that the tensors Φ and Ψ form a basis for two irreducible representations of the unitary group in tensor space of second rank. The irreducible representations of the unitary group may therefore also be characterized by a Young scheme.

This occurs in the general case. It is found that the irreducible representations of the unitary group U_n in tensor space are characterized by the Young schemes

$$[\lambda] \equiv [\lambda_1, \lambda_2, \ .\ .\ .\ , \lambda_n],$$

where

$$\lambda_1 \geqslant \lambda_2 \geqslant \ldots \geqslant \lambda_n; \quad \sum_{i=1}^{n} \lambda_i = k,$$

k being the rank of the tensor space. Consequently, the irreducible representations of the unitary group are determined by giving n numbers of λ_i.

Similarly, it is possible to consider the irreducible representations of orthogonal and simplectic groups in tensor space. It is found that for these groups, a knowledge of l or $(2j+1)/2$ numbers is sufficient for the complete determination of irreducible representations. Such results may be obtained by considering tensors having a definite symmetry character, and then eliminating from all the tensors of this class, subspaces having a definite number of traces (Spuren) equal to zero.

In the example of second rank tensors, this process appears as follows. We take the symmetrical part of the tensor, which is transformed in unitary transformations according to the representation [2]. We form its trace

$$\mathrm{Sp}\,\Phi = \sum_m (-1)^{l-m} \varphi_m^{(l)}(1)\, \varphi_{-m}^{(l)}(2). \tag{15}$$

The symmetrical tensor now has the trace

$$\Phi_{m_1 m_2} = \Phi_{m_1 m_2}^{(0)} - \frac{1}{2l+1} \delta_{m_1 m_2} \mathrm{Sp}\Phi.$$

The tensor $\Phi^{(0)}$ then has a trace equal to zero, and we separate any symmetrical tensor into a part with a trace equal to zero and a part with a metric tensor. It is not difficult to see that this separation is invariant with respect to orthogonal transformations. It is found that this separation already gives irreducible spaces. In the general case, the invariant subspaces are characterized by the type of symmetry and the number of invariant pairs (15). The considerations are similar for the simplectic group.

Classification of States in the Shell Model. Interaction Energy of One Shell

We shall examine the first fundamental problem formulated in the introduction. We assume that we have k particles in one shell (ls-coupling), i.e., the configuration

$$l^k.$$

The basis system then consists of all tensors of k-th rank in the vector space of single-particle functions

$$\varphi_m^{(l)} \chi_{m_\sigma}^{(\sigma)} \chi_{m_\tau}^{\tau}.$$

The dimensionality of the vector space is $4(2l+1)$. We assume in this space unitary transformations from the group $SU_{4(2l+1)}$. The tensors are then transformed according to a k-tuple direct product

$$[SU_{4(2l+1)}]^k.$$

The physically interesting part of these functions obviously consists of the class of antisymmetric tensors, which are transformed according to the irreducible representation

$$SU_{4(2l+1)}\,[1^k]. \tag{16}$$

We now select a system of classification of the wave functions such that this corresponds to reduction by subgroup

$$SU_{2l+1} \otimes SU_4 \tag{17}$$

in a space of individual particles. The irreducible representation (16) breaks down on reduction into subgroups (17)

$$SU_{+(2l+1),\,[1^k]} \to \sum_{\lambda} SU_{2l+1,\,[\lambda]} \otimes SU_{4,\,[\tilde{\lambda}]}\,.$$

At the same time, [λ] passes through all the Young schemes of k boxes with $n < 2l + 1$ rows; $[\tilde{\lambda}]$ denotes the Young scheme transposed to [λ]. The number of rows in $[\tilde{\lambda}]$ cannot be greater than four, therefore [λ] cannot have more than four columns. The subsequent types of reduction into subgroups are now clear. We shall examine the successive steps of the reduction:

$$SU_{2l+1} \to O_{2l+1} \to O_3,$$
$$SU_4 \to SU_2 \otimes SU_2.$$

As result, we obtain the new classification of states in the form

$$|\,(l^k)\,[1^k]\,[\lambda]\,(\sigma)\,LM_LSM_STM_T>. \tag{18}$$

Since, if Serber forces are used, the symmetry group for interaction has the form

$$O_3^{(r)} \otimes SU_4,$$

it may be stated that the energy of the system is diagonal according to [λ] and L, and does not depend on S and T (Wigner supermultiplet). In the general case, the energy will not be diagonal according to (σ). We put to ourselves the question: What form should the interaction have so that (σ) is also an exact quantum number? For this, it is sufficient if the interaction commutes with infinitesimal operators of the group O_{2l+1}. It is found that such an interaction may be constructed from infinitesimal operators of the group, called Casimir operators.

For the group O_{2l+1} such an operator has the form

$$G(O_{2l+1}) = \frac{1}{2} \sum_{i<j} \sum_{k=1}^{2l} [1-(-1)^k]\,(2k+1)\,u^k(i) \otimes u^k \cong \sum_{i<j} (P_{ij}^{(r)} - g_{ij}),$$

where $P_{ij}^{(r)}$ is the Majorana operator and g_{ij} is the Racah operator.

$$<l^2,\, LM\,|\,g_{ij}\,|\,l^2,\, LM> \approx \delta_{LO}$$

is the pairing operator.

If now the Wigner interaction is expanded according to multipole fields

$$V_W = \sum_{i<j} \sum_{\lambda\,\text{even}} <\lambda 0 l 0\,|\,l 0>^2 (2l+1)\,F^{\lambda} u(i) \otimes u^{\lambda}(j)$$

and

$$F^{\lambda} = \frac{2\lambda+1}{5} \cdot \frac{<2 0 l 0\,|\,l 0>^2}{<\lambda 0 l 0\,|\,l 0>^2}\,F^2$$

(the boundary case of short-range forces) is required for the Slater integrals, the expression for the interaction energy may be written by means of Casimir operators

$$<[\lambda](\sigma)L|V_W|[\lambda](\sigma)L> = \frac{k(k-1)}{2}F^0 + \frac{l(l+1)(2l+1)}{10(2l-1)(2l+3)}F^2\Big[\lambda_1(\lambda_1-1)+\lambda_2(\lambda_2-3)+\ldots+$$

$$+\frac{1}{2}(\nu_1(\nu_1+2l-1)+\ldots+\nu_l(\nu_l+1)) - \frac{k[k-(2l^2+l+1)]}{2l+1}\Big].$$

A similar expression may be written for the Majorana forces. The considerations for jj-coupling are similar. Of course, it must not be expected that any accurate results will be obtained from the spectra of light nuclei on these assumptions. However, it will be possible to detect the trend of experimental spectra. These discussions will therefore be regarded only as initial in character.

Construction of the Wave Functions

After classification, the next problem is the construction of the wave functions. The problem is solved in quite an elegant fashion by means of the group theory. We assume that we have found the wave function of a system of (k−1) particles on the l shell. These functions have the form

$$|(l^{k-1})[1^{k-1}][\lambda'](\sigma')L'S'T'>$$

(the projections of the moments have been omitted). These functions are transformed according to the irreducible representation

$$SU_{4(2l+1),[1^{k-1}]}.$$

The single-particle functions of the k'th particle are transformed according to the representation

$$SU_{4(2l+1),[1]}.$$

The product of two of these functions is then transformed according to the direct product

$$[1^{k-1}]\otimes[1] = [1^k] + \ldots.$$

This direct product is reducible in the general case. Among the other irreducible components, the representation $[l^k]$, which we require, appears once. The correct function of the k particles therefore ought to be represented by linear combinations of products of (k−1)-particle functions and single-particle functions.

$$|(l^k)[1^k].[\lambda](\sigma)LST> = \sum_{\lambda',\sigma',L',S',T'} <[1^{k-1}][\lambda'](\sigma')L'S'T',$$

$$lst|[1^k][\lambda](\sigma)LST>|(l^{k-1})[1^{k-1}][\lambda'](\sigma')L'S'T';\ lst;\ LST>.$$

The coefficients of this linear combination are called fractional parentage coefficients for the separation of one particle. These coefficients represent a certain generalization of Clebsch-Gordan coefficients, by means of which the direct product of two representations of the three-dimensional orthogonal group O_3 is reduced. The same is required here for the unitary group, a definite scheme of reduction according to subgroups being in addition necessary. We shall consider the more general problem: The construction of k-particle wave functions from the products of the k_1- and k_2-particle functions ($k=k_1+k_2$). This problem is solved by means of multiparticle parentage coefficients obtained from the reduction of the direct product. The calculation of parentage coefficients is a very difficult problem which is solved in the general form only by group-theory methods. The problem has been solved completely only for the p shell and for $j = 3/2$, $5/2$ shells in j–j coupling. Solutions are known in some other special cases (for l equal to 2, 3 or $j = 7/2$). Many physicists have recently given this problem serious attention. What gives us a knowledge of parentage coefficients?

It is obvious, knowing the parentage coefficients, that the matrix of the interaction energy may be written for various types of interaction, and this matrix may be diagonalized. We then calculate other values (probability of transitions, etc.), i.e, we obtain necessary information about the shell.

General Results of the Theory and New Ideas

We shall examine in detail what the theory has provided as applied to light nuclei.

We know that in most cases, spin-orbital interaction leads to a mixture of states with different principal oscillator quantum numbers. This leads to the fact that nuclei with $4 < A < 16$ ought to be mainly described by one configuration P^k, it being necessary to take into account the existence of both central interactions and spin-orbital interactions (intermediate coupling). On the whole, the results of the theory in this region are quite satisfactory. One circumstance should still be mentioned. If the formula for the energy of the p shell is rewritten, we get

$$\langle E\rangle_{[\lambda],L} = \frac{k(k-1)}{2} F^0 + \frac{3}{50} F^2 \left[\lambda_1(\lambda_1 - 1) + \ldots + \frac{1}{2} L(L+1) - \frac{k(k-4)}{3}\right],$$

i.e., for the given supermultiplet, the energy has the structure of a (cut-off) rotational band.

This result is distorted by high values of the spin-orbit forces. Some extreme case exists for nuclei of the $f_{7/2}$ shell. Obviously, the spin-orbit forces are so strong that the application of j–j coupling appears reasonable ($A \sim 50$). Seniority should be regarded as a good quantum number (especially for nuclei with Z or N equal to 28). This means that here the pairing forces have considerable significance. The Casimir operator for a symmetry group (it may be written in a way similar to the Casimir operator for the orthogonal group) serves as prototype for such forces.

It is clear, however, that in the general case, a more detailed description is possible only if the configuration interaction is taken into account, i.e., only under certain boundary conditions is it possible to expect that a configuration already gives sufficiently good results. This follows from the fact that in recent years a large number of collective effects have been found in light nuclei (nuclei from O^{16} to Mg^{24}). Consequently, the ground state is already correlated, i.e., there are admixtures of higher configurations. All the configurations of an oscillator shell must be taken into account. At the same time, it is already known that the second oscillator shell (i.e., the p shell) discloses a rotational band.

It is found that similar effects are possible also in other oscillator shells. To find them, we shall examine the properties of the harmonic oscillator.

We know that the operator of a Hamiltonian for the three-dimensional harmonic oscillator in the language of second quantization has the form

$$H = \hbar\omega \sum_{\alpha=1}^{3} a_\alpha^+ a_\alpha,$$

where

$$a_\alpha^+ = \sqrt{\frac{\varepsilon}{2h}}\, x_\alpha - \frac{i}{\sqrt{2\hbar\varepsilon}} P_\alpha, \quad \alpha = 1, 2, 3, \quad \varepsilon = \varkappa m$$

are the boson creation operators. From the mathematical point of view, H is the Hermitian bilinear form which does not alter in unitary transformations. Since the commutation ratios do not vary, the group U_3 or SU_3 is the symmetry group of the three-dimensional oscillator. This finds its expression in the form of the spectrum:

$$E_N = \hbar\omega\,(N + {}^3/_2),$$
$$l = N, \quad N-2, \ldots, 1 \quad \text{or} \quad 0.$$

The spectrum is degenerate not only according to m but also according to l. This symmetry results in the fact that simultaneously with the superseniority scheme, which uses the reduction

$$SU_{2l_1+1+2l_2+1+\ldots} \to 0_{2l_1+1+2l_2+1+\ldots} \to 0_3,$$

it is also possible to consider the scheme

$$SU_{2l_1+1+2l_2+1+\ldots} \to SU_3 \to 0_3.$$

If for l_1, l_2, ... we take all the values of the oscillator shell, rotational bands (cut-off) again appear in the spectrum. To understand this, let us examine the scheme of classification according to SU_3 for the example of the (s–d) shell (here N=2). The initial configuration has the form

$$(2s\ 1d)^k.$$

It is obvious that we now have six basic functions in the orbital space and therefore the space part of the space function is characterized by the Young scheme [λ], i.e., by continuous representations of the group SU_6. It is subsequently necessary to consider the reduction of these representations according to the subgroup SU_3. The irreducible representations of the group SU_3 are characterized by the two numbers (ν, μ). Actually, the irreducible representations of the group U_3 are characterized by a Young scheme with three rows

$$[\lambda_1, \lambda_2, \lambda_3].$$

If the group SU_3 is taken instead of the group U_3 the irreducible representations remain irreducible, but the representations

$$[\lambda_1, \lambda_2, \lambda_3] \text{ and } [\lambda_1 - \lambda_3, \lambda_2 - \lambda_3, 0]$$

become equivalent, i.e., equal except for the similarity transformation. It is therefore possible to obtain all the irreducible representations of the group SU_3 already if the two numbers $[\lambda_1, \lambda_2]$ are known. We then take as characteristic of the irreducible representations of the group SU_3 the numbers

$$\nu \equiv \lambda_1 - \lambda_2,$$
$$\mu = \lambda_2,$$

i.e.,

$$[\lambda_1, \lambda_2] \equiv (\nu, \mu).$$

With these remarks for the simplest cases we get the following sets of numbers:

k	$[\lambda]$	$(\nu\mu)$
1	$[1]$	$(2, 0)$
2	$[2]$	$(40)(02)$
	$[1^2]$	(21)

After carrying out this reduction we expand the irreducible representations of the group SU_3 according to the irreducible representations of the group U_3, i.e., we develop the L structure of the irreducible representations of the group SU_3. This process may be represented in a general form and the following results may be obtained:

$$(\nu, \mu) \to (D_\nu + D_{\nu-2} + \ldots) \otimes (D_\mu + D_{\mu-2} + \ldots) - (D_{\nu-1} + D_{\nu-3} + \ldots) \otimes (D_{\mu-1} + D_{\mu-3} + \ldots).$$

For the case of $\mu = 0$, for example, we get

$$(\nu, 0) \to (D_\nu + D_{\nu-2} + \ldots) \otimes D_0 = D_\nu + D_{\nu-2} + \ldots,$$

i.e., the values of $L = \nu, \nu-2, \ldots$ are contained in $(\nu, 0)$.

For $\mu = 1$, we have

$$(\nu, 1) \to (D_\nu + D_{\nu-2} + \ldots) \otimes (D_1) - (D_{\nu-1} + \ldots) \otimes$$
$$\otimes D_0 = D_{\nu+1} + D_\nu + D_{\nu-1} + \ldots + D_2 + D_1.$$

In the general case the L structure of the irreducible representation $(0, \mu)$ may be classified in the following way: They are rotational bands with

$$K = \mu, \mu - 2, \ldots, 1 \text{ and } 0$$

and each of them contains the values of $L = K, K+1, \ldots, K+\nu$. If $K = 0$, $L = \nu, \nu-2, \ldots, 1$ or 0. In this scheme of classification, the states are characterized by the following quantum numbers:

$$|(2sld)^k [1^k] [\lambda] (\nu, \mu) KLM_L SM_S TM_T\rangle.$$

So far all this has been very formal. We now ascertain what interaction forces are diagonal in this scheme. We find the Casimir operator for the group SU_3. For this purpose, we first of all write down the infinitesimal operators of the group U_3. This is done best of all in the form of spherical tensors (in doing this it is necessary to take into account the fact that the space in which U_3 acts is now a space of the basis operators a_α^+). In this representation, the nine infinitesimal operators have the form

$$H_0 = \frac{1}{2} b^2 p^2 + \frac{1}{2b^2} r^2,$$

$$l_x, l_y, b_z \quad (\text{or } l_+, l_- \text{ and } l_0),$$

$$Q_q^{(2)} = \sqrt{\frac{4\pi}{5}} \left\{ \frac{1}{b^2} r^2 Y_q^{(2)}(\theta_r, \varphi_r) + p^2 b^2 Y_q^{(2)}(\theta_p, \varphi_p) \right\}.$$

The operator H_0 is scalar and is the Hamiltonian of the oscillator. The remaining eight operators are infinitesimal operators of the group SU_3, the first three operators being angular momenta, while the last five form a spherical tensor of second rank.

We write the Casimir operator of the group SU_3 in the form

$$G(SU_3) = \frac{1}{4} \sum_{i<j} [3(l_i \otimes l_j) + (Q^{(2)}(i) \otimes Q^{(2)}(j))].$$

Its matrix elements in the scheme of SU_3 are of course diagonal and have the values

$$\langle(\nu\mu) L | G(SO_3) | (\nu\mu) L\rangle = (\nu^2 + \nu\mu + N^2) + 3(\nu + \mu).$$

Since, however, the operator $\sum_{i<j} l(i) \times l(j)$ is also diagonal in the given SU_3-multiplet, this means that the second operator also is diagonal, in which case

$$<(\nu\mu) L | Q \otimes Q | (\nu\mu) L> = 4(\nu^2 + \nu\mu + \nu^2) + 12(\nu + \mu) - 3L(L+1).$$

That is to say, not only is the operator $Q \otimes Q$ diagonal, but its eigenvalues depend on L according to the law L(L+1). This means that not only does classification according to the SU_3 scheme lead to classification of the levels according to rotational bands, but also the fact that two-particle interactions of the type $(Q \otimes Q)$ are taken into account gives a characteristic dependence on L. All this is easy to understand if the form of the operator Q is examined in greater detail. We write $Q_q^{(2)}$ (i) in the form

$$Q_q^{(2)}(i) = \sqrt{\frac{16\pi}{5}}\, r^2(i)\, Y_q^{(2)}(\theta_r(i), \varphi_r(i)).$$

This is possible since the momentum p and the coordinate of r are equivalent in the language of the harmonic oscillator. The two-particle interaction then has the form

$$Q^{(2)}(i) \otimes Q^{(2)}(j) = \frac{16\pi}{5} r^2(i) \cdot r^2(j) \left(Y^{(2)}(i) \otimes Y^{(2)}(j)\right) = 8r^2(i)\, r^2(j)\, p_2(\cos\theta_{ij}).$$

If the two-particle potential is expanded according to powers of the distance between particles, we get

$$V(r_{ij}) = a + br_{ij}^2 + cr_{ij}^4 + \ldots.$$

The constant a does not contribute to the splitting of the levels, the second term r_{ij}^2 is taken into account by the general oscillator field, the term r_{ij}^4, giving the splitting of the levels, has the form $Q \otimes Q$. We shall not discuss here in detail the problems connected with the application of this scheme to specific nuclei. Papers on these lines have lately appeared fairly frequently. Agreement between theoretical and experimental data is sometimes good and sometimes not. However, the following conclusion is important: The shell model permits the explanation of some general relations in the properties of light nuclei. It is therefore possible to understand why some light nuclei are deformed or why the seniorities in other cases are good quantum numbers. It is obvious, however, that the shell model is too rough, or at least the conditions by which we are limited for various considerations are too rough.

The main problem is to take into account certain features in the interaction between nucleons. Firstly, it is necessary to take into account spin-orbit interaction, and secondly the central interaction, in which both a force of type p_2 and evidently a pairing force play a part. The problem can be solved exactly for only one type, since taking the other type subsequently into account involves disturbance of the symmetry, and the problem is solved only by diagonalization of the energy matrix. The influence of higher configurations on low levels in light nuclei is in principle not yet clear. At least, in addition to the basic configuration, it is necessary to take into account the contribution of certain subsequent configurations, as is done in Elliott's scheme.

LITERATURE CITED

1. Hamermesh, M., Group Theory and Its Application to Physical Problems, Addison-Wesley, 1962.
2. Flach, G., and Reif, R., Gruppentheoretische Methoden im Schalenmodell der Kerne.
3. Elliott, J. P., The Nuclear Shell Model in Selected Topics in Nuclear Theory, IAEA, Vienna, 1963.

INVESTIGATION OF THE STRUCTURE OF THE NUCLEUS BY MEANS OF REACTIONS PRODUCED BY H^3 NUCLEI

O. Nathan

Denmark

Elementary Review of the Generalized Model. Empirical Tendencies in the Spectra of Heavy Nuclei

We first of all consider some important tendencies in nuclear spectra and then examine some fundamental points of the scheme with pair coupling, the ordered coupling scheme and the competition between these schemes.

We shall examine the doubly magic nucleus Pb^{208}. One particle or one hole moving outside the core of Pb^{208} may be described in terms of the simple shell-model, the motion of the particle in a static spherical well being considered. In these nuclei, there are also still more complex excited states at energies above ~2.14 MeV, due to the rupture or transition of pairs in the core of Pb^{208}. For the present, however, we shall disregard these states.

We consider the well-studied Pb^{207}. The values of the quantum numbers and the positions of the levels have been taken from data on β decay and stripping reactions, and have been found to be in very good agreement with the results of calculations of the motion of a particle in a potential of Saxon-Woods type. We can thus use these states with sufficient reliability for the construction of simple states in lighter isotopes of lead having two or more neutron holes. For this purpose, we make use of the limit shell model (LSM) without residual interactions.

We shall examine Pb^{206}; the lowest configuration will be $(p^{-1}{}_{1/2})^2$, corresponding to the 0^+ state. The next higher configuration $(p^{-1}{}_{1/2} f^{-1}{}_{5/2})$ gives 2^+ and 3^+ states. However, whereas for the 0^+ state all the particles move in pairs with a total moment $I = 0$, for the 2^+ or 3^+ states, the angular momentum of a pair of particles is not equal to zero, i.e. they have the seniority $\nu = 2$. As is well known, this classification (senority scheme) is often useful in nuclear spectroscopy.

The following conclusions may be drawn from a comparison of the predictions of the limit-shell model (LSM) for the spectum of Pb^{206} with data obtained mainly from the decay of Bi^{206}:

1. Close to the ground state, there are fewer states than follow from the LSM; the ground state is evidently lower by approximately 1 MeV than most of the other states of the spectrum.

2. The first excited state, which is the 2^+ state, has a greater probability of E2 transition to the ground state than this ought to be for a single-particle transition

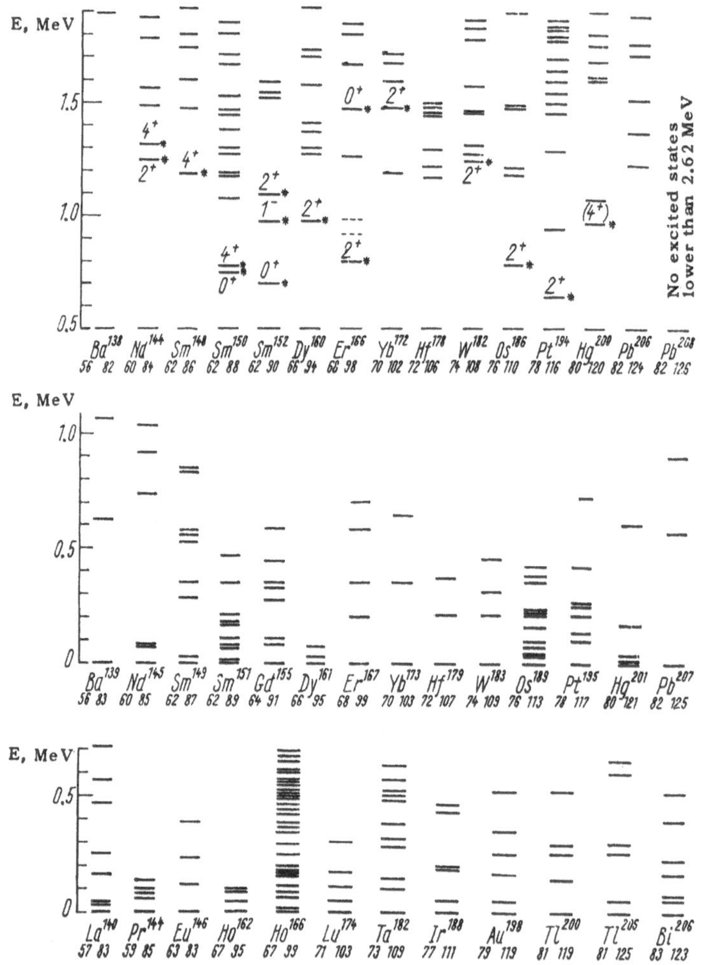

Fig. 1

$$(p^{-1}\,{}^{1}/_{2}\, f^{-1}\,{}^{5}/_{2}) \rightarrow (p^{-1}\,{}^{1}/_{2})^2.$$

It should be noted that many values of spins and parities, predicted by the simple model, are actually found in empirical spectra, although the energy agreement is not always exact. If we consider the lighter isotopes of lead, the above-mentioned features of the spectra become clear. The limit-shell model predicts a large number of states close to the ground state of Pb^{204} and Pb^{202}, but due to the lowering of the ground state all these states are clearly at energies higher than 1MeV. The above-mentioned effect is called the even-even energy gap. It ought also to be noted that the first 2^+ state is evidently of special significance; It is situated relatively low and has an increased value of reduced transition probability.

We shall examine the case for nuclei heavier than lead. Figure 1 shows some characteristic spectra of internal excitations in the region from Ba to Pb. The states of clearly collective character are low; in even systems an energy gap exists, while it is nonexistant in odd systems. Figure 2 gives the single-particle values of the B(E2)/B(E3) ratios, which are single-particle for $2^+ \rightarrow 0^+$ transitions. This ratio is always greater than unity and attains values of ~ 200 in

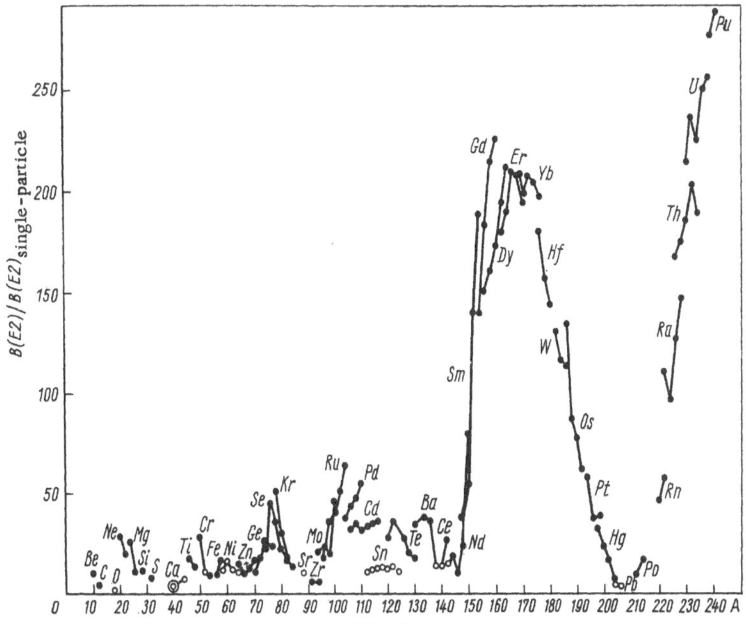

Fig. 2

some regions, in particular for $150 < A < 190$ and $A > 224$. In these regions a new feature of the energy spectra is observed, the lowest states of even-even nuclei have the spin sequence: 0^+, 2^+, 4^+... and the energy of these states ought to be quite close to the rule $I(I+1)$, i.e., we find band spectra, similar to the spectra of molecular rotation. At the same time, the energy of the 2^+ state is reduced to 40-100 keV.

To understand the energy gap and the collective 2^+-states, we introduce the residual interactions between nucleons, in addition to the potential of the shell model. For heavy nuclei, remote from closed shells, it is impossible to perform the calculations without introducing quite schematic forces. In practice, the division of the residual forces into short-range and long-range components is widely used. It is quite rational to assume that short-radius forces result in the occurrence of an energy gap in the spectra of even systems. In the ground state of an even system, the particles move predominantly in pairs. The states in which the particles forming a pair are situated are associated with time reversal. In these states, the particles are situated close to each other. The short-range forces are consequently very important in such a system and if the forces are attractive, this may lead to a definite correlation between the pairs of particles which, in its turn, leads to a lowering of the ground state.

At the same time, a considerable increase in the probability of a $2^+ \rightarrow 0^+$ transition expresses the coherence of many-particle motion which may be caused by forces of larger radius. When the effect of long-range forces becomes particularly important compared with the influence of the short-range component, the nuclear field may assume constant deformation at which rotational motion becomes possible.

Short-Range Forces. Scheme with Pair Coupling

Figure 3a shows the spectrum of a system of two identical particles in a state with definite j and with δ-form forces of attraction. The spectrum consists of a system of almost (but not wholly) degenerate states with values of spin I^π equal to 2^+, 4^+, 6^+, ... $s=2$, and also states with $I^\pi = 0^+$ and $s=0$, for which the interaction is particularly strong; an energy gap is present.

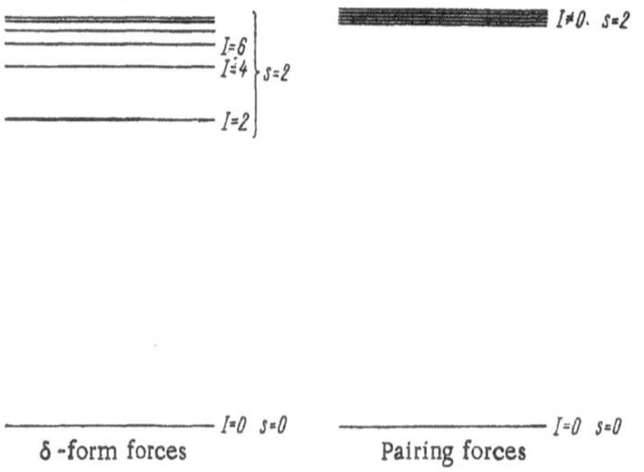

Fig. 3

The wave function of the 0^+ state contains all the states of (jm, j−m) pair with equal weight and definite phases.

It is often more convenient to work with other model spins, called pair spins. By definition, these spins correspond to isotropic scattering of nucleon pairs. Figure 3b shows the spectrum of a pair for this case. All states with seniority s = 2 are degenerate and interaction occurs only in the ground state. The wave function of the 0^+ state again contains all the states of pairs with equal weights, but the phase relationships are different from the δ-forces.

The pair forces are better taken into account in second quantization formalism, which we shall not consider. We shall examine the solution for the more complex case where there is more than one pair of particles, and states with several values of j may be realized, while pair forces act between the particles. For this case, there is a simple variational solution proposed by Bardeen, Cooper and Shrieffer (BCS). The corresponding wave function is shown in Fig. 4. The particles fill the levels only in pairs; the level ν is empty (with the probability u_ν^2) or filled (with the probability v_ν^2). Figure 4 shows the situation for different values of the force of pair interaction G, which are compared with the mean distance ρ^{-1} between the single-particle levels.

The wave function is characterized by two parameters: the position of the Fermi surface λ and the diffusivity of the Fermi surface Δ. The value of 2Δ is approximately equal to the energy gap. For the BCS solution, we have:

$$v_\nu^2 = \frac{1}{2}\left(1 - \frac{\varepsilon_\nu - \lambda}{E_\nu}\right),$$

$$E_\nu = \sqrt{(\varepsilon_\nu - \lambda)^2 + \Delta^2},$$

$$u_\nu^2 + v_\nu^2 = 1, \quad \Delta = G\sum_\nu v_\nu u_\nu,$$

where for given single-particle energy ε_ν and given number of particles N, the values of u_ν and v_ν depend only on the pair interaction G. For $G \ll \rho^{-1}$, the single-particle levels are filled to a definite level, depending on the number of particles. For $G \gg \rho^{-1}$, all the levels are filled uniformly. For $G \sim \rho^{-1}$, we obtain the BCS pair distribution. The use of the BCS solution leads to

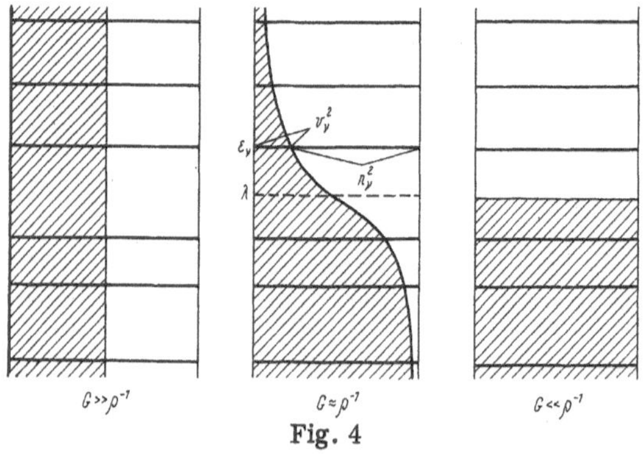

Fig. 4

nonconservation of the number of particles. On the average, the wave function contains the correct number of particles, but there are fluctuations which lead to definite complications, especially for excited states.

In discussing the excited states of such a system, it is convenient to pass to quasiparticles, using the canonical Bogolyubov-Valatin transformation. Compared with the states of the limit single-particle model (residual interaction is absent, the Fermi distribution is stepped), quasiparticle states have supernormalized properties (energy, quadrupole moment, etc.). At the same time, the quasiparticle, associated with only one single-particle state, has the same I^{π}'s as that state. The transition to quasiparticles results in a variation in the probability value of the filling of a given state. The quasiparticle thus has certain properties of a particle and a hole simultaneously. Very much above the Fermi surface, it is almost a particle, very much below, it is a hole.

The ground state of an even system corresponds to a quasiparticle vacuum. The energy of the single-quasiparticle state relative to a vacuum is E_{ν}; such states correspond to the ground state and excited states in the odd system. The energy difference of the two states in the odd system is equal to $E_{\nu'} - E_{\nu}$. In even–even systems, the lowest states are two-quasiparticle states, the energy of which relative to the ground state is $E_{\nu'} + E_{\nu} \geq 2\Delta$. Thus, there are no "internal" excitations which would be situated below the energy gap 2Δ for particles of the kind under consideration.

Long-Range Forces. Ordered Scheme of Coupling

Pair forces correspond to isotropic scattering of nucleon pairs. Long-range forces lead to predominantly forward scattering, that is to say, interaction rarely occurs between particles outside the plane of their orbit; this is indeed the characteristic of the field. In the multipole expansion of such fields, the single-pole term gives the spherical field of the shell model. The dipole term connects the particles of different shells and describes the shift of the center of mass of the nucleus.

Thus, we find that the first important term is not represented in the LSM, this is the term representing the quadrupole component of the field. If there are sufficiently many particles outside a filled shell, it may be shown that the quadrupole component of the forces (P_2 forces) results in the deformation of the nuclear field which may or may not have axial symmetry. For nuclei closer to doubly magic nuclei, the field produced by forces of quadrupole type is not strong enough to result in stable deformation. However, these forces weaken the resistance of the nucleus to deformation, and the result is that low-frequency fluctuations of the shape of the nucleus are produced. The low-lying 2^+-states in even–even nuclei may be regarded as the manifestation of these quadrupole fluctuations.

Competition Between Coupling Schemes

Pair and quadrupole coupling schemes compete with each other. For a small number of particles outside the filled shells, the pair forces predominate and are taken into consideration by transition to quasiparticles, and the quadrupole part of the Hamiltonian is expressed by quasiparticle operators. However, with increase in the number of particles outside the filled shells, the relative influence of the quadrupole forces increases. This is basically due to the fact that, with respect to long-radius forces, all the nucleons make a coherent contribution. Thus, if there are n nucleons in addition to a filled shell, the long-radius forces give an energy of interaction proportional to n^2; pair forces, however, only act between two nucleons and consequently their energy contribution is proportional to n. In fact, stable nuclear deformation is found empirically only for nuclei having shells which are approximately half filled (by protons or neutrons). For nuclei having a stable deformation, the principal effect of pair forces is the energy gap, the value of which is close to the value of the gap for spherical nuclei.

The single-particle states, over which the particles are distributed, are states having a deformed Nilsson potential. It has been shown that, on the whole, the formalism developed for considering pair correlations in the case of spherical nuclei may also be used for deformed nuclei.

The situation with coupling schemes is at present understood best of all in two limit regions: where the nuclear field is most stable in form, i.e., close to doubly closed shells, and where there is strong deformation. For nuclei situated between these two limit cases (and most of the nuclei studied actually come within this category), the situation is very complicated due to the coupling of single-particle motion with low-lying vibrations. Progress in the detailed understanding of the spectra of these nuclei is very slow.

Rotational Motion, Moments of Inertia

Phenomenological Treatment of Rotational Motion

In the preceding sections, some fundamental consequences of the introduction of short-range residual interactions were discussed. We shall consider the nuclear moments of inertia as example of the application of the model with pairing. First of all, however, I should like to enumerate some basic characteristics of the rotational motion of nuclei.

It is well known that in deformed even-even nuclei, the rotational bands of the ground states appear with surprising regularity. In deformed odd-mass nuclei at low energies, some rotational bonds are found, each of which is based on its internal (Nilsson) state. Consequently, the complete nuclear wave function may be represented in good approximation in the form of the product of the rotational part (D function), which describes the rotation of the system as a whole), and the internal wave function, which describes the motion of a particle in the deformed field. In other words, it is possible to use an adiabatic approximation. Nonadiabatic effects are considered separately as small perturbations. It should be noted that compliance with the rule $I(I+1)$ for the distance between the levels of the rotational bands is compatible with the assumption of the axial symmetry of the nucleus. Originally, Bohr and Mottelson developed the theory of rotational spectra within the framework of adiabaticity, symmetry with respect to reflection and axial symmetry. As is well known, this early model with strong coupling was successful in classifying scattered data on energy spectra, moments and probabilities of transitions in nuclei.

Moments of Inertia

Whereas the phenomenological model with strong coupling is able to describe quite satisfactorily many features of rotational spectra, the basic parameters of the collective motion cannot be calculated without considering the details of the internal motion of the individual nucleons. Thus, for example, the moment of inertia J is contained in the phenomenological model as a parameter which must be determined empirically by means of the known relationship

$$E_{\rm rot} = \frac{\hbar^2}{2J} I(I+1).$$

At first, the classical model of the rotational motion of a liquid drop was used. The corresponding moment of inertia, however, was much less than the observed value. A more realistic approximation is the cranking model, introduced by Inglis:

$$J_c = 2\hbar^2 \sum_i \frac{|\langle i | J_c | 0 \rangle|^2}{\varepsilon_i - \varepsilon_0}.$$

Here J_c is the component of the angular moment operator along the x axis, perdicular to the axis of symmetry. Summation is taken over i particles having energies ε_i. The formula for

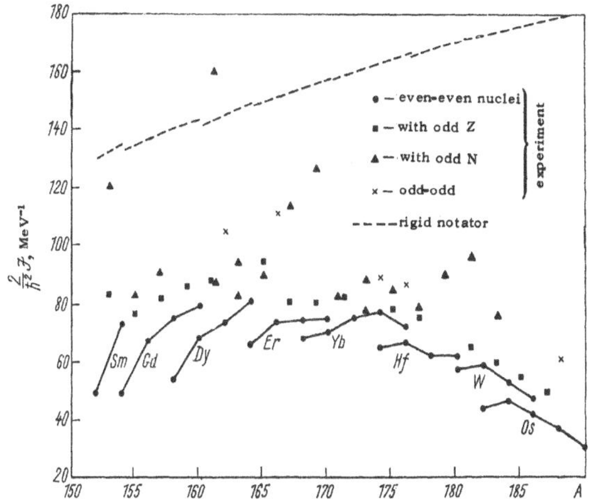

Fig. 5

the moment of inertia given above may be deduced, for example, by considering the motion of nucleons in a field which is subject to forced rotation. The additional energy which is necessary for the particles to follow the field adiabatically is calculated. By making assumptions concerning the nature of the potential and the interparticle forces, it is possible to use the formula under discussion as starting point for a microscopic description of the collective motion, i.e., a description in which the parameters of motion are examined in terms of single-particle motion.

In the case of the rotation of independent particles, however, the cranking model gives a moment of inertia approximately equal to the solid state moment $J_S \approx {}^2/_5\,AMR_0^2$, which is known to be much greater than the observed value. Figure 5 shows the results of calculations for even–even and odd–odd nuclei of the rare-earth region. This problem became more intelligible when pair interaction was taken into account. In the quasiparticle approximation two modifications appeared in the Inglis formula for the moment of inertia.

1. For even–even nuclei, the denominator $\varepsilon_1 - \varepsilon_0$ is replaced by the sum of the two quasiparticle energies $E_{\nu'} + E_{\nu}$, which is always greater than 2Δ.

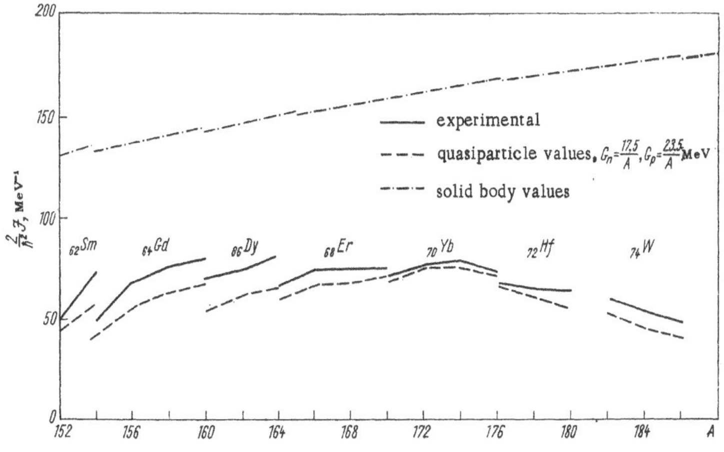

Fig. 6

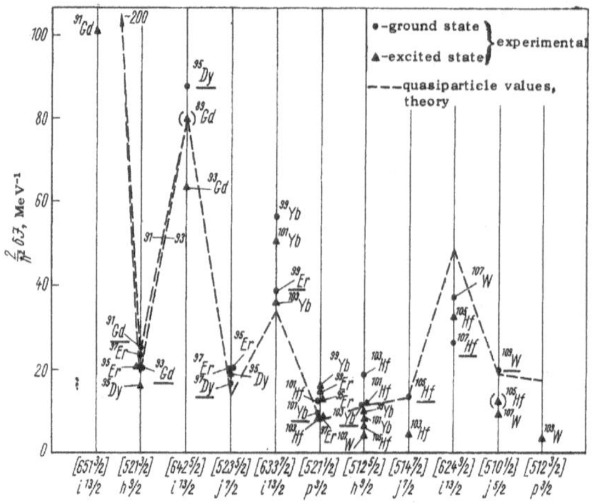

Fig. 7

2. The factor $(uv_{\nu'} - v_{\nu} u_{\nu'})^2$ appears in the numerator. Both modifications result in a decrease in the moment of inertia. The results of calculations for even–even rare-earth nuclei (carried out by Nilsson and Prior) are shown In Fig. 6. Agreement with experiment was now much better and even quite fine empirical tendencies are reproduced in detail. The calculations were based on the known internal states of the axially symmetrical system (Nilsson orbits, which are not discussed here) with deformation parameters selected from measured probabilities of $0^+ \rightarrow 2^+$ transitions, and with a force of pair interaction G consistent with the other properties of nuclei of this region (difference in mass between odd and even nuclei). There are no other free parameters in the calculation.

These results for the moments of inertia of inertia of even–even nuclei, although 5-20% below the experimental values, nevertheless provide convincing evidence of the truth of the basic assumptions of the model, ordered coupling scheme and pair-interaction forces. Further confirmation of the agreement of these approximations is obtained in the calculation of nuclei with odd A. Figure 6 shows that the moments of inertia of nuclei with odd A lie above the corresponding values for even–even nuclei, and the values for odd–odd nuclei are on the average still higher. The increase in J is due to the fact that a single-quasiparticle state may be connected directly to another single-quasiparticle state by the operator J_x without rupture of the pair. The corresponding energy denominator is small. The resulting contribution more than compensates the decrease in value of J_x, due to the fact that the level occupied by the odd particle drops out of the overall summation. Some Nilsson states of spherical subshells $h^{11}/_2$, $i^{13}/_2$, and $j^{15}/_2$ have particularly large matrix elements of the angular momentum operator for states of one subshell. The corresponding even particles consequently have large moments of inertia. Figure 7 shows the excess of moments of inertia of nuclei with odd A, δJ for nuclei with an odd number of neutrons in the rare-earth region, together with Prior's calculations. The highest values of δJ were obtained for three states of the subshell $i^{13}/_2$. Comparison of the theoretical and experimental values of the moments of inertia of deformed nuclei may be used for identification of the states.

Vibrations of Spherical Nuclei

We have already discussed the definite consequences of pair interaction for rotational motion in deformed nuclei. We shall now examine the influence of long-range residual forces in the structure of spherical nuclei, and in particular on the energy system of vibrational 2^+- and 3^--states.

Just as for rotations, we can distinguish two main tendencies in describing vibrations: the phenomenological and microscopic approaches. In the former, we assume adiabaticity and consider the vibrations in the first place according to the low amplitude of harmonic vibrations, ignoring the shell structure of the nucleus.

Mass parameter and coefficient of rigidity are determined experimentally. In the microscopic description, these parameters are calculated on the basis of data concerning the positions of single-particle levels and the force of the residual interaction. We shall first of all consider the general structure of an ideal harmonic-quadrupole spectrum:

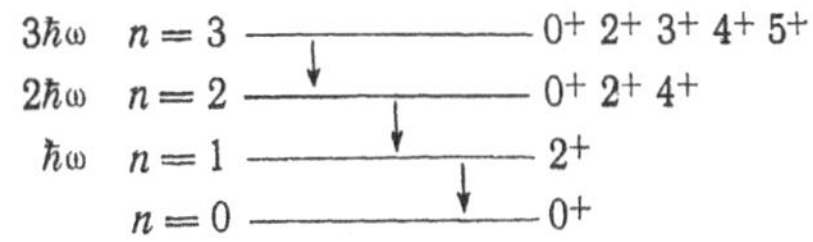

It consists of a system of equidistant states formed by the composition of vibrational quanta. The first excited state is 2^+ followed by the degenerate triplet $0^+2^+4^+$, and so forth. The γ decay of these states proceeds mainly by means of E2 transitions between successive phonon states ($\Delta n = 1$) with intensities characterized by branching rules. Some empirical data are given in Fig. 8 showing the energies of the first excited 0^+- and 4^+-states and the second excited 2^+-states. Unit energy is the energy of the first excited 2^+-states. Qualitatively, this pattern confirms the vibrational interpretation in the mass range $60 < A < 140$, where the energy ratios on the average have a value of about 2. It is clear, however, that the fluctuations in these ratios are much greater than the fluctuations of the energy ratios of the $4^+/2^+$-states in deformed regions. We thus come to the conclusion that adiabatic approximation is not so good for vibrational excitations as it was for rotational excitations. On the whole, this pecularity is due to the fact that the vibrational frequency is fairly high, approaching values corresponding to internal excitations. Thus, there is an important connection between vibrational and internal motions, which partly disturbs the harmonicity of the vibrations. This conclusion has subsequently been confirmed by data on the ratio of the intensities in γ transitions in vibrational cascades. As example, we consider the case of Se^{76} (see Fig. 8) where an almost degenerate two-phonon triplet was identified. The numbers against the arrows denote the relative values of $B(E2)\downarrow$, measured by the Aldermaston group. In the phenomenological harmonic model, the values of $B(E2;\, 4^+ \to 2^+)$, $B(E2;\, 2^+ \to 2^+)$, and $B(E2;\, 0^+ \to 2^+)$ ought to be twice the value of $B(E2;\, 2^+ \to 0^+)$.

In Se^{76} this prediction is satisfied only for the $4^+ \to 2^+$ transition, while for the other two-phonon transitions, the position is less hopeful.

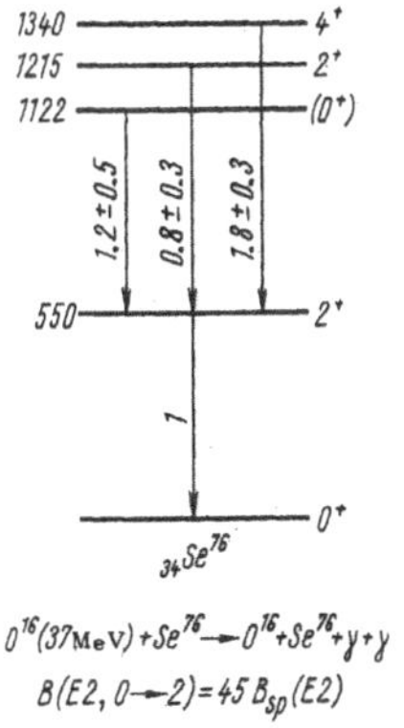

Fig. 8

In conclusion, it may be said that the phenomenological model describes the position of low-lying vibrational bands qualitatively rather than quantitatively. Well developed three-phonon and four-phonon vibrational bands have not yet been definitely identified, and the influence of anharmonicity already affects the two-phonon states; it is therefore found to be impossible to describe the vibrational states without involving details of single-particle representation.

Connection between Collective Vibrational States and Shell Structure in the Nucleus

The connection between quadrupole vibrational states and shell structure appears on examination of the first excited 2^+-states. The falling curve undoubtedly has energy scatter, the energies of the

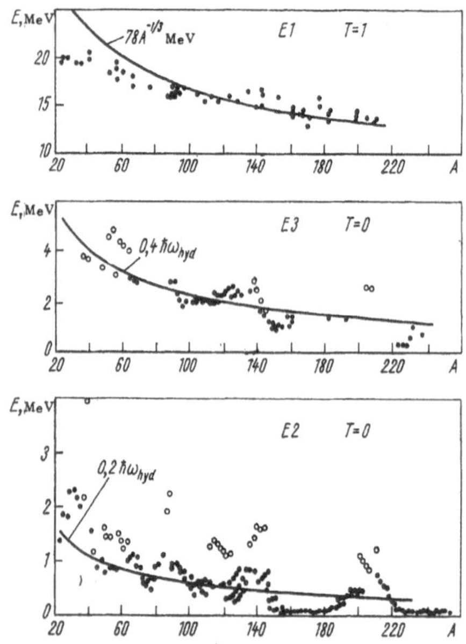

Fig. 9

vibrational states being appreciably higher near the closed shells. However the collective 3⁻-states have somewhat smaller energy fluctuations, and the curve of the giant dipole resonance shows a smooth dependence on energy (Fig. 9).

To obtain some qualitative idea of this problem, let us consider the instantaneous fluctuations in systems with pairing. In the ground state of even nuclei having a spherical form, the particles move in pairs, the contribution of which to the moment of the nucleus is equal to zero. The nucleus thus possesses spherical symmetry. In the absence of long-range forces of residual interaction, the fluctuations of the pairs bringing the nucleus out of spherical symmetry are independent and incoherent. In the presence of long-range forces, the fluctuations of one pair produce a small increase in the mean field of the nucleus, varying with time and acting on all the pairs in the nucleus. There thus appears a system of connected oscillators and a collective form of vibrations may arise, including the coherent motion of many nucleons. Corresponding excited states of a vibrational character may be regarded as a combination of quasiparticle states, the frequencies of the vibrational states differing from the frequencies of the quasiparticle states owing to coupling. The magnitude and sign of the frequency variation depend on the type of interaction. Some approximate methods, developed for the solution of this problem, lead to considerable calculation difficulties. The following remarks, however, may be useful to experimenters.

Collective vibrational states may be approximately classified according to isotopic spin. Vibrations in which the neutrons and protons move in the phase T = 0, are shifted towards low frequencies owing to the attractive character of the nuclear forces in this isotopic state; low-lying 2^+- and 3^--states are an example of vibrational states with T = 0. Vibrations, in which the protons and neutrons move in opposite directions (T = 1) are energetically disadvantageous, owing to the symmetry of the energy of the nucleus. The giant dipole resonance is an example of such a high-frequency type. The relations between the shell model and the collective vibration model are given in Fig. 9, showing the frequency distribution of quasiparticle transitions of type E1, E2, and E3 for protons of hypothetical even nuclei, arbitrarily selected in the selenium region. From Fig. 9, it is impossible to say anything about the influence of long-range nuclear forces. If these forces are taken into consideration, the intensity of the transitions to many closely situated states has a tendency to concentrate in one state, which is the corresponding vibrational state. Low-lying octupole states are included mainly in quasiparticle transitions between N and N + 1 shells, having the unperturbed energy $\sim \hbar\omega_{osc}$. In the range of medium and heavy nuclei, however, spin-orbital coupling results in a small number of low-energy quasiparticle transitions of E3 type. In Fig. 10, these transitions are referred to as $\Delta N = 1$ inside the shell. Thus, the spectrum of collective octupole vibrations may contain one or more states with isotopic spin T = 0, depending on the detailed examination of the unperturbed internal spectrum. The unperturbed spectrum of type E1 transitions consists of transitions with $\Delta N = 1$ and $\Delta N = 3$, but the potential of the harmonic oscillator is related to the special rule of selection $\Delta N = 1$, therefore transitions with the variation $\Delta N = 3$ may be expected to be weak in a real nucleus.

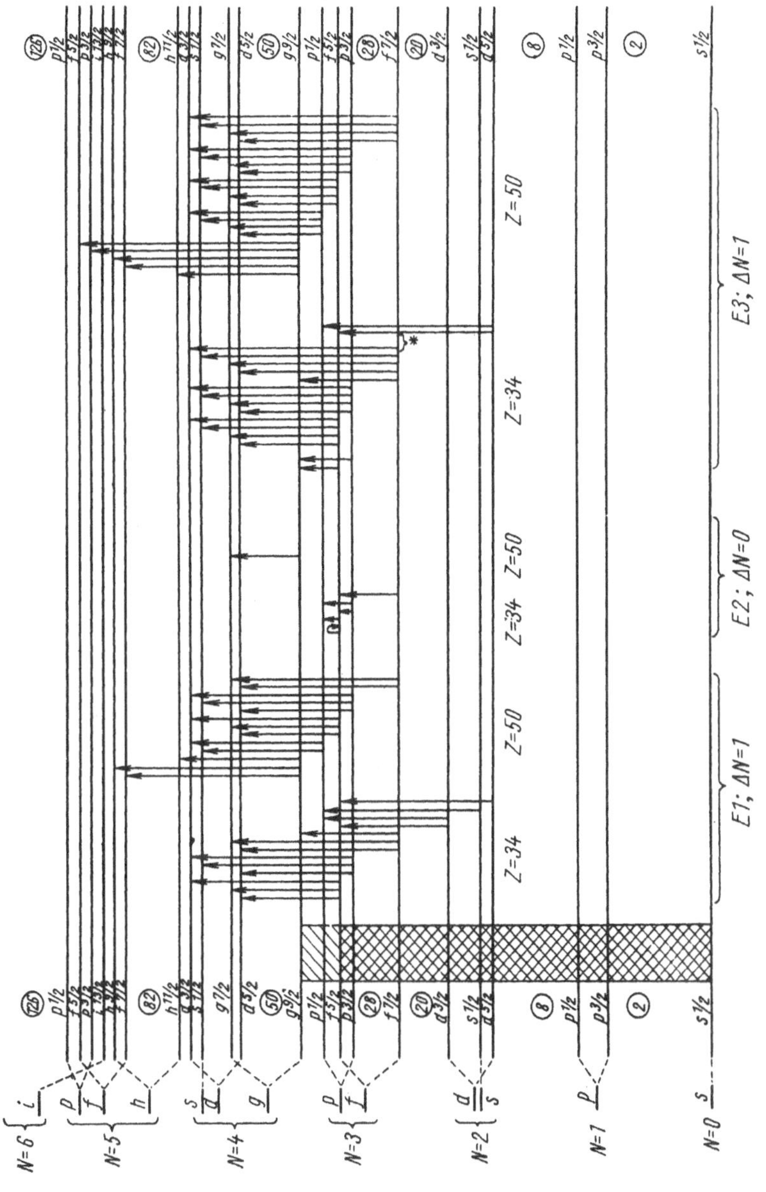

N=6
N=5
N=4
N=3
N=2
N=1
N=0
i
p
f
h
s
d
g
p
f
d
s
p
s
126
82
50
28
20
8
2
p 1/2
f 5/2
p 3/2
i 13/2
h 9/2
f 7/2
h 11/2
d 3/2
s 1/2
g 7/2
d 5/2
g 9/2
p 1/2
f 5/2
p 3/2
f 7/2
d 3/2
s 1/2
d 5/2
p 1/2
p 3/2
s 1/2
Z=34
Z=50
Z=34
Z=50
Z=34
Z=50
E1; ΔN=1
E2; ΔN=0
E3; ΔN=1
*

Fig. 10

We shall consider the unperturbed spectrum in filling of the shell $Z = 50$. The spectrum of the dipole states closely resembles the spectrum shown in Fig. 10, and it is therefore not surprising that the energy of the large dipole resonance for lead is scarcely distinguished from the energy of the giant dipole resonance for selenium. However, in the case of type E2 transitions, none of the low-energy proton transitions shown in Fig. 9 occurs at $Z = 50$, and only the transition $g\ ^9/_2 \rightarrow g\ ^7/_2$, $\Delta N = 0$ remains. This is the fundamental consideration explaining why lead isotopes have higher excitation energies. In the general case, type-E2 oscillations depend very strongly on the degree of filling of the different subshells. A relative maximum of type-E2 states has been found in Sn^{114}. This effect is due to the relative filling of the neutron subshells $d\ ^5/_2$ and $g\ ^7/_2$. For octupole energies, the expected behavior is described by means of slow variations, due to transitions between shells having small fluctuation superpositions caused by transitions inside the shell. These comments on single-phonon energies show the close connection between collective excitations and shell structure. We wished to obtain quantitative agreement. It is therefore not clear whether the details of the classification of the energies of the 2^+ and 3^- states in spherical nuclei can be satisfactorily described in the framework of microscopic calculations without the use of additional parameters.

It should be noted that the discussion of the octupole states has a tentative character owing to the lack of information on the classification of states situated above the 3^--states and to the obscure situation in the $A \approx 200$ region.

LITERATURE CITED

Nathan, O., and Nilsson, S. G., Alpha, Beta, and Gamma Ray Spectroscopy, K. Siegbahn, Editor, North Holland Publ., Amsterdam, 1964.

(t, p) - STRIPPING IN HEAVY NUCLEI†

As is known, the processes of stripping and capture, including transitions of an individual nucleon, are very common methods of research in nuclear spectroscopy. The importance of these reactions is that they determine the degree of filling of the energy levels by individual nucleons. We shall examine the related process including the stripping of a neutron pair from an impinging triton. This and the converse process of the (p, t) reaction have already been studied for several years on light nuclei, but the (t, p) process on heavy targets had not been systematically studied up to 1964.

The details of the (t, p) process depend on the spatial correlation of the two captured neutrons. In a triton, both neutrons are in the singlet state with a relative angular momentum equal to zero, and comparatively large cross sections of the (t, p) reaction are possible for those states in which the neutron pair is transmitted as a whole. Consequently, it may be assumed that the (t, p) reaction will provide information on the state of the pairs. If we had a really good theory of the reactions, including the finite radius of interaction and the finite dimension of triton, it would be possible to study in more detail the short-range spatial correlations between the nucleons in heavy nuclei. Such a theory, however, has not yet been developed. Experiments with pair stripping and capture in heavy nuclei, which have so far been made, are very rare. The group in Minnesota (Hintz et al. [1]) made the first experiments on the (p, t) reaction at 40 MeV using a linear accelerator, but unfortunately at low resolution (300-400 keV). The variable energy cyclotron in Boulder was also used for the (p, t) reaction. Work at high resolution was carried out by the Aldermaston group in England (Hinds and Middleton) using a tandem. We shall enumerate the reasons why so few experimental groups study the (t, p), (p, t), (He^3, n), and (n , He^3) reactions:

† Most of the experimental data presented in this lecture are unpublished and provisional.

1. In (p, t) reactions, Q is equal to from −3 to −10 MeV. Consequently, such experiments cannot be carried out by means of tandems, at least not with heavy nuclei.

2. (t, p) reactions may easily be studied by means of tandems, but work with tritium is dangerous.

3. The angular distributions often show a sharp maximum and minimum. High resolution and simultaneous recording at many angles are important features of proton detectors.

4. (He^3, n) or (n, He^3) reactions are difficult for study with good resolution by present techniques.

The first work on the (t, p) reaction in Aldermaston with the Hinds-Middleton spectrograph [2] was done on light targets. Recently the Aldermaston group was joined by a group from the Niels Bohr Institute in Copenhagen. In the experimental process, the combined group used a multispectrograph [2] containing 24 channels. The particles were recorded photographically. The angular range was from 5 to 175° in 7.5° steps. In the experimental process a resolution of 15-20 keV was obtained (target thickness 100-300 mg/cm^3).

The inspection of 20 m of film takes 500-1000 h, depending on the number of lines. The choice of the nucleus targets was determined for two reasons:

1) A desire to obtain information on nuclear structure for different types of nuclei.

2) The need to produce empirical rules for the mechanism of the reaction and for studying the dependence on angular distribution in a sufficiently wide mass range.

We shall first of all consider the selection rules and spectroscopic amplitudes in (t, p) reactions. We represent pair stripping as a direct single-act process, resembling in many respects single-particle stripping, but in a two-nucleon process there are important differences, due to the correlation between the transmitted particles.

The table shows the selection rules. We denote the total and orbital moments, as well as the spin of the transmitted pair by J, L, S, respectively; J_i and J_f are the total moments in the initial and final states of the nuclei; l_1 and l_2 are the orbital moments of the states to which the pair passes; Λ is the orbital moment of the pair in the mass center system, and λ is the relative orbital moment; ν is the seniority quantum number. The approximate rules are based on the simplifying assumptions:

1. The wave function of triton is a purely S-state.

2. The neutron pair is in a purely relative S-state ($\lambda = 0$).

It is considered that these assumptions are satisfied by 90-95%.

In single-particle stripping, the cross section may be factored as follows:

$$\sigma \sim S\Phi(l, j, Q, E_i, \theta)$$

1. Statistical factor.

2. Spectroscopic factor, resulting from overlapping of the initial and final nuclear wave functions and containing information on the nuclear structure.

3. Reduced cross section comprising distorted waves and energy coupling effects.

In single-particle stripping, the factor S may be calculated with good accuracy by the distorted-waves method, l is its principal characteristic; the dependence on E_i and Q is weak. It may be noted that the cross section in the (d, p) reaction decreases with l approximately as 2^{-l}.

$\mathbf{L} = \mathbf{l}_1 + \mathbf{l}_2 = \boldsymbol{\Lambda} + \lambda_\lambda$ – relative motion of the pair;
$\boldsymbol{\Lambda}$ – motion of the pair in the mass center system.

Selection Rules

Exact	Approximate
$\lvert\Gamma_i - \Gamma_f\rvert \leqslant \Gamma \leqslant \Gamma_i + \Gamma_f$ $\Delta\pi = (-1)^{l_1+l_2} = (-1)^{\lambda+\Lambda}$ $\Delta v = 0, \pm 2$	$\lvert\Gamma_i - \Gamma_f\rvert \leqslant L \leqslant \Gamma_i + \Gamma_f$ $\Delta\pi = (-1)^L$

For even–even targets we have (approximate rule)

$$L = \Gamma_f, \quad \Delta\pi = (-1)^{\Gamma_f}.$$

The angular distributions possibly depend basically on L (equal to J_f in the even–even case).

We now return to the (t, p) reaction, taking Sn^{118}(t, p) as example. The target and final ground state of the nucleus represent the mixture

$$a\left[(3s^{1/2})^2\right]_{0^+} + b\left[(2d^{3/2})^2\right]_{0^+} + c\left[(1h^{11/2})^2\right]_{0^+} + \ldots.$$

In the (t, p) reaction, the final ground state may be formed by filling any of these orbits by pairs, the process being coherent and capable of resulting in either strengthening or weakening. This is quite a common feature of two-nucleon stripping; some terms may be included coherently in the transition with the same value. Consequently,

$$\sigma(\theta) \sim \Big|\sum_{\substack{l_1 j_1 \\ l_2 j_2}} [S_L(l_1 j_1 l_2 j_2)]^{1/2} [\Phi_L(l_1 \delta_1 l_2 j_2, E\theta)]^{1/2}\Big|^2,$$

i.e., we cannot generally factor $\sigma\theta$ into a spectroscopic factor and a factor allowing for the distorted-wave effect (with the exception of those cases where the states of the target and final nucleus differ only by one neutron pair). The angular distributions depend on the nuclear wavefunctions and can tell us only about the spectroscopic amplitudes. It should be noted that the reduced probabilities of the reactions contain the product of two functions, and consequently will be extremely sensitive to the selected coupling energies.

The empirical data obtained show that σ decreases with increase in L approximately as follows:

$$\sigma_{L=4} : \sigma_{L=2} : \sigma_{L=0} \approx 1 : 6 : 40, \quad \text{or} \quad \sigma \approx 2.5^{-L}.$$

As additional example of single and double neutron stripping we consider the reactions

$$Ca^{43}(d, t)Ca^{42} \text{ and } Ca^{40}(t, p)Ca^{42}.$$

Bjerregaard et al. [3] assumed that the first and second excited 2^+ states of Ca^{42} have respectively the following approximate wave functions:

$$2_1^+ : \frac{1}{\sqrt{2}} \{[(f^7/_2)^2]_{2^+} + [f^7/_2 p^3/_2]_{2^+}\},$$

$$2_2^+ : \frac{1}{\sqrt{2}} \{[(f^7/_2)^2]_{2^+} - [f^7/_2 p^3/_2]_{2^+}\}.$$

If the ground state of Ca^{43} is quite purely $(f\ ^7/_2)^3$; the (d, p) reaction takes place only through $(f\ ^7/_2)^2$ as part of the wave function of Ca^{42}, and no difference is found between the two 2^+-states. But the (t, p) reaction from Ca^{40} may strengthen one state and suppress the other, as was also found by Middleton and Hinds.

We shall consider in more detail some reactions of

$$Sn^{118}(t,\ p)\,Sn^{120}.$$

Here, the pair forces predominate and the ground state is strongly mixed. The sign in the mixture are the same and we observe an intensification of the transition for the ground state. According to Yoshida, pairing intensifies transition to the ground state compared with transition to the two-particle 0^+-state, formed by means of the j^2 configuration, by a factor of ~ 10 in the lead region, namely

$$\frac{(2\Delta|G)^2}{(j + {}^1/_2)^2}.$$

For 2^+ oscillations, it is sometimes possible to obtain an intensification due to long-range forces. However, for Sn^{118}(t, p) Sn^{120} Yoshida predicted that the vibrational 2^+-state would be formed with approximately the same cross section as the ordinary state.

It is probably possible to identify some weak 0^+-states from the position of the maximum and minimum. It should be noted that some of these excited 0^+-states were previously identified in the (d, p) reaction by Cohen and others. Above 1.17 MeV in Sn^{120} there is no state having in the (t, p) reaction an angular distribution like that which the 2^+-state has at 1.17 MeV.

It is clear from the cited experiments that due to the lack of a well-founded theory for two-nucleon stripping reactions, it is impossible to determine the spin and parity of weak highly excited states. Even groups of levels having LT = 0 show some individuality in angular distribution, and the corresponding determination of the spin may consequently be regarded as merely provisional.

The case of Sm^{144}(t, p) Sm^{146} qualitatively differs considerably from Sn. The distributions for the ground and first 2^+-states are similar to the case for Sn, but the $0^+/2^+$ ratio is quite different, indicating that the nucleus of Sm^{146} has less pairing than Sn. We know that the state at 1.39 MeV is a doublet [5] (with splitting of ~ 1 keV), consisting of 3^- and 4^+ states. In the given experiment it is impossible to distinguish the proportions of these states. The state at 2.61 MeV is quite analogous to the 0^+ state, excited 3.3 times more weakly than the ground state.

The angular distributions for the stronger groups of Sm^{148}(t, p) Sm^{150} are shown in Fig. 11. The ground state of Sm^{150} is very similar to the ground state in Sm^{144}(t, p), but the 2^+ state has a very smooth maximum at $\theta \approx 42°$. The intensification of the ground state relative to the first 2^+-state is less than that found in tin, but is definitely stronger than in Sm^{144}(t, p). We know [6] that the state at 0.75 MeV is 0^+ (two phonon); it is excited 4 times more weakly than the ground state. No other 0^+-states have been identified in this reaction.

We shall now consider Sm^{150}(t, p) Sm^{152}; in this process, however, the situation is very complex since, evidently, both the intial and final states have an ill-defined form. It is assumed that the ground state of Sm^{150} is spherical, while it is deformed in the nucleus Sm^{152}, but part of

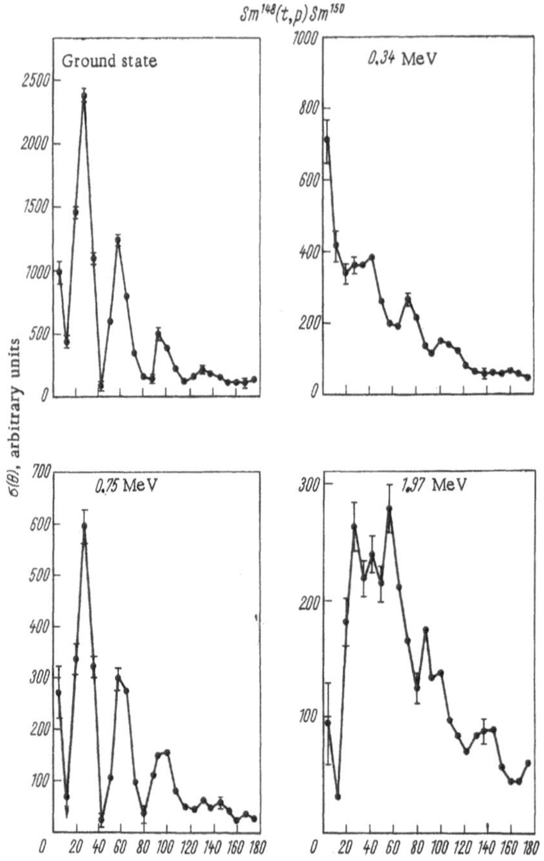

Fig. 11

the time, these two states may have other forms. Consequently $Sm^{150}+2n$ may have a considerable overlap with the ground state of Sm^{152}, as well as with excited states of different forms. Mention should be made of the existence of three strong groups of levels with $L=0$: the ground state, β-vibrational state, and an unknown 0^+-state. Since the generalized model does not give the second excited 0^+ state at 1.9 MeV, it has been provisionally assumed to have a spherical form [7]. It should be noted that a 2^+ state is possible, situated at 0.2 MeV above 0^+. Such an interpretation by means of states of unstable form is confirmed by mass data indicating that the energy obtained in the deformation of the nucleus, is ~ 0.8 MeV on transition from 88 to 90 neutrons.

Beta vibrations are not necessarily strong in the (t, p) process. Actually

	$\sigma_{max}\beta\ 0^+/\sigma_{max}$(ground state)
$Sm^{150}(t, p)$	0.7
$Sm^{152}(t, p)$	0.09.

Gamma vibrations have not been observed in any of these cases.

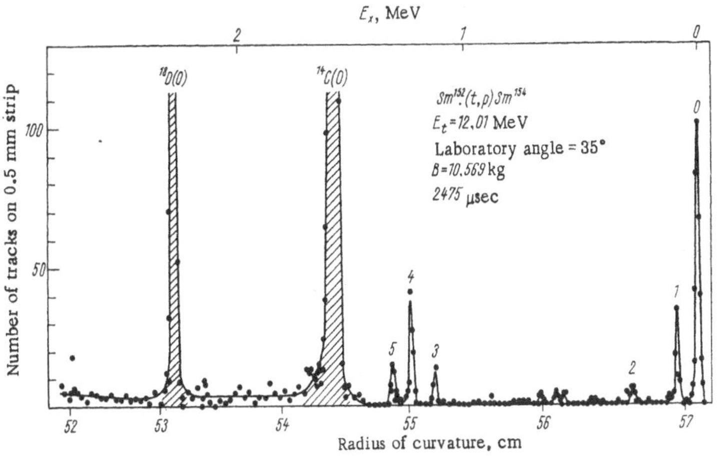

Fig. 12

The spectrum of Sm^{152}(t, p) Sm^{154} is shown in Fig. 12. It contains several excited states, but there are no transitions as strong as in the ground state. The angular distribution shows the existence of possible levels with L = 0 at 1.2 MeV. This state is excited 30% more weakly than the ground state, but does not appear at Coulomb excitation and consequently is not β vibrational This may be a state with less deformation or a collective state including pair correlation.

The Sm^{150}(t, p) Sm^{156} reaction has not been studied in detail. It contains many more (mainly weak) excited states than the two preceding cases, although evidently Sm^{156} and Sm^{154} ought to have a similar structure. The 4^+ state of Sm^{156} distinctly occurs; the corresponding angular distribution decreases smoothly with θ. The state at 1.07 MeV is probably the 0^+ state; it is excited 6% more weakly. The ground state is excited approximately in the same way as β-vibrational states. If there are other excited 0^+-states, their cross section is less than for the 1.07 MeV state.

Further experiments have been carried out with Pb^{204}, Pb^{206} targets, and some light nuclei in the Ca – Cu range.

LITERATURE CITED

1. Hintz, N. M., Nucl. Spectroscopy with Direct Reactions, 11:425, (ANL-6878), March 1964.
2. Middleton, R., and Hinds, S., Nucl. Phys., 34:404 (1962).
3. Bjerregaard, J., et al., Phys. Rev., 136:B1348 (1964).
4. Yoshida, Nucl. Phys., 33:685 (1962).
5. Proc. of Minsk Conf., 1965.
6. Nuclear Data Sheets.
7. Hinds, S., et al., Phys. Rev. Lett, 14:48 (1965).
8. Elbek, B., Private communication.
9. Bes'y, D. Private communication.
10. Hintz, N., and Mottelson, B., Private communication.
11. Bes, D., Private communication.

DEFORMED STATES OF NUCLEI

B. S. Dzhelepov

USSR

Introduction

The concept of "deformed nucleus" has become so firmly rooted in modern physics that it would not appear to be in need of any definition or improvement. Five years ago it was assumed that there were three regions of "complete" deformation with clearly pronounced boundaries: a region of light deformed nuclei ($A \cong 24$), medium deformed nuclei ($150 < A < 192$) and heavy deformed nuclei ($A > 226$). In the succeeding years, however, an ever increasing number of nuclear states have been included in deformed states; this process is being intensified and will evidently be still further accelerated. At the same time, the inclusion of nuclei in the category of deformed nuclei is done on the basis of very diverse and sometimes debatable criteria; we shall later attempt to classify them but at first we shall consider the direction in which the number of deformed nuclear states is being or may be extended.

There are five such directions:

1) New regions of deformed nuclei;
2) Boundary regions;
3) Separate islets or individual nuclei;
4) Different states of one nucleus;
5) Mixed spherical and nonspherical states.

Let us examine these more closely.

1. New Regions of Deformed Nuclei

In 1961, Sheline et al. [1] discovered a new region of deformed nuclei and predicted the existence of yet one or two other regions. At that time, it was assumed that for stable deformation to occur, it was necessary that there should be a sufficiently large number of nucleons of both sorts (or the corresponding holes), in addition to a filled shell – not less than five or six nucleons or holes of each sort (it is now known that even two nucleons suffice in light nuclei; ${}_{10}Ne^{20}_{10}$ is deformed).

We place all the deformed nuclei in an N;Z diagram (Fig. 1). A net corresponding to filled shells of neutrons and protons may be applied to the diagram. Nuclei situated in the dashed-line regions have a sufficient number of nucleons (or corresponding holes) in addition to a filled shell, and are therefore deformed. The regions 4-7 are the assumed new regions of complete deformation. A new (fifth) region has practically been found, and already eight nuclei are known in it: Ba^{124}, $Ba^{126}-Ba^{130(131)}$, Cs^{125}, Cs^{127}, Cs^{129}. In his first paper, Sheline demonstrated the

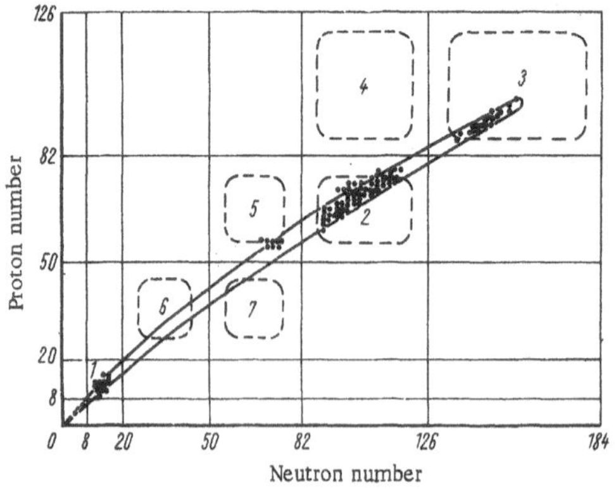

Fig. 1. Diagram of deformed nuclei.

presence of deformation only by the low excitation-energy of the first 2^+-levels. Then L. K. Peker succeeded in showing from spectroscopic data that Ba^{129} (and possibly Ba^{131}), Cs^{125}, Cs^{127}, Cs^{129} were deformed. Recently, Clarkson succeeded in observing rotational bands down to $I = 10^+$ in Ba^{124} and Ba^{126}; to confirm the correctness of spin assignment, the angular correlation in the $(4 \rightarrow 2)$ $(2 \rightarrow 0)$ cascade was studied. In addition, it was found possible to measure the half-life of one of the 2_1^+ states, and to show that the quadrupole moment Q_0 was high. All the work mentioned demonstrates convincingly that nuclei in the fifth region are actually deformed. The existence of the other regions 4, 6, and 7 (see Fig. 1) appears to be quite probable.

2. Boundary Regions

Discussions as to whether there are slightly deformed nuclei situated close to the boundaries of regions of complete deformation are still continuing.

The generally accepted view is that the curve $\beta = f(A)$ twice reaches the axis of abscissas, outlining strict regions of complete deformation; these views are based on the following consideration. The shape of the nucleus is determined by the minimum of total energy; surface tension forces tend to make the nucleus spherical; nucleons in excess of a filled shell tend to cause the nucleus to be deformed. Depending on which forces predominate, the nucleus will be spherical or deformed. Since the deforming forces depend on the number of particles (or holes) outside the filled shell, it is unlikely that over a long A range, there will be almost exact compensation of the surface forces and deforming forces and the $\beta(A)$ curve will approach the axis of abscissas smoothly. These are, of course, only general considerations and the solution belongs to experiment; all the available deformation criteria will have to be applied to nuclei situated in a transition region. In the region around Nd–Sm, among the even–even nuclei, the first to be deformed are Nd^{150}, Sm^{152}, and Gd^{154}. Among nuclei having an odd number of neutrons, the first to be deformed are Sm^{151}, and Gd^{153}. For nuclei having an odd number of protons, the situation is not so clear. Eu^{153} nuclei are undoubtedly deformed. É. E. Berlovich has collected the arguments in support of the nuclei Eu^{147}, Eu^{149}, Eu^{151} being also slightly deformed ($\beta \cong 0.1$). These arguments are, however, debatable, and the main argument – the rotational band – has not yet been found for these nuclei, although it ought to have been under the conditions of experiments already made.

A situation, which is still less clear, is found for the other edge of the region of deformed nuclei in the region of isotopes of Os and Pt. The $\beta = f(A)$ curve approaches the axis of abscissas more smoothly. The rotational bands are mixed strongly with vibrational bands, and it is evidently impossible to separate these forms of motion. The nuclei Os^{190}, Os^{191}, Pt^{189} are deformed, but the conclusions are indefinite with regard to Os^{192}, Pt^{190}, and Pt^{192}.

3. Islets of Deformed Nuclei

It is clear from Fig. 1 that, for stable deformation, a minimum of 5 or 6 nucleons of each sort in excess of the shell is necessary, which almost precludes the possibility of the occurrence of deformed light nuclei, especially if the subshells, containing 28 and 40 nucleons, have to be considered on the same basis as a shell. All these small islets may remain; thus, for example, for the nuclei $_{34}Se^{75}_{41}$ and $_{35}Br^{76}_{41}$, the above-mentioned requirement is satisfied and there are facts which indicate that these nuclei are really deformed.

It is possible that such islets also exist in other places (perhaps some isotopes of Ag are deformed). L. K. Peker has suggested that conditions for stable deformations are more favorable in odd nuclei than in even–even nuclei; in the islets there ought to be more odd–odd and odd nuclei than even–even nuclei.

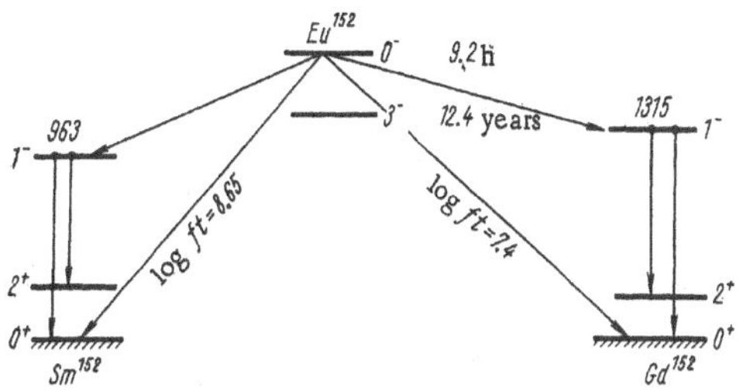

Fig. 2. Elements of the Eu^{152} decay schemes.

4. Different States of One Nucleus

In 1957, L. K. Peker [2] advanced the hypothesis that in the nucleus Eu^{152}, the ground state ($I=3^-$, $T_{1/2}=12.4$ years) is deformed and the first isomeric state ($I=0^-$, $T_{1/2}=9.2$ h) is spherical. This hypothesis is confirmed by two arguments.

a. Decays (Fig. 2)

$$0^- \; Eu^{152} \rightarrow 0^+ \; Sm^{152}$$
$$0^- \; Eu^{152} \rightarrow 0^+ \; Gd^{152}$$

are equally forbidden according to spin and parity. In such cases log ft usually differs by 0.2–0.3. Here, however, for the first disintegration log $ft = 8.65$ and for the second log $ft = 7.4$; the probabilities differ by a factor of nearly 20. The difference in log ft may be due to the fact that the variation in the shape of the nucleus produces an additional forbiddenness.

b. The transition between the isomeric and ground states of Eu^{152} is of type M3 with an energy of about 50 keV. Such isomeric transitions ought to have a half-time of the order of one second; the corresponding conversion lines ought to be observed in the disintegration of the 0^- state. However, no one has been able to find them. It is possible that the transition occurs between the spherical and deformed states, and this imposes additional prohibition on the isomeric transition.

It has been assumed in recent years that different states of the same nucleus may have different equilibrium shapes.

Recently, Bjerregaard et al. [3] measured the proton spectrum in the reaction Sm^{150}(t, p) Sm^{152}. This reaction provides almost unique possibilities; the ground state in Sm^{150} is evidently spherical (see Section 5) and in Sm^{152} it is deformed. It was found that in the above-mentioned reaction with considerable cross section a previously unknown 0^+-level, 1091 keV, of Sm^{152} was excited. The authors assumed this was the spherical state of Sm^{152}. It is not clear, however, why this state is not excited in the decay of the also spherical 0^- state of Eu^{152}.

At the Gordon conference in 1965, Mottelson pointed out that among the excited states of the nuclei Tl^{193}–Tl^{201}, a state of $9/2^-$ type is found, which can scarcely be spherical, since the Mayer scheme does not give such characteristics; at the same time, the ground states of these nuclei are spherical.

5. Mixed Spherical + Nonspherical States

Let us assume that the same nuclear state may be both spherical and at the same time deformed. Such an assumption, as far as the writer is aware, has not appeared in the literature.

However, it obviously does not contradict any general principles. To verify this, it is merely necessary to recall the deuteron theory; the ground state of the deuteron is a "mixture" of 96% S+4% D, and if the S state is spherical, the D state does not possess spherical symmetry, or in modern parlance it is deformed (evidence of this is the quadrupole moment of the deuteron. Assumptions concerning such mixing of states have not been made, probably because there were no experimental facts which required such a hypothesis for their explanation. One such fact has been found, and perhaps other data are to be found in the literature. In the disintegration of the same $Eu^{152}(0^-)$, the 1^-, 963 keV level of Sm^{152} and 1^-, 1315 keV level of Gd^{152} are excited. Each of them is discharged to the 0^+ and 2^+ states of its own nucleus. On discharge of the 1^-, 963 keV Sm^{152} state to the levels of the ground-state band of the deformed nucleus Sm^{152}, the Alaga rules ought to be observed. These are observed for $K_i = 0$,

$$\frac{B(E1,\ 1- \to 0+)}{B(E1,\ 1- \to 2+)} = \frac{1}{2}, \text{ and experiment gives } 0.55 \pm 0.02.$$

For the discharge of the 1^-, 1315 keV state of Gd^{152}, there are no Alaga rules, since the nucleus Gd^{152} is spherical. However, the experimental ratio

$$\frac{B(E1,\ 1- \to 0+)}{B(E1,\ 1- \to 2+)} = \frac{I_\gamma(1- \to 0+)}{I_\gamma(1- \to 2+)}\left(\frac{E_{1\to 2}}{E_{1\to 0}}\right)^3 = 0.48 \pm 0.06$$

is in fact one which the Alaga rules require. There are two views on this coincidence: 1) it may be accidental, and it is then meaningless to seek the cause, or 2) it may not be accidental and then the cause should be sought. This cause may be the following: In the 0^+ and 2^+ states of Gd^{152}, the ground-state component corresponds to the spherical-equilibrium shape, but there is a small component with a deformed-equilibrium shape. If the 1^- state is also deformed, its disintegration will occur predominantly to deformed components of the 0^+ and 2^+ states of Gd^{152}, and this results in the observed compliance with the Alaga rules. This, however, is only assumption. Evidently similar cases are encountered in heavy nuclei on the disintegration of type-1^- states.

6. Deformation Criteria

The diversity of the criteria currently applied in solving the problem as to whether or not one state or another is deformed calls for some classification of these criteria. Evidently, there are two criteria which ought to be considered as the principal ones. A nuclear state is deformed if the internal quadrupole moment exceeds 4 barns (we are not considering deformed light nuclei), or if there is a rotational band in the composition of at least two excited states, which agree in spin and parity, and the energies of which are in satisfactory agreement with the relationships existing in this atomic weight region.

A third criterion may be added for even–even nuclei; if in an even–even nucleus, the first 2^+-type level has an energy of less than 200 keV, the ground state of the nucleus is deformed. For the first criterion, a complete theory of quadrupole moments is not available, and if for example the quadrupole moment is equal to 2 barns, we do not know what this means. Movement of a proton along the equator of a nucleus should give rise to the quadrupole moment

$$Q_1 = -\frac{2j-1}{2j+1} < r^2 >,$$

where j is the moment of the proton. Whatever j may be, Q_1 can never exceed 1 barn. It is possible that during its movement, the proton attracts other protons and this increases Q_0, and the dynamic part of Q_0 is produced during the vibrations of the nuclear surface; conversely, in all deformed nuclei, if Q_0 is measured, it is greater than 4 barns. The second criterion, the presence of a rotational band, is more accessible to analysis of excited states, since the coefficients of the rotational formula differ considerably at different points of the deformed regions.

The third criterion is a simplified modification of the old Bohr and Mottelson criterion [4]: A nucleus is deformed if an excited level is found which may be assumed to be the first rotational level with an excitation energy $E(2^+) < 13\hbar^2/J_s$. These three criteria reflect the data already acquired; they are satisfied by all nuclei which is customary to regard as deformed; at the same time, none of the criteria permits us to treat as deformed any nucleus for which the experimental facts definitely point to the spherical shape.

In addition to these principal criteria of deformation, we shall also consider the auxiliary criteria:

1) The presence of quantum characteristics agreeing with the predictions of Nilsson's scheme but absent from Mayer's scheme;

2) The high value of $B(E2)\uparrow$ for Coulomb excitation, indicating the collective character of the excited state;

3) An increase in the probability of the reaction in which the deformed nucleus is the starting point;

4) Compliance with the "direct" or "reverse" Alaga rules.

5) Facilitated β-decay from deformed states and hindered decay from spherical states.

7. Value of the Deformation Parameter β

Practically the most reliable method of determining β is to calculate β from Q_0 according to the formula

$$Q_0 = \frac{3}{(5\pi)^{1/2}} Z R_0^2 \beta (1 + 0.16\beta).$$

In even–even nuclei, the quadrupole moment Q_0 is determined from the lifetime of the rotational levels, in odd and odd–odd nuclei by the optical method.

The results relative to even–even deformed nuclei of medium weight are shown in Table 1 and in Fig. 3, prepared by V. D. Vitman.

It should be noted that the value of β^2 may also be determined from the isotope shift of the $S_{1/2}$ terms [5]. The same values of β are then obtained as in those determined from quadrupole moments.

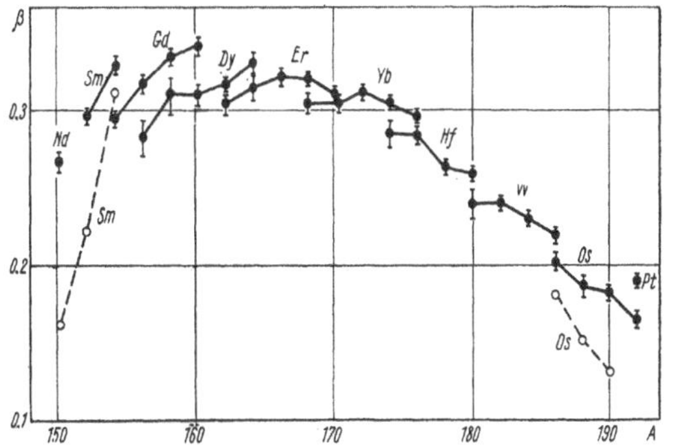

Fig. 3. Deformation parameters of even–even nuclei calculated from $T_{1/2}$ of (2_1^+).

TABLE 1. Values of $T_{1/2}$, Q_0, and β for the First 2^+-Levels of Deformed Even–Even Nuclei 150 < A < 192

Nucleus	$T_{1/2}$, nsec (measurement of lifetime)	$T_{1/2}$, nsec (Coulomb excitation)	$\overline{T}_{1/2}$	Q_0 (measurement of lifetime)	Q_0 (Coulomb excitation)	$\overline{Q}_0$ (barn)	β	$B(E2)\uparrow$	X
Nd^{150}	1.54±0.07	1.49±0,06	1.51±0,05	5.04±0.13	5.16±0.09	5.11±0.13	0.266±0.06	2.65±0.10	1.04±0.05
Sm^{152}	1.408±0.024	1.25±0.06	1.37±0.04	5.88±0.07	6.24±0.13	5.96±0.11	0.296±0.05	3.86±0.15	1.13±0.05
	—	1.39±0.07	1.40±0.02	—	5.91±0.11	5.89±0.06	0.293±0.03	3.47±0.13	1.01±0.05
Sm^{154}	2.74±0.24	2.77±0.21	2.76±0.16	6.77±0.38	6.73±0.09	6.73±0.09	0.329±0.04	4.50±0.17	0.99±0.12
Gd^{154}	1.18±0.02	1.33±0.13	1.18±0.02	6.24±0.12	5.87±0.26	6.18±0.11	0.294±0.05	3.43±0.30	0.89±0.09
Gd^{156}	2.17±0.03	2.25±0.18	2.17±0.03	6.81±0.23	6.70±0.16	6.74±0.13	0.317±0.06	4.47±0.21	0.96±0.08
Gd^{158}	2.44±0.09	2.33±0.19	2.42±0.08	7.18±0.29	7.26±0.14	7.24±0.13	0.337±0.06	5.26±0.21	1.05±0.09
Gd^{160}	2.52±0.14	2.37±0.20	2.47±0.12	7.35±0.34	7.57±0.14	7.54±0.13	0.342±0.06	5.70±0.21	1.06±0.11
Dy^{156}	—	0.81±0.07	0.81±0.07	—	6.18±0.25	6.18±0.25	0.282±0.11	3.79±0.30	—
Dy^{158}	—	1.64±0.17	1.64±0.17	—	6.85±0.30	6.85±0.30	0.310±0.13	4.67±0.40	—
Dy^{160}	2.02±0.05	2.25±0.21	2.03±0.05	7.14±0.26	6.70±0.22	6.88±0.17	0.310±0.07	4.46±0.30	0.91±0.10
Dy^{162}	2.24±0.06	2.22±0.18	2.24±0.06	7.09±0.28	7.12±0.09	7.12±0.09	0.317±0.04	5.05±0.13	1.01±0.08
Dy^{164}	2.44±0.24	2.33±0.21	2.37±0.16	7.38±0.48	7.51±0.17	7.50±0.16	0.331±0.07	5.64±0.25	1.04±0.15
Er^{162}	—	1.40±0.11	1.40±0.11	—	7.02±0.18	7.02±0.18	0.304±0.07	4.89±0.25	—
Er^{164}	1.64±0.09	1.56±0.15	1.62±0.08	6.94±0.30	7.13±0.25	7.05±0.20	0.303±0.08	5.04±0.35	1.05±0.12
Er^{166}	1.85±0.03	1.75±0.16	1.85±0.03	7.41±0.30	7.62±0.16	7.58±0.14	0.322±0.06	5.87±0.24	1.06±0.09
Er^{168}	1.91±0.03	1.82±0.16	1.91±0.03	7.42±0.30	7.64±0.12	7.61±0.11	0.321±0.04	5.80±0.19	1.05±0.09
Er^{170}	—	1.94±0.16	1.94±0.16	—	7.43±0.10	7.43±0.10	0.311±0.04	5.51±0.14	—
Yb^{168}	—	1.58±0.14	1.58±0.14	—	7.43±0.18	7.43±0.18	0.305±0.07	5.43±0.25	—
Yb^{170}	1.56±0.03	1.61±0.15	1.56±0.03	7.58±0.31	7.45±0.17	7.48±0.15	0.305±0.06	5.53±0.25	0.97±0.09
Yb^{172}	1.55±0.08	1.67±0.15	1.58±0.07	7.98±0.42	7.70±0.13	7.73±0.13	0.312±0.05	5.89±0.20	0.93±0.10
Yb^{174}	1.91±0.20	1.75±0.16	1.81±0.12	7.45±0.51	7.62±0.11	7.61±0.11	0.305±0.04	5.78±0.17	1.09±0.16
Yb^{176}	2.00±0.14	1.68±0.15	1.84±0.12	6.83±0.44	7.47±0.11	7.45±0.11	0.297±0.04	5.57±0.16	1.19±0.16
Hf^{174}	—	1.40±0.16	1.40±0.16	—	7.28±0.24	7.28±0.24	0.285±0.09	5.26±0.35	—
Hf^{176}	1.38±0.04	1.46±0.14	1.39±0.04	7.50±0.32	7.28±0.17	7.33±0.15	0.284±0.06	5.27±0.25	0.95±0.09
Hf^{178}	1.49±0.04	1.53±0.14	1.49±0.04	6.87±0.28	6.79±0.12	6.80±0.11	0.263±0.04	4.57±0.16	0.98±0.10
Hf^{180}	1.52±0.03	1.55±0.13	1.52±0.03	6.80±0.27	6.73±0.13	6.74±0.12	0.259±0.04	4.51±0.17	0.98±0.09
W^{180}	1.30±0.03	—	1.30±0.03	6.39±0.24	—	6.39±0.24	0.239±0.09	—	—
W^{182}	1.37±0.01	1.42±0.12	1.37±0.01	6.51±0.24	6.44±0.13	6.46±0.11	0.240±0.04	4.12±0.17	0.97±0.08
W^{184}	1.26±0.03	1.16±0.08	1.24±0.03	6.03±0.25	6.28±0.12	6.23±0.11	0.230±0.04	3.94±0.15	1.08±0.09
W^{186}	1.035±0.035	1.02±0.07	1.03±0.03	5.94±0.18	5.96±0.15	5.95±0.12	0.219±0.05	3.56±0.18	1.01±0.08
Os^{186}	0.83±0.02	0.58±0.12	0.82±0.05	5.54±0.16	6.78±0.67	5.61±0.16	0.202±0.06	4.6 ±0.9	1.43±0.30
Os^{188}	0.712±0.018	0.52±0.04	0.67±0.05	4.95±0.11	5.86±0.17	5.22±0.20	0.186±0.07	3.40±0.20	1.36±0.10
Os^{190}	0.50±0.20	0.328±0.023	0.33±0.02	4.2±0.9	5.18±0.16	5.14±0.16	0.182±0.05	2.68±0.16	1.52±0.61
Os^{192}	—	0.275±0.018	0.275±0.018	—	4.64±0.15	4.64±0.15	0.164±0.05	2.15±0.14	—
Pt^{192}	0.0272±0.0011	—	0.0272±0.0011	5.51±11	—	5.51±0.11	0.189±0.04	—	—

Notes:

1. For Sm^{152}, the lower row gives the values when the values of Sheline et al. [Nucl. Phys. 16:518 (1960)] are excluded from the data on Coulomb excitation.
2. In the calculation of $T_{1/2}$ from the Coulomb excitation and Q_0 from $T_{1/2}$, the theoretical values of KVK were used, with errors $\Delta\alpha_K/\alpha_K = 2\%$; $\Delta\alpha_L/\alpha_L = 5\%$ and $\Delta\alpha_M/\alpha_M = 30\%$.

It is very difficult to determine the value of β theoretically. It is necessary to find the minimum of the total energy of the nucleus, and to take into account all the nucleons, pair interactions, superfluidity, etc. Such calculations have been performed by Mottelson and Nilsson (6), Marshalek, Person and Sheline [7], Bes and Szymanski and others. They obtain good agreement between theoretical and experimental data within the limits of 10%. At the 1965 Gordon conference, Beranger and Kümmer communicated the results of calculations of the equilibrium deformation for some nuclei. The calculations were made on the basis of the Hamiltonian $H = H_{\text{mean pot.}} + H_{\text{pair}} + H_{QQ}$. The mean potential was assumed to be the Nilsson potential, also a potential differing somewhat from the latter; the constants used for H_{pair} and H_{QQ} were the generally accepted ones. The following results were obtained:

Sm^{149}	$\beta_0 = 0$	$\gamma_0 = 0$	Os^{186}	$\beta_0 = 0.18$	$\gamma_0 = 0$
Sm^{150}	0.16	0	Os^{188}	0.15	0
Sm^{152}	0.22	0	Os^{190}	0.13	0
Sm^{154}	0.31	0	Os^{192}	0	0
			W^{186}	0	0

It should be noted that γ_0 is everywhere equal to 0. For Sm^{150}, $\beta_0 = 0.16$ was obtained, although actually this nucleus is obviously spherical.

LITERATURE CITED

1. Sheline, R. K., et al., Phys. Rev. Lett., 7:446 (1961).
2. Peker, L. K. Zh. Éksp. Teor. Fiz., 33, 291 (1957).
3. Hinds, S., et al., Phys. Rev. Lett., 14:48 (1965)
4. Alder, K., et al., In the collection "Deformation of Atomic Nuclei" [Russian Translation], Moscow, Izd. in lit. 1958.
5. Ionesco-Pallas, N., Phys. Rev. Lett., 6:93 (1963).
6. Mottelson, B., and Nilsson, S., Phys. Rev., 99:1615 (1955).
7. Marshalek et al., Rev. Mod. Phys., 35:108 (1963).

ROTATIONAL BANDS OF THE GROUND STATES OF EVEN - EVEN NUCLEI

Basic Facts

There are currently many models of the atomic nucleus. The predictions of the various models are particularly convenient to verify from the rotational bands of the ground states, since in these bands it is possible to excite the maximum number of levels; the ground states of even–even nuclei are very remote from all the nonrotational excited states (the "gap" is about 1 MeV); it is possible that perturbations, due to other forms of motion, will not be large in any of the rotational bands of the ground states.

In β processes, only the lower levels of the rotational bands are excited, most often only the 2_1^+ and 4_1^+ levels, more rarely the 6_1^+ levels and only in five cases the 8_1^+ levels. This is due to the fact that β-active states having a higher spin than 8 are rarely encountered, and β processes are very strict with regard to spin variation.

Rotational levels with high spin have also not been excited in nuclear reactions at low energies; a slow particle is unable to transmit a high rotational momentum to the nucleus.

The situation changed when the experimenters began to use fast heavy particles. In the grazing collision of a proton having an energy of 10 MeV with a nucleus of moderate weight, a rotational momentum of MUT = ~5ħ may be transmitted to the nucleus. It would appear that by increasing the energy of the protons to 100 MeV, 16ħ may easily be attained. This is not so, however: firstly, at higher energies, the nucleus becomes semitransparent to nucleons, and secondly, a cascade of fast nucleons carries off at once a large proportion of energy and momentum, and as a rule only a small proportion of the momentum of the impinging particle is transmitted to the nucleus.

It is found to be practically meaningless to increase v, and therefore we increase m, while retaining the optimum value for v (energy about 10 MeV per nucleon).

In the last few years, in (α, xn) reactions for Coulomb excitation, and nuclear reactions with heavy ions (C^{12}, N^{14}, etc.), it has been possible to excite much higher rotational levels than before. Stephens et al. were able to excite levels with a spin of up to 16^+ in five nuclei, and with less certainty 18^+-levels in Hf^{170}, Hf^{172}, and W^{182}, and even a 20^+ level in Hf^{170}.

Table 2 gives all the information on rotational bands which the author has succeeded in collecting up to July 1, 1965.

Such abundant experimental material enables comparison with the predictions of the various models to be made much more carefully than previously.

We shall refrain here from making a comparison for all the nuclei, but shall discuss one for which the band is known quite reliably up to the level $I = 16^+$, the nucleus Hf^{170}. Comparison with the models was made by N. A. Voinova (Figs. 3, 4; Table 3).

Bohr-Mottelson Model

Even in a first note on deformed nuclei, Bohr and Mottelson remarked that the simple rotational formula $E_I = AI(I+1)$ gives deviations from the experimental data. They pointed out that agreement was improved by adding to it a second term: $+ BI^2(I+)^2$.

TABLE 2. Energy of Ground-State Band Levels
a. State with Spins up to 8^+

Nucleus	$E(2^+)$	$E(4^+)$	$E(6^+)$	$E(8^+)$	$\frac{100}{A}$
Nd^{150}	131±1	391±2	—	—	4.38
Sm^{152}	121.784±0.007	366.45±0.08	—	—	4.73
Sm^{154}	81.99±0.02	266±3	—	—	7.24
Gd^{154}	123.06±0.03	371.14±0.16	—	—	4.68
Gd^{156}	88.97±0.01	288.16±0.03	584.5±0.5	—	6.66
Gd^{158}	79.5104±0.0018	261.448±0.005	539.031±0.018	898.3±0.4	7.50
Gd^{160}	75.26±0.01	245±5	—	—	7.89
Dy^{156}	138.0±0.2	404.4±0.4	771.1±0.7	1212*	4.13
Dy^{158}	99.10±0.15	317.6±0.3	641*	1037**	5.95
Dy^{160}	86.8±0.2	283.8±0.3	581.5±0.5	972**	6.86
Dy^{162}	80.7±0.1	265.9±0.4	549.1±0.8	923**	7.40
Dy^{164}	73.39±0.01	248±10	468±15	—	8.22
Er^{162}	102±2	338±5	—	—	5.86
Er^{164}	91.5±0.1	301.0±1.5	617*	1028*	6.52
Er^{166}	80.57±0.02	265.0±0.1	545.4±0.3	905±10	7.40
Er^{168}	79.80±0.02	264.3±0.4	548.9±0.8	—	7.50
Er^{170}	79.31±0.02	—	—	—	(7.57)
Yb^{164}	122.5±0.4	384.0±1.2	758±3	1219±4	4.78
Yb^{166}	101.8±0.3	329.7±1.0	667±2	1097±3	5.82
Yb^{168}	87.54±0.07	289*	591*	978*	6.83
Yb^{170}	84.26±0.02	277.8±0.2	—	—	7.09
Yb^{172}	78.70±0.01	260.32±0.10	540.0±0.4	—	7.60
Yb^{174}	76.46±0.01	252±5	—	—	7.81
Yb^{176}	82.13±0.02	—	—	—	(7.30)
Hf^{166}	158.7±0.5	470.7±1.4	898±3	1407±4	3.61
Hf^{168}	123.9±0.4	385.0±1.1	756±3	1212±4	4.70
Hf^{170}	100.0±0.3	320.6±1.0	641±2	1041±3	5.90
Hf^{172}	94.5±0.3	307.9±0.9	627.0±1.9	1036±3	6.29
Hf^{174}	90.9±0.5	297.6±1.0	608.6*	1009.8*	6.54
Hf^{176}	88.35±0.01	290.3±0.3	601±7	—	6.75
Hf^{178}	93.14±0.02	306.8±0.2	632.6±0.4	1059.6±0.6	6.41
Hf^{180}	93.33±0.02	308.58±0.13	641.1±0.3	1084.9±0.7	6.41
W^{172}	122.9±0.4	376.9±1.1	727±2	1147±3	4.72
W^{174}	111.9±0.4	355.0±1.1	704±2	1137±3	5.26
W^{176}	108.7±0.4	348.5±1.1	699±2	1140±3	5.43
W^{178}	108	346	701	1156	5.46
W^{180}	103.6	337.6	688.3	1139.2	5.74
W^{182}	100.092±0.012	329.36±0.05	680.4±0.2	—	5.96
W^{184}	111.13±0.06	363.97±0.12	—	—	5.36
W^{186}	122.48±0.08	400±5	—	—	4.86
Os^{184}	119.8±0.3	383.8±1.2	775.0±1.6	—	4.93
Os^{186}	137.03	433.79	868.62	—	4.29
Os^{188}	154.93±0.06	478.4±0.5	—	—	3.75
Os^{190}	186.7±0.1	547.7±0.5	1047±3	1661±5	3.06
Os^{192}	205.80±0.02	489.15±0.22	—	—	2.60
Pt^{190}	295	—	—	—	—
Pt^{192}	316.51±0.02	784.60±0.06	—	—	1.71

The value of A was calculated from the Bohr-Mottelson two-term formula $A = \frac{5}{2}E(2^+) - \frac{3}{140}E(4^+)$; 100/A for Er^{170} and Yb^{176} are in brackets, since only $E(2^+)$ is known for them the value of $E(4^+)$ has been obtained by extrapolation from neighboring isotopes.

*Data taken from the schemes presented by Élbek at the Conference in Tbilis, 1964.

**Data obtained from the (α, xn) reaction.

TABLE 2 (continued)

b. States with Spins from 10^+ to 20^+

Nucleus	$E(10^+)$	$E(12^+)$	$E(14^+)$	$E(16^+)$	$E(18^+)$	$E(20^+)$
Dy^{158}	1512	2037	—	—	—	—
Dy^{160}	1442	—	—	—	—	—
Yb^{164}	1748 ± 5	2322 ± 7	2928 ± 9	—	—	—
Yb^{166}	1604 ± 5	2172 ± 7	2774 ± 8	3402 ± 10	—	—
Hf^{166}	1971 ± 6	2565 ± 8	3178 ± 10	—	—	—
Hf^{168}	1734 ± 5	2304 ± 7	2910 ± 9	—	—	—
Hf^{170}	1503 ± 5	2013 ± 6	2564 ± 8	3147 ± 9	(3762 ± 12)	(4414 ± 13)
Hf^{172}	1519 ± 5	2063 ± 6	2651 ± 8	3273 ± 10	3915 ± 12	—
W^{172}	1616 ± 5	2129 ± 7	2577 ± 8	3253 ± 10	3850 ± 12	—
W^{174}	1635 ± 5	2186 ± 7	2780 ± 8	—	—	—
W^{176}	1648 ± 5	2206 ± 7	2801 ± 8	3425 ± 11	—	—
W^{178}	1691					

Much earlier in the study of the rotational spectra of diatomic molecules terms of this type had been introduced to take into account the elongation of molecules under the effect of rotation. If on rotation a nucleus is deformed slightly, and its moment of inertia varies insignificantly, this may be allowed for by the inclusion of a second term. There may be, however, also other causes for the occurrence of such a term (variation of superconducting properties during rotation, etc.).

Comparison of the formula with the data in Table 2 shows that B is negative and almost 1000 times less than the value of A; the moment of inertia increases only slightly on rotation of the nucleus. The two-term formula with A and B gives good agreement with experiment only for low rotational energies. For $I \geq 8$, the second negative term is too large, the increase in the energy of the levels becomes slower at first, and then this energy begins to decrease. This has never been observed experimentally.

Figure 3 shows the levels of Hf^{170} calculated from the Bohr-Mottelson single-term, two-term and three-term formulas. The constant A in the one-term formula was determined according to $E(2_1^+)$, the constants A and B in the two-term formula were determined according to $E(2_1^+)$ and $E(4_1^+)$. The observed deviations from experiment are still more evident in Fig. 4 giving the dependences on I; the numerical values are given in Table 3. The appreciable lack of agreement with experiment calls for further refinement in the rotational formula.

Bohr and Mottelson [1] advanced arguments in support of the introduction of a third term in the form of $CI^3(I+1)^3$. Such a term is based on the assumption that, on rotation, the moment of inertia of the nucleus increases as the quadratic function ω^2 (ω being the angular velocity of rotation of the nucleus.

The introduction of the third term and the choice of the coefficients A, B and C according to the three levels (2_1^+, 4_1^+ and 6_1^+) naturally improves agreement in the level of the 8_1^+ level, but at higher spins, the discrepancy rapidly increases. This is well shown in Fig. 4 and Table 3.

Of course, the region of discrepancy may be shifted by the introduction of terms of type $DI^4(I=1)^4$ and so forth. Whatever form the function $E=f(I)$ may actually have, it may be represented as a polynomial in powers of $I(I+1)$. But if it is necessary to use many terms of the series, it means that the law $E=f(1)$ has not been divined well. The impression is created that

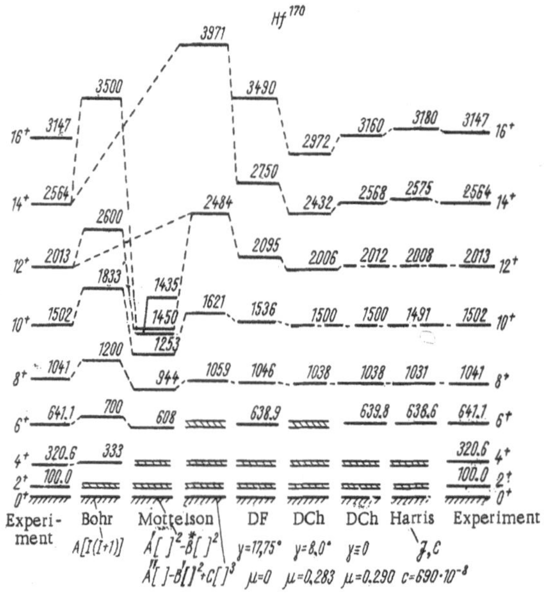

Fig. 3. Comparison of the experimental data on the ground-state rotational band of Hf^{170} with the predictions of various models: with the Bohr-Mottelson, Davydov-Filippov, Davydov-Chaban formulas and the Harris formula. For each model, when determining the parameters, use was made of the lower levels shown by double lines (one, two, or three, depending on the number of parameters in the model). The Davydov-Chabon model (for $\gamma = 0$) and the Harris model give the best agreement with experiment.

the Bohr-Mottelson formula is too formal and ought to be replaced by methods in which the form of the function $E = f(I)$ has been defined better.

Davydov-Filippov Model

In 1958, A. S. Davydov and G. F. Filippov [2] assumed the nucleus to be a nonaxially symmetrical body. To describe the shape of the nucleus, two parameters were introduced, the deformation β_0 and nonaxiality γ_0. For calculating the energies of the rotational band of such a nucleus, two simplifying assumptions had to be made: 1) that the parameters β_0 and γ_0 did not vary during rotation, i.e., the nucleus rotated without altering its shape, and 2) it was necessary to adopt some simple expression for the dependence of the principal moments of inertia of the nucleus on the parameters β_0 and γ_0.

The first assumption is natural in the initial stage of development of the model, while the second is more complex. If the nucleus is a solid or rigid body, expression of the three moments of inertia in terms of β_0 and γ_0 will not be difficult. If, however, calculation of the energy of the rotational levels is carried to the end, obvious contradictions with experiment become noticeable. The nucleus is a nonrigid triaxial ellipsoid. It thus becomes necessary to assign to the nucleus properties typical of the hydrodynamics of an ideal liquid or of a partly super-

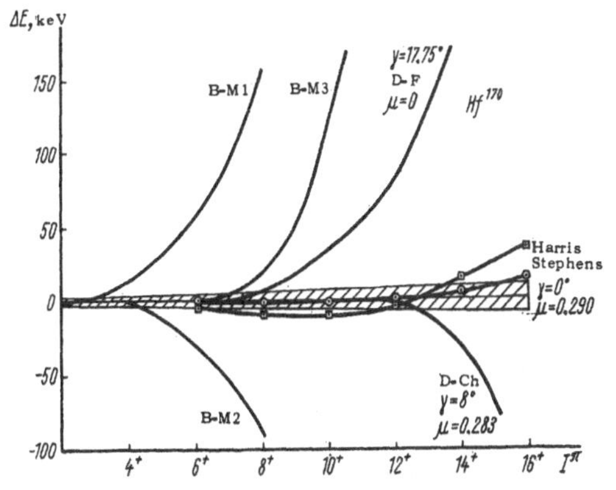

Fig. 4. Divergence of predictions of various models from the experimental energies of the levels (data the same as in Fig. 3). The divergent shaded strip is the error of the experimental values of the energies of the levels (0.3%). Agreement is best in the Stephens model (Davydov-Chaban for $\gamma = 0$) and Harris models. The Davydov-Chaban model (for $\gamma = 8°$) gives appreciable discrepancies for $I \geq 14$.

fluid material. A. S. Davydov and G. F. Filippov assumed that the moments of inertia depended on β_0 and γ_0 in the way in which this is prescribed by the hydrodynamic model of the nucleus. After the two assumptions had been made, the problem was finally solved. It was found that the rotation consisted of two parts, a normal part (with the sequence of spins $0_1, 2_1, 4_1, 6_1, 8_1, 10_1, \ldots$), and an anomalous part comprising bands with I equal to $2_2, 3_2, 4_2, 5_2, 6_2, \ldots$; $I = 6_4, \ldots$ The results were then compared with experiment. It was necessary to select two levels for determining the parameters β_0 and γ_0 (or J_0 and γ_0), and compare with experiment the energies of the remaining levels. A. S. Davydov and G. F. Filippov selected the levels 2_1^+ and 2_2^+ for determining the parameters. This choice was the simplest, but evidently not very happy, since there were always doubts as to whether a type-2^+ level belonging to the anomalous rotation band had actually been chosen, or some other type-2^+ level, by chance turning out to be low.

Such a choice of levels for determining the parameters J_0 and γ_0 is not obligatory. In making the calculations for the data of Fig. 1, we did not want to go beyond the limits of the principal rotational band, and used $E(2_1^+)$ and $E(4_1^+)$ for the determination of J_0 and γ_0. The results are given in Figs. 3 and 4 and Table 3. Discrepancies are observed at high values of I. It becomes necessary to modify the initial assumptions, to reject the model or to complicate it.

TABLE 3. Comparison of Experimental Values of the Energies of Levels of Hf^{170} Calculated from Different Models

Experiment		Bohr and Mottelson [1]			Davydov and Filipov [2] for γ = 17.75° (μ = 0)	Davydov and Chaban [3] for γ = 8° (μ = 0.284)	Davydov and Chaban [3] for γ = 0° (μ = 0.290)	Harris [5] for $J_0 = 2.952 \times 10^{-2}$ $C = 6.90 \times 10^{-8}$
Level	E, keV	$E = AI(I+1) = A[\,]$	$E = A'[\,] + B'[\,]^2$	$E = A''[\,] + B'[\,]^2 + C[\,]^3$				
2+	100.0	100.0 Assumed	100.0 Assumed	100.0 Assumed	100.0 Assumed	100.0 Assumed	100.0 Assumed	100.0 Assumed
4+	320.6	333	320.6 Assumed	320.6 Assumed	320.6 Assumed	320.6 Assumed	320.6 Assumed	320.6 Assumed
6+	641.1	700	608	641.1 Assumed	638.9	641.1 Assumed	639.8	636.6
8+	1041	1200	944	1059	1046	1039	1038	1031
10+	1502	1833	1253	1621	1536	1500	1500	1491
12+	2013	2600	1450	2484	2095	2006	2012	2008
14+	2564	3500	1435	3971	2750	2432	2568	2575
16+	3147	4533	1092	6656	3492	2972	3160	3180

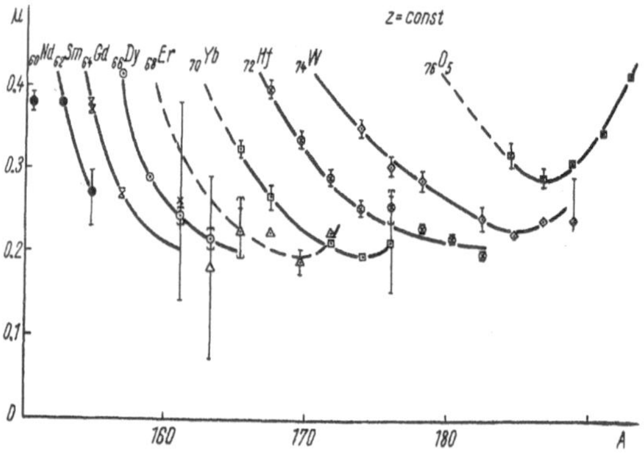

Fig. 5. Values of softness parameter μ for even–even nuclei calculated according to $E(2_1^+)$ and $E(4_1^+)$ (for $\gamma = 0$).

Davydov-Chaban Model

In 1960, A. S. Davydov and A. A. Chaban [3] proposed to modify the previous model of A. S. Davydov and G. F. Filippov and to allow for the possibility of β vibrations of the nucleus (vibrations along the major axis of the ellipsoid), when the effective value of β could be different in different states. Another parameter μ, the softness of the nucleus, was introduced for describing the properties of the nucleus. The assumptions about γ_0 and about the moment of inertia remained the same as in the Davydov-Filippov model, but it was proposed to assume that γ is the "effective" value in γ vibrations. It is possible to calculate the rotational band in this case also, but the final formulas are found too complicated for calculations. Sufficiently complete comparison with this model has therefore not been made.

For determining the parameters J_0, γ and μ, the energy of three levels must be used. It is natural to use the three lower levels 2_1^+, 4_1^+, and 6_1^+. Figures 3 and 4 and Table 3 give the results for Hf^{170}. Agreement of the results is very good up to $I = 12^+$ and any substantial deviation is observed only in levels with $I = 14^+$ and $I = 16^+$.

Stephens Modification

Stephens et al. [4] compared the large rotational bands they had found in nine nuclei with the various models. They discovered that the best agreement was obtained with the Davydov-Chaban model for $\gamma = 0$. As will be seen from Fig. 4, this agreement for Hf^{170} is really excellent; it is in the limits of 0.3 up to $1 = 16^+$; it is nearly as good also for the other eight nuclei they investigated.

For $\gamma = 0$, for determining J_0 and μ, it is sufficient to know $E(2_1^+)$ and $E(4_1^+)$. This information is already available for 39 nuclei (see Table 2). N. A. Voinova has calculated μ for these nuclei. The results are given in Fig. 5; the errors in the experimental points are due to errors in $E(2_1^+)$ and $E(4_1^+)$.

The assumption of $\gamma = 0$ gives rise to some difficulties; the anomalous band (γ-vibration levels) extends to infinity. Therefore, although agreement with experiment in the case of Hf^{170} is better for $\gamma = 0$ than for γ and μ taken from the energies of the 2_1^+, 4_1^+, 6_1^+ levels, it seems to us that the last choice of parameters is to be preferred.

Harris Formula

At the end of 1964, Harris [5] proposed a new approach to the calculation of rotational bands. He obtained surprisingly good agreement with the experimental bands with a very limited number of assumptions.

In classical mechanics, the energy and momentum of a rotating body are determined by the formulas

$$E = \frac{1}{2} J\omega^2, \quad P = J\omega,$$

where ω is the angular velocity and J is the moment of inertia. In quantum mechanics, P is quantized and is equal to $\hbar\sqrt{I(I+1)}$; if, now, we exclude velocity from the equations for E and P, the usual rotational formula $E = (\hbar^2/2J)\, I(I+1)$ is obtained. It may, however, be assumed that for some reason or other E and P contain terms with higher powers of ω:

$$\begin{cases} E = \frac{1}{2} J\omega^2 + \alpha\omega^4, \\ P = \hbar\sqrt{I(I+1)} = J\omega + \beta\omega^3. \end{cases}$$

Harris showed that the constants α and β in second approximation in the Inglis model may be reduced to one constant

$$\begin{cases} E = \frac{1}{2} J\omega^2 + \frac{3}{2} C\omega^4, \\ P = \hbar\sqrt{I(I+1)} = J\omega + 2C\omega^3. \end{cases}$$

Eliminating ω from these two equations, we get a cubic equation for the energy of the rotational band levels:

$$\frac{128C^2E^3}{J^3\hbar^6 I(I+1)} - \frac{32CE^2}{J\hbar^4 I(I+1)} + \frac{2JE}{\hbar^2 I(I+1)} + \\ + \frac{72CE}{J^2\hbar^2} - 1 - \frac{27CI(I+1)}{J^3} = 0. \quad (1)$$

Putting

$$\frac{C}{J^3} I(I+1) = x,$$

we have

$$E = \frac{\hbar^2}{2J} I(I+1)\{1 - x + 4x^2 - 24x^3 + \ldots\}.$$

Thus, if $C = 0$ or if $C \neq 0$, but x is very small, the original Bohr-Mottelson formula is obtained; if the term with x is taken into account, this is equivalent to the introduction of the second Bohr-Mottelson term $BI^2(I+1)^2$; if the term $4x^2$ is taken into account, this is equivalent to the introduction of $CI^3(I+1)^3$. Thus, at first sight, we arrive at the Bohr-Mottelson formula. Actually, however, for high values of I, the terms of the series increase and the solution of the cubic equation (1) must be used. For describing the whole of a rotational band, Harris requires only two parameters J and C, which he determines by the method of least squares from the energy of all the known levels. After this, for light nuclei, the mean deviation for the levels up to 12^+ is approximately 0.4% with an error in the experimental data of 0.3%.

The constants J and C may, of course, be determined not from all the levels but, for example, from the two low levels, $E(2_1^+)$ and $E(4_1^+)$. The results are given in Figs. 3 and 4 and in Table 3.

Conclusions Regarding Rotational Formulas

The following conclusions may be drawn from the comparison of all the models with experimental data.

1. The Bohr-Mottelson rotational formulas with one, two and three terms do not describe well the experimental facts.

2. The Davydov-Filippov model is also in poor agreement with experiment.

3. The Davydov-Chaban model and the Harris formula are suitable for further analysis.

A special analysis is required for the choice of the optimum modification in the Davydov-Chaban model. In the Davydov-Chaban model, use is made of physically intelligible parameters, while in the Harris formula, the parameter C still has a formal value. A number of properties of the β band, and if $\gamma \neq 0$ also of the γ band follow from the Davydov-Chaban model; the Harris method provides no information on these bands.

Physical Characteristics of the Ground States of Even – Even Nuclei

At the present time nuclear states may be characterized by two overall characteristics, mass and charge, and three characteristics reflecting the distribution of the masses, charges, and currents.

J – Moment of inertia reflecting the distribution of the masses;

Q_0 – the electrical quadrupole momentum, reflecting the distribution of the charges;

g_R – gyromagnetic ratio, reflecting the distribution of currents during rotation.

We shall commence the discussion with the quadrupole moments.

Electrical Quadrupole Moment Q_0

The quadrupole moment Q_0 mainly reflects the statistics – the deviation in the distribution of the charges from the spherical. This means principally that electrostatics alone does not contribute to the experimental value of Q_0. Thus, for example, if the charges vibrate about a spherically symmetrical equilibrium shape (for example, β vibrations), the mean value of $|Q_\beta|$ with time may differ from zero.

It is known that the quadrupole moments of spherical odd nuclei sometimes attain 2 barns, while the quadrupole moment caused by the orbital motion of one odd nucleon $q = -\frac{2j-1}{2j+1} \times \langle \bar{r}^2 \rangle$, i.e., for ordinary small values of j, has a value of about 0.4 barn and never exceeds 1 barn. It is not clear what produces the excess quadrupole moment in spherical nuclei, whether it is the protons which attract each other and dynamically deform the nucleus (travelling wave) or whether it is the overall vibrations of the nucleus.

In even–even nuclei, the effective quadrupole moment Q is connected to Q_0 by the formula:

$$Q = \frac{I_0(2I_0 - 1)}{(I_0 + 1)(2I_0 + 3)} Q_0,$$

always equal to zero, since $I_0 = 0$. Optical measurements are therefore precluded and only one method of measuring Q_0 remains: the measurement of the lifetime of the rotational levels. We shall therefore examine these experimental data and then the values of Q_0 given in Fig. 9 and Table 2.

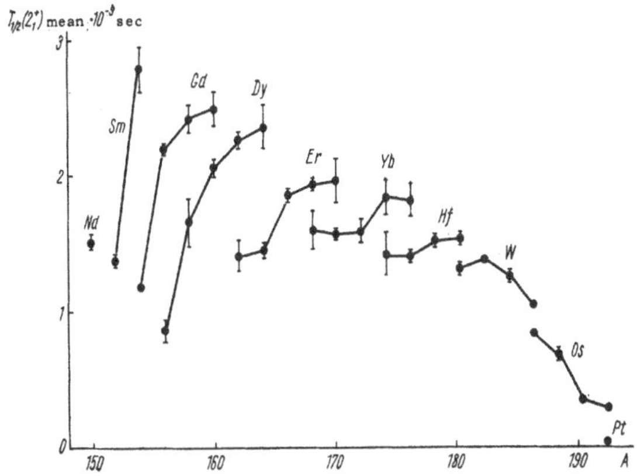

Fig. 6. Experimental data on half-life values of the 2_1^+ levels; the results of all methods have been taken into account.

Half-Life Values of Rotational Levels

$T_{1/2}(2_1^+)$. Experiments

At the present time, experimental data are available for almost all the 2_1^+ levels (Fig. 6, Table 2).

These data have been obtained by two methods:

1) From the delay in $\gamma-\gamma$ or $\beta-\gamma$ coincidences, in which the $2_1^+ \to 0_1^+$ transition is the second;

2) from the delay in γ radiation after the formation of the 2_1^+ state in a nuclear reaction or in Coulomb excitation by a pulsating beam;

3) by the method of Coulomb excitation, in which $\varepsilon_\gamma B(E2)$ is determined (from the γ rays) or $\varepsilon_k B(E2)$ is determined (from conversion electrons);

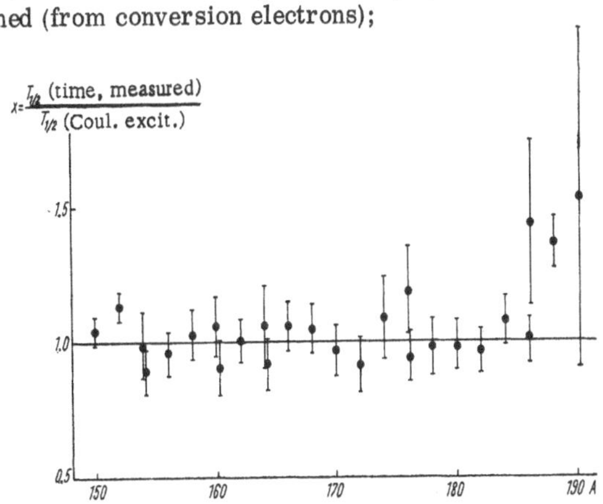

Fig. 7. Graph showing that the two methods, direct measurement of lifetime and Coulomb excitation, give results agreeing within the limits of experimental errors.

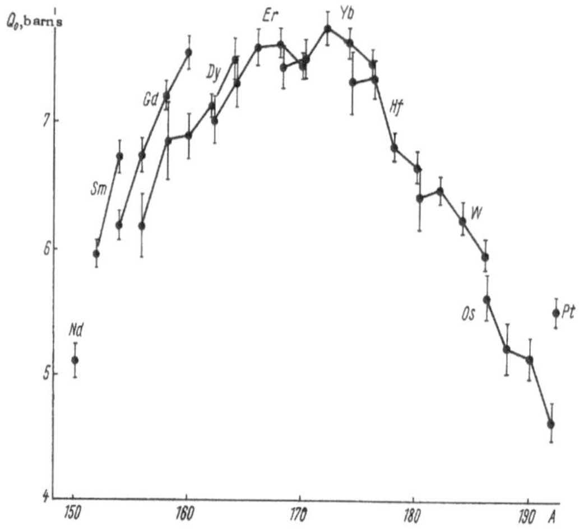

Fig. 8. Comparison of $t_{exp} = [T_{1/2}(4_1^+)/T_{1/2}(2_1^+)]_{exp}$ and $t_{theor} = [T_{1/2}(4_1^+)/T_{1/2}(2_1^+)]_{calc}$; the $4^+ \to 2^+$ transition is ~10% faster than predicted by the Bohr-Mottelson theory. The introduction of the Davydov-Ovcharenko correction reduces the discrepancy.

4) by means of the Mössbauer effect for one nucleus only – Yb^{170}; $T_{1/2} = (1.5 \pm 0.2)$ nsec.

In the first and second methods, the luminescence (deexcitation) time was measured, and in the third method, it was calculated from the formula

$$T_{1/2}(2_1^+) = \frac{283}{(E^{\bullet}_{2+})^5_{keV} B(E2, 0+ \to 2_1^+)(1 + \Sigma\alpha)} \text{ sec},$$

where $\Sigma\alpha$ is the total conversion coefficient for E2 transitions.

Since the methods of direct time measurement and probability of Coulomb excitation are quite different in both technique and theory, it is necessary to make sure that they give the same results. The data in Fig. 7 show that there is no systematic disagreement in the limits of the errors of measurement given by the authors. This renders the two methods equivalent and confirms that $\Sigma\alpha$ is given correctly by the tables in the limits of 5%.

The mean values of $T_{1/2}(2_1^+)$ from all the data are given in Table 2 and Fig. 6.

The value of $T_{1/2}(2_1^+)$ is not a smooth function of A since the values of $E(2_1^+)$ (and consequently $\Sigma\alpha$) and β (and consequently Q_0) are not smooth functions of A. Since the values of $E(2_1^+)$ are known with much greater accuracy than Q_0, the experimental data on $T_{1/2}(2_1^+)$ are used primarily for the determination of Q_0^2 and β.

$T_{1/2}$. Theory

The value of Q_0^2 is determined from the experimental values of $T_{1/2}(2_1^+)$ or $B(E2, 0 \to 2)$ from the formulas

$$Q_0^2 = \frac{2840}{E^5_{keV}(T_{1/2})_{sec}(1 + \Sigma\alpha)} \text{ barns}^2,$$

$$Q_0^2 = \frac{3.2\pi B\,(E2,\ 0 \to 2)_{e^2\,\mathrm{barns}^2}}{e^2} = \frac{10.04}{e^2} B\,(E2,\ \ 0 \to 2)_{e^2\,\mathrm{barns}^2}\ \mathrm{barns}^2$$

The results of calculations of Q_0 from these formulas are given in Fig. 8.

Both formulas are based on the assumption that the nucleus rotates without any change in shape or internal structure, Q_0 being an intrinsic constant of this structure, and the probabilities of all rotational transitions are determined by it and by the values of E and $\Sigma\alpha$.

In the Bohr-Mottelson and Davydov-Filippov models, the structure of the nucleus is not considered, and the assumption is made that it is the same in all rotational states. The formulas mentioned are therefore fully applicable. In models which take into account the structure of the nucleus or its variation by the action of rotation, the formulas mentioned are not sufficient and require refinement. Thus, for example, the Bohr-Mottelson model assumes that, in the 2_1^+, 4_1^+, etc., states, there is added to the wave function with K = 0 the wave function with K = 2 (admixture of β-vibrational band). Such mixing may lead to a slight change in Q_0, β, and $T_{1/2}$ ($T_{1/2}$ would be slightly increased). These corrections have not yet been calculated.

In the Davydov-Chaban model, the difference from zero of the softness parameter μ ought to alter B(E2) slightly, and consequently also $T_{1/2}$ (a paper by A. S. Davydov and V. I. Ovcharenko deals with this point).

In $4^+ \to 2^+$, $6^+ \to 4^+$, etc., transitions, the components M3 and E4 are not forbidden; if they exist, they may also alter $T_{1/2}$. These corrections also have not been considered.

The variations in $T_{1/2}$ are not large but they are different in $T_{1/2}(2_1^+)$, $T_{1/2}(4_1^+)$, etc., and therefore it may be possible to discover them with further improvement in the accuracy of measurements of $T_{1/2}(4_1^+)/T_{1/2}(2_1^+)$, etc.

$T_{1/2}(4_1^+)$. Experimental Data

The 4_1^+ states have a much shorter lifetime than the 2_1^+ states, less than 2×10^{-10} sec. Such short intervals of time are at the limit of the possibilities of the method of delayed coincidences; however, most of the information has been obtained by this method. Coulomb excitation is only possible in multiple form and this creates additional difficulties. All the results and ratios obtained are given in Table 4.

$$t_{\mathrm{exp}} = \frac{T_{1/2}(4_1^+)_{\mathrm{exp}}}{T_{1/2}(2_1^+)_{\mathrm{exp}}}.$$

TABLE 4. Half-life $T_{1/2}(4_1^+)$

Nucleus	Energy of $4^+ \to 2^+$ transition, keV	$T_{1/2}(4_1^+)$, nsec	$100\, t_{\mathrm{exp}} = 100\left[T_{1/2}(4+)/T_{1/2}(2+)\right]_{\mathrm{exp}}$	$100\, t_{\mathrm{theor}}$	$\frac{t_{\mathrm{theor}}}{t_{\mathrm{exp}}}$
Gd^{154}	248.08	3.9 ± 0.5	3.3 ± 0.4	4.15 ± 0.16	1.25 ± 0.17
Gd^{156}	199.19	10 ± 2	4.6 ± 0.9	5.1 ± 0.3	1.10 ± 0.23
Dy^{160}	197.0	8.3 ± 0.9	4.1 ± 0.5	5.4 ± 0.4	1.33 ± 0.18
Dy^{162}	185.2	13.2 ± 0.8	5.9 ± 0.4	6.2 ± 0.5	1.05 ± 0.11
Er^{166}	184.4	11.9 ± 0.8	6.5 ± 0.5	6.9 ± 0.6	1.07 ± 0.11
Er^{168}	184.5	12.1 ± 0.8	6.3 ± 0.4	6.8 ± 0.6	1.07 ± 0.11
Hf^{180}	215.25	7.1 ± 0.9	4.7 ± 0.6	5.0 ± 0.4	1.07 ± 0.16
Os^{186}	296.76	3.5 ± 2.0	4.9 ± 2.8	3.10 ± 0.16	$0.63^{+0.8}_{-0.2}$
Os^{188}	323.5	12.5 ± 0.4	1.9 ± 0.6	2.57 ± 0.08	$1.38^{+0.6}_{-0.3}$
Os^{190}	361.0	2.8 ± 1.4	8.5 ± 4	2.59 ± 0.06	$0.30^{+0.33}_{-0.10}$
Os^{192}	283.4	3.5 ± 0.5	12.7 ± 2.0	17.4 ± 0.04	1.37 ± 0.22

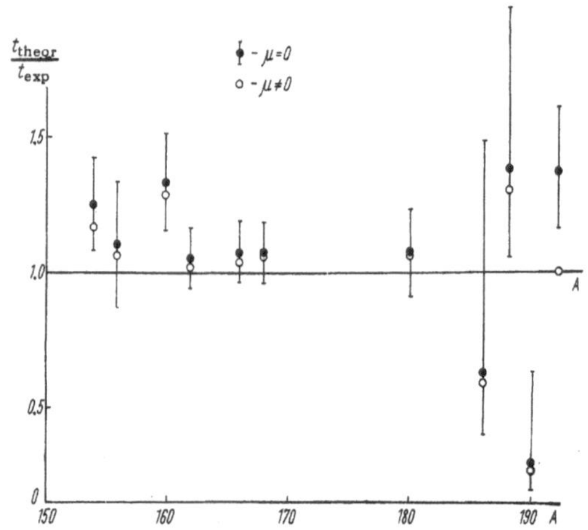

Fig. 9. Quadrupole moments of deformed even–even nuclei (graph constructed from the data of Fig. 6).

$T_{1/2}(4_1^+)$. Comparison with Theory

The theoretical value of t is obtained for the rotating invariable nucleus, for pure E2 transitions:

$$t_{theor}=\left(\frac{E_{2\to0}}{E_{4\to2}}\right)^5\frac{1+\Sigma\alpha_{4\to2}}{1+\Sigma\alpha_{2\to0}}\left(\frac{<_{2\to0}>}{<_{4\to2}>}\right)^2$$

($<2\to0>$ is the abbreviated symbol for the Clebsch-Gordan coefficient). Table 4 gives the values of t_{theor} and the ratio t_{theor}/t_{exp}. The latter values are also given in Fig. 9. The results, which relate to different nuclei, are not in bad agreement; only the points for Os^{181} and Os^{188} do not come within the overall relationship and require verification. The impression is gained that the ratio t_{theor}/t_{exp} is only 5 to 10% greater than unity: the $4^+\to2^+$ transition is faster than predicted by theory.

The causes which may give rise to these slight discrepancies have already been enumerated.

One of the causes, variation in shape during rotation, may be taken into consideration by using the Davydov-Chaban theory. A. S. Davydov and V. I. Ovcharenko calculated the correction for the softness of the nucleus in the form of an additional factor $S(I, \mu)$. For values of μ obtained from the energetics of the ground-state band, this correction is several percent; with the necessary sign, it also brings the experimental and calculated points closer together (see Fig. 8).

Moments of Inertia of Ground States of Even – Even Nuclei

The value of the moment of inertia J_0 may be determined approximately from the energy of the first excited level by means of the formula

$$E=\frac{\hbar^2}{2J_0}I(I+1)=AI(I+1),$$

whence

$$J_0=\frac{\hbar^2}{2A}.$$

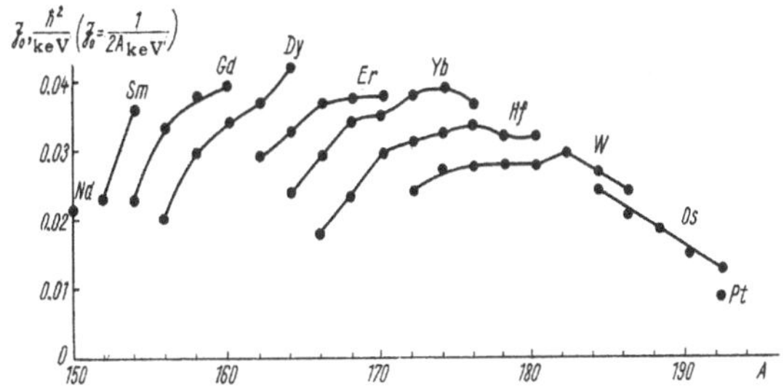

Fig. 10. Moments of inertia of ground states of deformed even–even nuclei; calculated according to the two term Bohr-Mottelson formula.

Often, J_0 is given not in grams per square centimeter, but in the units $\hbar^2$ per kiloelectron-volt ($1\hbar^2/\text{keV} = 6.940 \times 10^{-46}\ \text{g} \cdot \text{cm}^2$); in these units $J_0 = 1/2A$ keV.

With the introduction of complicated rotational formulas, the value of A varies somewhat (for example, with the introduction of the term $BI^2(I+1)^2$, A is increased by 0-4%), and therefore we agree upon the formula which may be used.

The last column of Table 3 gives the values of 100/A keV obtained by means of the Bohr-Mottelson two-term formula, i.e., determination of the parameters A and B from the energy of the 2_1^+ and 4_1^+ levels (C scarcely affects J_0).

The value of J_0 depends on Z and A. However, they are all less by a factor of 2-6 than the moment of inertia of a uniform ellipsoid of revolution with the same volume, mass and deformation parameter. This fact has long been known and many articles have dealt with its explanation.

The moment of inertia of an ordinary classical body is equal to $\int r^2 dm$ and is independent of the state of the substance and chemical bonds. If the observed moment of inertia is less than the calculated value, this means that the mass is concentrated closer to the axis of rotation, or that not all the mass is participating in the rotation. It is difficult to reject the assumption that the density of nuclear matter is almost uniform over the radius of the nucleus. It is just as impossible to imagine how there can be any portion of the nucleus which does not participate in the rotation; all the nucleons are always moving about the entire nucleus. To explain the low moments of inertia, therefore, it is necessary to resort to the concept of an ideal liquid or the superfluidity of nuclear matter. The concept of an ideal liquid leads to a moment of inertia which is too low, while concepts of superfluidity enable one to explain the low moments of inertia and even the strange shape of the curves in Fig. 10. This has been done by Nilsson and Prior [8]. The mean difference between theory and experiment is about 10% which is typical of calculations connected with superfluidity.

Values of g_R for Even – Even Nuclei

As is well known, the magnetic moments of the ground states of even–even nuclei are equal to zero; the nucleons move as if all their moments, magnetic and mechanical, orbital and spin, were compensated pairwise. On rotation of the nucleus, however, the magnetic and mechanical moments differ from zero; in quantum mechanics, as distinct from classical physics, it is necessary for this that the body should not be spherically symmetrical.

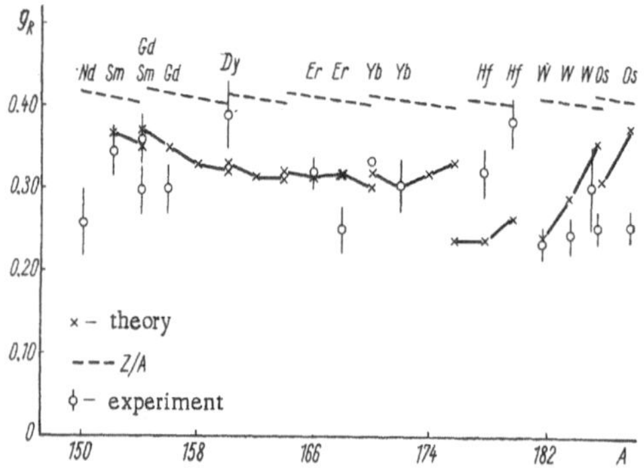

Fig. 11. Gyromagnetic ratios g_R for even–even nuclei.

Thus, in the 2_1^+, 4_1^+,... states, nuclei have a magnetic moment. Brady and Deutsch [6] were the first to point out the possibility of measuring these moments from the variation in the angular $\gamma-\gamma$ correlation in a magnetic field; if the first γ-quantum results in the formation of an excited state and the second in its disappearance, a certain time elapses between these acts; the nucleus, precessing in a strong magnetic field, succeeds in turning slightly, which affects the angular correlation. In the past few years, many modifications and refinements of this method have appeared. A detailed description of the method is given in [7].

The principle of the method, however, has not changed. Experimental results are usually expressed in the form of values of g_R, the gyromagnetic ratio.

In classical electrodynamics, it is shown that for any rotation of a system of charges having the same value of e/m_0

$$g_R = \frac{\mu}{P} = -\frac{1}{2c} \cdot \frac{e}{m_0}.$$

This formula is also retained in quantum mechanics. It may be written in more convenient form by expressing the mechanical moments in the units $\hbar$ and the magnetic moments in corresponding Bohr magnetons:

$$g_R = \frac{1}{2c} \cdot \frac{e}{m_0} \cdot \frac{\hbar}{\frac{\hbar e}{2m_0 c}} = 1.$$

In considering the rotation of a nucleus, the following complications are possible.

1. On rotation of the nucleus, the neutrons naturally do not contribute to the orbital moment, but to the mechanical moment, as if they made the protons heavier. It is therefore possible for g_R to have the value $Z/A = 0.4$.

2. The protons and neutrons have their own magnetic moments and spins; it is impossible to say how this affects g_R in the general form. However, by considering only the rotational states of the ground-state band, it is possible to ignore these factors: In the ground state, all moments are compensated. In a relatively slow rotation, this compensation ought not to be disturbed.

3. If quantum mechanical superfluidity exists in the nucleus, some of the protons and neutrons will not, so to speak, participate in the rotation.

If equal parts (for example, 50%) of the protons and neutrons do not participate in the rotation, μ and P will be diminished, but g_R will retain its previous value. Consequently, the experimental value of g_R, its difference from Z/A, reflects the different relative functions of the participation of protons and neutrons in the collective motion.

Let us examine the experimental data.

Table 5 gives all the experimental results obtained up to May 1, 1965. The weighted average values of g_R are given in Fig. 11. Corrections on account of the refinement of $T_{1/2}(2_1^+)$

TABLE 5. Experimental Values of g_R for Even–Even Nuclei

Isotope	Authors	Method	Type of source	g_R as given by the authors	Constants used by the authors: G_2 and G_4	β	$T_{1/2}$, nsec	Constants corrected by the authors: β	$T_{1/2}$, nsec	Refinement of g_R
Nd^{150}	Goldring et al.	$p_c, \gamma(H)$	Liquid	0.22 ± 0.04	$G_2=1\pm0.035$	2.3	1.65	2.095	1.51	0.26 ± 0.04
Sm^{152}	Goldring et al.	$p_c, \gamma(H)$	»	0.21 ± 0.04	$G_2=1\pm0.035$	1.7	1.40	1.121	1.37	0.31 ± 0.06
	Manning et al.	$\gamma, \gamma(H)$	»	0.28 ± 0.07	$G_2=1\pm0.13$	1.15	1.4	1.121	1.37	0.28 ± 0.07
	Sugimoto et al.	$p_c, \gamma(H)$	Solid	0.34 ± 0.16 (78)	—	1.07	—	1.121	1.37	0.32 ± 0.15
	Bauer et al.	$\gamma, \gamma(H, t)$	Liquid	0.35 ± 0.03	$G_2=1\pm0.03$	1.042	1.42	1.121	1.37	0.34 ± 0.030
	Debrunner et al.	$\gamma, \gamma(H)$	»	0.115 ± 0.075	$G_2=1.09\pm0.14$ $G_4=0.97\pm0.23$	—	1.40	1.121	1.37	0.114 ± 0.074*
										$\bar{g}_R=0.34\pm0.030$
Sm^{154}	Goldring et al.	$p_c, \gamma(H)$	»	0.21 ± 0.04	$G_2=1.10\pm0.10$	1.7	2.63	1.121	2.76	0.30 ± 0.03
Gd^{154}	Stiening et al.	$\gamma, \gamma(H)$	»	0.36 ± 0.06	$G_2=0.82\pm0.02$	—	—	1.121	1.18	0.36 ± 0.06
	Manning et al.	$\gamma, \gamma(H)$	—	0.40 ± 0.05	—	1.2	1.2	1.121	1.18	0.40 ± 0.05
	Bauer et al.	$\gamma, \gamma(H, t)$	»	0.367 ± 0.03	$G_2=0.93\pm0.01$	1.042	1.18	1.121	1.18	0.34 ± 0.03
										$\bar{g}_R=0.36\pm0.03$
Gd^{156}	Bauer et al.	$\gamma, \gamma(H, t)$	»	0.32 ± 0.03	$G_2=1\pm0.03$	1.042	2.22	1.121	2.16	0.305 ± 0.030
Dy^{160}	Manning et al.	$\gamma, \gamma(H)$	—	0.28 ± 0.08	$G_2=0.8\pm0.12$	6.0	1.7	6.37	2.03	0.22 ± 0.06
	Debrunner et al.	$\gamma, \gamma(H, t)$	Liquid Solid	0.18 ± 0.08	—	—	1.8	6.37	2.03	0.16 ± 0.06*
	Künding	$\gamma, \gamma(H, t^{\circ})$	Liquid Solid	0.46 ± 0.05	$G_2=0.84\pm0.06$ $G_4=0.62\pm0.04$	—	1.8	6.37	2.03	0.41 ± 0.04
	Cohen et al.	$\gamma, \gamma(H)$	Liquid	0.32 ± 0.10	$G_2=0.58\pm0.11$	—	1.8	6.37	2.03	0.28 ± 0.09
	Cohen et al.	Mössbauer effect	—	0.37	—	—	—	—	—	0.37
	Bauminter et al.	—	—	0.48 ± 0.06	—	—	—	6.37	2.03	0.48 ± 0.06
										$\bar{g}_R=0.39\pm0.04$
Er^{166}	Manning et al.	$\gamma, \gamma(H)$	Liquid	0.31 ± 0.06	$G_2=0.78\pm0.12$ $G_4=0.55\pm0.05$	7.7	1.66	7.26	1.85	0.30 ± 0.06
	Künding	$\gamma, \gamma(H)$	»	0.36 ± 0.06	$G_2=0.63\pm0.06$ $G_4=0.62\pm0.05$	—	—	7.26	1.85	0.36 ± 0.06
	Bodenstedt et al.	$\gamma, \gamma(H)$	»	0.260 ± 0.034	$G_2=0.80\pm0.08$ $G_4=0.56\pm0.04$	8.1	1.98	7.26	1.85	0.32 ± 0.02
										$\bar{g}_R=0.32\pm0.02$
Er^{168}	Bodenstedt et al.	$\gamma, \gamma(H)$	»	0.25 ± 0.03	$G_2=0.86\pm0.07$	7.26	1.92	7.26	1.91	0.25 ± 0.03
Yb^{170}	Hrynkiewicz	Mössbauer effect	»	0.334 ± 0.005	—	—	1.6	2.677	1.56	0.334 ± 0.005
Yb^{172}	Bodenstedt	—	—	0.304 ± 0.034	—	—	1.57	2.677	1.58	0.304 ± 0.034
Hf^{176}	Karlsson et al.	$\gamma, \gamma(H)$	»	0.29 ± 0.02	$G_2=0.74\pm0.09$ $G_4=0.818\pm0.046$	—	1.50	7.26	1.49	0.29 ± 0.02
Hf^{178}	Bodenstedt et al.	$\gamma, \gamma(H)$	Liquid	0.356 ± 0.035	$G_2=0.541\pm0.068$ $G_4=0.573\pm0.032$	—	1.49	7.26	1.49	0.356 ± 0.035
										$\bar{g}_R=0.32\pm0.03$
Hf^{180}	Bodenstedt et al.	$\gamma, \gamma(H)$	»	0.371 ± 0.032	$G_2=0.86\pm0.05$	—	1.49	—	1.52	0.380 ± 0.032
	Deutsch et al.	—	—	0.29 ± 0.04	—	—	—	—	—	0.29 ± 0.04
										$\bar{g}_R=0.341\pm0.032$
W^{182}	Kegel	$p_c, \gamma(H)$	—	0.404 ± 0.027	—	—	—	—	1.37	0.404 ± 0.027*
	Koerner et al.	$\gamma, \gamma(H, t)$	»	0.336 ± 0.044	—	—	1.41	—	1.37	0.346 ± 0.045*
	Goldring et al.	$p_c, \gamma(H)$	Solid	0.193 ± 0.018	$G_2=0.925\pm0.030$ $G_4=1$	—	1.43	—	1.37	0.201 ± 0.019
	Scharenberg et al.	$\gamma, \gamma(H, t)$	»	0.239 ± 0.020	—	—	1.37	—	1.37	0.239 ± 0.020
	Klyucharev et al.	$\gamma, \gamma(H)$	Metallic	0.247 ± 0.037	$G_2=0.82$ $G_4=1$	1	—	—	1.37	0.247 ± 0.037
	Klyucharev et al.	$\gamma, \gamma(H)$	Liquid	0.323 ± 0.048	$G_2=1$ $G_4=1$	1	—	—	1.37	0.323 ± 0.048
										$\bar{g}_R=0.23\pm0.02$
W^{184}	Goldring et al.	$p_c, \gamma(H)$	Solid	0.207 ± 0.016	$G_2=0.940\pm0.015$ $G_4=1$	—	1.24	—	1.24	0.207 ± 0.016
	Bodenstedt et al.	—	Liquid	0.38 ± 0.05	$G_2=0.54$ $G_4=0.67$	1	1.28	—	1.24	0.39 ± 0.05
	Scharenberg et al.	$p_c, \gamma(H, t)$	Metallic	0.275 ± 0.025	—	—	1.28	—	1.24	0.279 ± 0.025
										$\bar{g}_R=0.239\pm0.024$
W^{186}	Goldring et al.	$p_c, \gamma(H)$	Solid	0.292 ± 0.027	$G_2=0.940\pm0.015$ $G_4=1$	—	1.01	—	1.03	0.286 ± 0.026
	Scharenberg et al.	—	—	0.368 ± 0.042	—	—	1.01	—	1.03	0.36 ± 0.04
										$\bar{g}_R=0.308\pm0.034$
W**	Klein-Tebble et al.	$p_c, \gamma(H)$	Solid	0.264 ± 0.023	$G_2=0.966\pm0.049$ $G_4=1.07\pm0.15$	1	—	—	—	0.264 ± 0.023
Os^{186}	Bodenstedt et al.	$\gamma, \gamma(H)$	Liquid	0.316 ± 0.028	$G_2=0.92\pm0.07$ $G_4=0.95\pm0.04$	—	0.84	—	0.82	0.316 ± 0.028
	Lerjefors et al.	$\gamma, \gamma(H)$	»	0.30 ± 0.08	$G_{4eff}=0.86\pm0.09$	—	0.70	—	0.82	0.25 ± 0.07
										$\bar{g}_R=0.30\pm0.05$
Os^{188}	Karlsson et al.	$\gamma, \gamma(H)$	»	0.36 ± 0.04	$G_2=0.78$ $G_4=0.87$	—	0.58	—	0.67	0.31 ± 0.03
	Goldring et al.	$p_c, \gamma(H)$	Metallic	0.20 ± 0.02	$G_2=1.025\pm0.030$ $G_4=0.94\pm0.12$	—	0.73	—	0.67	0.22 ± 0.02
	Shell et al.	—	The same	0.23 ± 0.03	$G_2=0.894\pm0.020$	—	0.71	—	0.67	0.24 ± 0.03
										$\bar{g}_R=0.25\pm0.02$

* Data not considered in the determination of $\bar{g}_R$.

** Measurements made on an unseparated mixture of isotopes of $W(W^{182}-W^{184}, W^{186})$.

and of the paramagnetic constant β have been made of all the results, following which the results were averaged in accordance with the errors as stated by the authors. The experimental values of g_R are less than Z/A. If the cause resides in pair correlations of superfluid type, they act more strongly on the protons than on the neutrons.

Calculation of the values of g_R can be made only by means of a model which takes superfluidity of nucleon matter into account. The results of such calculations by Nilsson and Prior are shown in continuous lines in Fig. 11. The agreement obtained is 10%.

LITERATURE CITED

1. Bohr, O., and Mottelson, B., At. Énerg., 14:41 (1963).
2. Davydov, A. S., and Filippov, G. F., Zh. Éksperim. i Teor. Fiz., 35:440 (1958).
3. Davydov, A. S., and Chaban, A. A., Nucl. Phys., 20:499 (1958).
4. Stephens, F., et al., Phys. Rev. Lett., 12:225 (1964); Diamond, R., et al., UCRL-11421; Stephens, F., Paper presented at the 15th All-Union Conference on Nuclear Spectroscopy, Minsk, January, 1965.
5. Harris, S., Phys. Rev. Lett., 13:663 (1964).
6. Brady, E., and Deutsch, M., Phys. Rev., 72:870 (1947).
7. Perturbed Angular Correlations, eds. E. Karlsson, E. Matthias, K. Siegbahn. North Holland Publ. Co., 1964.
8. Nilsson, S., and Prior, O., Mat. Fys. Medd. Dan. Vid. Selsk., Vol. 32, No. 16 (1960).

γ BANDS

Introduction

Almost all even–even nuclei have type-2^+ levels among the excited states. The nature of these levels will obviously vary.

We shall consider only deformed nuclei of medium weight and low excitation energies of up to 2 MeV. Even in this narrow range of nuclei and excitations, however, there are 2^+-levels, evidently of four types.

1. In almost all known nuclei, there are low-lying 2^+-levels having an excitation energy of 70–140 keV; these are the well-known first rotational levels of the 2_1^+ ground state. We shall not consider them at present.

2. In some nuclei type-2^+ levels are known, which belong to the β-vibrational bands (0_β^+, 2_β^+, 4_β^+, etc.). These bands are relatively easy to recognize from the presence of the 0^+ level. Available information about them will be given in the next lecture. In many nuclei, after the exclusion of groups 1 and 2, there remain 2^+ levels having an excitation energy of 600 to 1500 keV, on the average one level per nucleus, but in three nuclei (Hf^{178}, W^{182}, Os^{188}) two per nucleus are known. The energy of these levels is given in Table 6 and Fig. 12. There are a number of arguments in support of the levels shown in Fig. 12 belonging to at least two different types; we shall call them 2_ν^+ and 2_{gb}^+.

3. As long ago as 1956, Bohr, Mottelson, et al. [1] advanced the hypothesis that the majority of these levels, low states of γ-vibrational bands, states in which the vibrations disturb the axial symmetry of the nucleus, make the nucleus into a triaxial ellipsoid. These states must be of 2^+ type, and we shall henceforth call them 2_γ^+ states. Until 1956, it was possible merely to estimate the energy of such vibrations, on the basis of the general constants of nuclear matter (compressibility, elasticity, etc.). This energy was found to be about 1 MeV, corresponding to the average experimental values.

In 1958, A. S. Davydov and G. F. Filippov [2] advanced the hypothesis that a nucleus in the ground state has deformations, described by the parameters β_0 and γ_0 (an approximately triaxial

TABLE 6. Energy of 2^+ Levels (Excluding 2_1^+ and 2_β^+ Levels)

Nucleus	Energy, keV	Nucleus	Energy, keV
Nd^{150}	1070	Yb^{170}	(1482)
Sm^{152}	1086	Yb^{172}	1468
Sm^{154}	1440	Yb^{176}	1270
Gd^{154}	998	Hf^{178}	1164, 1483
Gd^{156}	1155	W^{180}	828
Gd^{158}	1185	W^{182}	1222, 1258
Gd^{160}	1020	W^{184}	904
Dy^{158}	947	W^{186}	730
Dy^{160}	966	Os^{186}	767
Dy^{162}	890	Os^{188}	633, 1461
Dy^{164}	770	Os^{190}	557
Er^{164}	870, 1308	Os^{192}	489
Er^{166}	787	Pt^{190}	598
Er^{168}	822	Pt^{192}	613
Er^{170}	930	Pt^{194}	618
Yb^{168}	(987)		

ellipsoid). Such a nucleus on rotation without change in shape ought to have among the excited states two type-2^+ levels, the levels 2_1^+ and 2_2^+; the first level belongs to the ordinary rotational band, and the second to the Davydov-Filippov anomalous band. The position of the 2_1^+ level depends little on the value of γ; the position of the 2_2^+ level, on the contrary, depends substantially on γ. For $\gamma \to 0$, the 2_2^+ level recedes into infinity.

At this stage, the Bohr-Mottelson and Davydov-Filippov concepts differed sharply. The former took into account the vibrations of the axially symmetrical nucleus, while the latter considered the rotation of the axially nonsymmetrical nucleus, of invariable shape. At that time, these concepts were often contrasted with each other, although quantitative comparisons could not be made, since the Bohr-Mottelson concepts were not expressed in any formula. Subsequently, however, the points of view began to approach each other. It was found that for explaining the observed ratios of the intensities of the transitions between the 2_γ^+ level and the ground state, it was necessary to assume the presence, in the 2_1^+, 4_1^+, etc., states of an admixture having K = 2 (Mottelson's Z correction [3]). Thus, only the ground state remained axially symmetrical (pure according to the quantum number K), while in the other cases nonaxiality was assumed. It was also found that equilibrium nonaxiality γ_0 in the Davydov-Filippov concept was equal to zero; there remained the γ vibrations with some effective $\tilde{\gamma}$ value, responsible for the position of the 2_2^+ level.

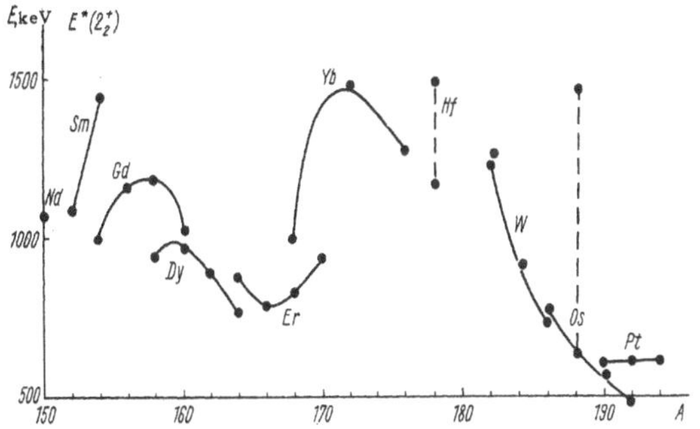

Fig. 12. Energy of type-2^+ levels (excluding the 2_1^+ and 2_β^+ levels).

Thus, it had already become more difficult to discriminate between these concepts; the one concerned vibrations (oscillations) and the other rotation. In both approaches however, it was assumed that the 2^+ levels had a collective nature. For the quantitative description of their properties, only the Davydov-Chaban model [4] was used, in which all the properties of the 2_2^+ states and the associated rotational states 3^+, 4^+, ..., necessarily followed from three parameters: moment of inertia j_γ, nonaxiality γ and softness of the nucleus μ. It should be mentioned that according to this model, there ought necessarily to be one 2_γ^+-level in each even–even nucleus. Judging by the values of γ, following from the energy of the 2_1^+, 4_1^+, and 6_1^+ levels, it should not exceed 2 MeV. It should also be mentioned that since we are speaking of collective levels, it is to be expected that their properties will vary smoothly on passing from one nucleus to the next.

4. The last type of levels, 2^+, are two-particle, resulting from the rupture of a nucleon pair and the distribution of the nucleons over previously unfilled levels. We shall denote them by 2_{gb}^+. The possibility of the existence of levels of this type was discussed long ago; it was clear that their excitation energy was of MeV order. However, a detailed study of the properties of these levels was made difficult by the fact that "superfluous" nucleons in the nucleus behave differently from free nucleons; they entrain other nucleons and polarize the remaining system of particles. Theoretically, they are already quasiparticles moving in an effective field. The study of the behavior of quasiparticles has received particular development with the appearance of the superfluid model of the nucleus. The properties of two-quasiparticle nonrotational levels have been discussed in detail in the lectures by V. G. Solov'ev. It should be noted that calculations made by V. G. Solov'ev [5, 6] predict on the average one state of 2_{gb}^+ type for the 1-2 MeV range. Consequently, in the 1-2 MeV range one may expect one collective 2^+-state and on the average one two-particle 2_{gb}^+-state. In fact, however, in the above-mentioned energy range, in most cases only one 2^+-state is found. There is undoubtedly a deficiency in the number of 2^+ states. It should be noted that the deficiency in the number of observed levels is found almost always on comparison of the two-particles levels of the superfluid model with experimentally observed levels; there are 15-20 theoretical levels in the 1-2 MeV range, and the observed number is one-third of this. Finally, in β decay, nuclear reactions or γ transitions, only those levels are observed which are close to the spin of the initial nucleus; however, the impression is created that one of these causes is insufficient to explain the disagreement. It is possible that in β decay, there are prohibitions connected with the classification according to ft; it is possible that some other prohibitions exist. It would be desirable to analyze the experimental data on the basis of a postulation of the following problem, an unusual one for experimenters: With what degree of reliability can it be confirmed that one level or another predicted by theory does not occur experimentally?

To simplify the terminology, we shall call the first type of 2^+ levels (after the exclusion of 2_1^+ and 2_β^+) γ-levels (2_γ^+), bearing in mind that possibly some of them will have to be transferred to the 2_{gb}^+ category.

Energies of γ Bands

2_2^+ Levels

Thirty type-2^+ levels are known, situated in the range from 450 to 1500 keV. The energy of the levels is shown in Table 6 and Fig. 12. It is not clear from the disposition of the points in Fig. 12 whether all the levels found are of one type. For classification, it is essential to have careful and diverse investigations of the physical properties of the levels: moments of inertia, quadrupole moments, gyromagnetic ratios, parameters of rotational fields, etc.

TABLE 7. Number of Levels in the 1-2 MeV Interval

Nucleus	Number of levels of 2^+_{gb} type in the 1-2 MeV interval	Number of other two-particle levels in the same interval
Dy^{160}	1	20
Dy^{162}	1	17
Er^{166}	2	13
Er^{168}	3	11
Yb^{172}	2	18
Hf^{178}	0	20
W^{182}	0	20

γ Bands

In the case of some nuclei, several levels in the γ band are already known, and in Er^{166} are six levels of the band, up to a level with spin 7. On the basis of the results of an analysis of the bands of the ground states, it would be possible to compare the energies of the levels with the Davydov-Chaban or Harris formulas. Knowing the energy of three levels, for example 2^+_γ, 3^+_γ, and 4^+_γ, it is possible to calculate the parameters j, γ, and μ relating directly to the γ band (we shall denote them by j_γ, γ_γ, and μ_γ). Just in the same way, from two levels, for example 2^+_γ and 3^+_γ, it is possible to calculate the Harris parameters j_γ and C_γ. There are no obstacles to such calculations, but they have not yet been done. In the meantime, it is quite possible that sets of j, γ, μ (or respectively j and C) for the ground-state β- and γ-bands will easily be distinguished. The Davydov-Chaban theory in its current state does not provide for any variation of the above-mentioned parameters over a band, or differences in the parameters in different bands. N. A. Voinova has calculated the energy of some type 3^+_γ- and 4^+_γ-levels for "compromise" parameters j, γ, and μ, determined from the energy of the 2^+_1, 4^+_1, and 2^+_γ levels. The results are given in Table 8. The energy of the 3^+_γ and 4^+_γ levels is less than the experimental values by 7-92 keV; at the same time, the energy of the 0^+_β, 2^+_β, and 4^+_β levels is greater than the experimental values by 60-220 keV. These results already show that sets of j, γ, and μ are different for different bands. Due to these uncertainties, attempts have been made to simplify this problem. Table 3 shows that in the ground-state band of Hf^{170}, the relative position of the first levels 0^+, 2^+, 4^+, 6^+ is described satisfactorily to an accuracy of 5% by a two-term formula of the type

$$E_I = E_0 + AI(I+1) + BI^2(I+1)^2.$$

TABLE 8. Energy of Rotational Levels of β and γ Bands of Some Nuclei

Nucleus	γ	μ	γ-band			β-band		
			γ and μ	Experiment	Δ	γ and μ	Experiment	Δ
Sm^{152}	11.5*	0.37*	2+ 1087	1087	0	0+ 762 0+	685	+83
			3+ 1143	1235	−92	2+ 891 2+	811	+80
			4+ 1243	1373	−130	1087 4+	1042	+135
			5+ 1363	1552	−189			
Gd^{154}	12*	0.37*	2+ 997.3	997.3	0	0+ 741	680.6	+60
			3+ 1084	1128.5	−44.5	2+ 895	815.7	+79
			4+ 1194	1265.3	−71.3	4+ 1194	1048	+146
Gd^{156}	10.5*	0.25*	2+ 1156.9	1156.9	0	0+ 1230	1010	+220
			3+ 1225	1251.0	−26			
			4+ 1326	1358.7	−36			
Er^{166}	12.7*	0.20*	2+ 787.6	787.6	0	0+ 1730	1460	+270
			3+ 832	860.6	−29			
			4+ 935	957.2	−22			
			5+ 1073.6	1073.6	−0.6			
			6+ 1225	1216	+9			
W^{182}	11.19	0.184	2+ 1221.8	1221.8	0			
			3+ 1324	1331	−5			
			4+ 1442	1437	−5			
W^{184}	13.77	0.170	2+ 904.4	904.4	0			
			3+ 1006.8	1006.8	0			
			4+ 1173.5	1134.7	+38.8			
W^{186}	15.73	0.222	2+ 730	730	0			
			3+ 859	830	+29			

Remark: The values of γ and μ are based on the energies of 2^+_1, 4^+_1, 2^+_γ levels.

*Data from the paper by Davydov; the other data calculated by N. A. Voinova.

For small values of I, the models of Davydov-Chaban and Harris reduce to this formula. We therefore attempt to apply the following analogous formula to the γ bands:

$$E_I = E_0 + A\{I(I+1) - I_0(I_0+1)\} + B\{I(I+1) - I_0(I_0+1)\}^2 = E_0 + Ax + Bx^2.$$

For all the γ bands we construct graphs of $\Delta E/(x_1 - x_2) = f(x_1 + x_2)$, where ΔE is the experimentally measured difference of the energies of two levels of the γ band, and x is as defined above. If this formula describes the experimental data well, the points lie on a straight line; the ordinate of the point of intersection of this straight line with the axis of ordinates determines the value of A_γ (and correspondingly $j_\gamma = 1/2A_\gamma$), and the slope determines B_γ.

The following conclusions may be drawn from an examination of Fig. 13, a and b.

1. There are no systematic deviations from the straight lines;

2. Individual points sometimes give deviations from the straight line of several keV. These may be experimental errors, and it is then necessary to redetermine some of the points, or they may be local displacements due to interaction of the levels. Such interactions in the γ band are to be feared, since there are many other levels in the energy region, in which the γ levels are situated. Practically almost nothing is known about these interactions and it would be desirable for a study of them to be made. It should be pointed out that in the bands of the ground states, such local displacements are very small and lie within the limits of the experimental errors.

Moments of Inertia of γ Bands

Figure 14 and Table 9 give the experimental values of the ratio J_γ/J_0. The moments of inertia J_γ are close to J_0; as a rule they differ by not more than 15%, sometimes by 25%. The old Davydov-Filippov theory predicted their exact equality [the rule is $E(3_\gamma) - E(2_\gamma) = E(2_1)$]. It is not yet known what the Davydov-Chaban formula gives: It is not clear what ought to be observed if the 2^+ level is two-particle. As is known, the addition of one nucleon to an even-even nucleus considerably increases the moment of inertia, the increase being from 6 to 50%, depending on the orbit of the added nucleon (see [7, 8]). This effect is not due to the addition of mass, since the subtraction of a nucleon from an even-even nucleus, results in an increase in the moment of inertia on the average by the same amount: evidently the odd nucleon to some extent impairs the superfluid properties of the even-even nucleus and causes the moment of inertia to approach its solid body value. The addition of two different nucleons to an even-even nucleus increases the moment of inertia still further: the additions of the two nucleons are combined (this is evidently expressed in the fact that by utilizing given odd nuclei, we take into account to some extent the transition from particles to quasiparticles). The statistics are not copious, the moment of inertia of only 15 odd-odd nuclei is known (see "Literature Cited" on p. 112), but agreement of the experimental and total additions is good. In view of this agreement, it may be expected that two-particle excited states of even-even nuclei will also have a moment of inertia greater than that of the ground state, and that it may be calculated by totalling the additions produced by the two nucleons. It is thus possible that the increase in the moment of inertia for two-particle levels will be found to be greater than the experimentally observed increase. From this point of view, all the 2^+ levels given in Table 14 are γ-vibrational 2_γ^+-levels. However, detailed calculations of the moments of inertia for two-particle systems are essential for the final assessment.

Lifetime of 2_γ^+ Levels

The 2_γ^+ levels have a lifetime of less than 10^{-11} sec, and therefore the direct measurement of the luminescence time is difficult to determine experimentally. Practically all the in-

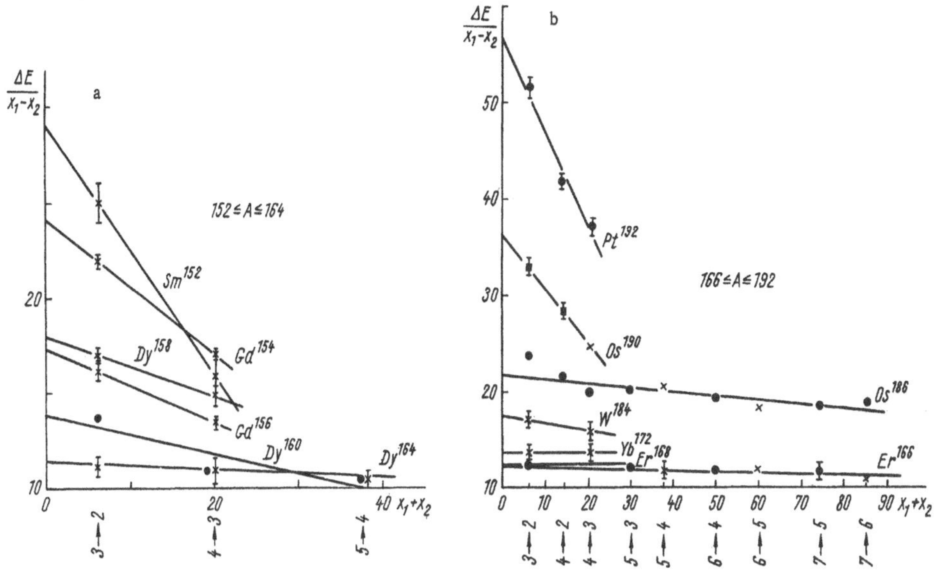

Fig. 13. Energies of bands based on 2^+-levels. ΔE) Energy difference of levels with spins I_1 and I_2, $x_i = I_i(I_i+1) - I_0(I_0+1)$. If the Bohr-Mottelson two-term formula satisfies experiment, the points ought to lie on a straight line. ×) ΔE transition was not observed, but energy levels are known between which it ought to take place; ●) ΔE measured directly.

formation is provided by the method of Coulomb excitation, the value of B(E2; $0^+ \rightarrow 2_\gamma^+$) being measured. Knowing this and the ratio of the two decay branches of the 2_γ^+ level, it is possible to calculate $T_{1/2}(2_\gamma^+)$; the corresponding values are given in Table 10. Since the distribution of the discharge intensities over the various channels, like the number of these channels, also depends on the individual properties of the nucleus, it is more convenient to consider the values of B(E2; $0^+ \rightarrow 2_\gamma^+$) directly. Known data are given in Fig. 15. The points are irregularly arranged. Points relative to Yb: for Yb^{176}, B(E2) is small, while for Yb^{172}, it has been possible to establish only the upper limit of B(E2; $0^+ \rightarrow 2^+$) < 0.04 barn. The anomalously low values of B(E2) permit one to assume that the states considered are not collective but two-particle states. It is curious that these are in fact the points for which the value of $E(2_2^+)$ is high; the collective 2^+ levels in the Yb isotopes are evidently still higher. At the 1964 Paris Conference it was stated that Hansen had succeeded in observing yet another 2^+-level in Yb^{172} with an energy of about 1700 keV; possibly, this is the 2_γ^+ level.

The Davydov-Filippov and Davydov-Chaban theories permit comparison of the values of B(E2; $0^+ \rightarrow 2_\gamma^+$)/B(E2; $0^+ \rightarrow 2_1^+$) with theory. Such a comparison was made by Artamonova for the

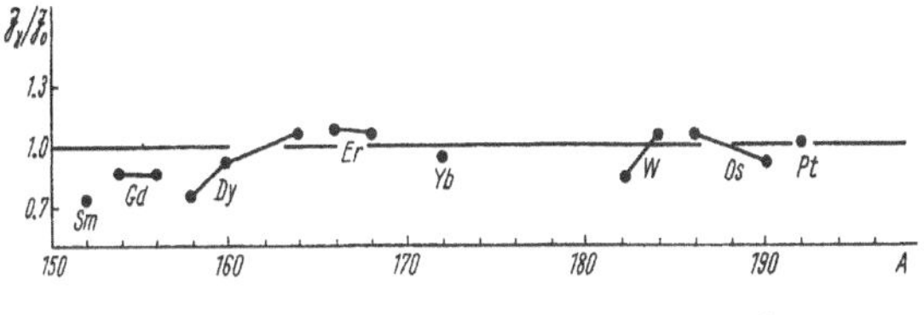

Fig. 14. Moments of inertia of γ bands.

TABLE 9. Moments of Inertia J_γ and J_0

Nucleus	γ Band A_γ, keV	Ground-state band A_0, keV	Ground-state band $J_\gamma/J_0 = A/A_\gamma$	Nucleus	γ Band A_γ, keV	Ground-state band A_0, keV	Ground-state band $J_\gamma/J_0 = A/A_\gamma$
Sm^{152}	29.1	21.14	0.726	Er^{168}	12.4	13.33	1.075
Gd^{154}	24.1	21.37	0.887	Yb^{172}	13.7	13.16	0.959
Gd^{156}	17.5	15.03	0.857	W^{182}	20.23	16.78	0.83
Dy^{158}	18.0	16.81	0.934	W^{184}	17.7	18.66	1.055
Dy^{160}	13.9	14.58	1.049	Os^{186}	21.9	23.31	1.06
Dy^{164}	11.2	12.2	1.09	Os^{190}	36.5	32.68	0.895
Er^{166}	12.4	13.51	1.09	Pt^{192}	56.9	58.48	1.03

Davydov-Filippov model in 1962 and showed satisfactory agreement between theory and experiment; no comparison has been made for the Davydov-Chaban model. There are evidently no calculations of this ratio for the superfluid model of the nucleus.

The values of $E(2^+_\gamma)$, J_γ/J_0, and $B(E2;\ 0^+ \to 2^+_\gamma)$ are all we have at present for the classification of the 2^+ states. This, however, is insufficient for drawing any confident conclusions.

Quadrupole Moments Q_γ and Deformation Parameter β_γ for Levels of the γ Band

For determining Q_γ it is necessary to know the partial probabilities of some E2 transitions inside the γ band. It is then possible to determine Q^2_γ by means of formulas of the type of

$$\lambda_{\gamma E2}(3^+_\gamma \to 2^+_\gamma) = cE^5_{3\gamma\to 2\gamma} Q^2_\gamma < 3220 \mid 3222 >^2,$$

$$\lambda_{\gamma, E2}(4^+_\gamma \to 2^+_\gamma) = cE^5_{4\gamma\to 2\gamma} Q^2_\gamma < 4220 \mid 4222 >^2$$

This would be the best method of determining Q^2_γ. It is possible to determine $\lambda_{\gamma E2}(3^+_\gamma \to 2^+_\gamma)$ if the lifetime of the 3^+_γ level and the relative intensities of all the discharge channels of this level are known.

TABLE 10. Value of $B(E2;\ 0^+_1 \to 2^+_2)$, barn

Nucleus	$E^*(2^+_\gamma)$	$B(E2;\ 0 \to 2_\gamma)$	Nucleus	$E^*(2^+_\gamma)$	$B(E2;\ 0 \to 2_\gamma)$
Nd^{150}	1070	0.068 ± 0.012	Er^{170}	930	0.10 ± 0.02
Sm^{152}	1086	0.12 ± 0.02*	Yb^{172}	1468	< 0.04
	1086	0.068 ± 0.012	Yb^{176}	1270	0.060 ± 0.015
Sm^{154}	1440	0.066 ± 0.015	W^{182}	1222	0.12
Gd^{154}	998	0.13 ± 0.05			0.26 ± 0.06*
Gd^{156}	1155	0.06 ± 0.02	W^{186}	730	0.17 ± 0.03*
Gd^{158}	1185	0.08			0.17 ± 0.05
Gd^{160}	1020	0.093 ± 0.015			0.20
Dy^{160}	966	0.069 ± 0.020*			0.24 ± 0.03
Dy^{162}	890	0.094 ± 0.018			0.06
Dy^{164}	770	0.13 ± 0.02	Os^{188}	633	0.33 ± 0.10
Er^{164}	870	0.18 ± 0.05			0.20
Er^{166}	787	0.11			0.20 ± 0.06*
	787	0.21 ± 0.04	Os^{190}	557	0.23
Er^{168}	822	0.17 ± 0.03			0.18 ± 0.04*

*Data obtained from the book by I. Lindskog et al., "α-β-γ-Spectroscopy," National Holland Publishing Company, Amsterdam, 1965; the other data are from the paper by I. Lindskog et al. presented at the Tbilisi Conference, 1964.

However, there are practically no reliable measurements of $T_{1/2}(3_\gamma^+)$ or $T_{1/2}(4_\gamma^+)$ and consequently it is impossible to determine the partial probabilities directly from experiment. There remains the indirect method: first to calculate the values of $\lambda_{\gamma,E2}(3_\gamma^+ \to 2_1^+)$ and $\lambda_{\gamma,E2}(3_\gamma^+ \to 4^+)$ from the experimental values of $B(E2;\ 0_1^+ \to 2_\gamma^+)$ using the formula

$$\left.\begin{aligned}
\lambda_{\gamma,E2}(3_\gamma \to 2_1) &= cE^5_{3_\gamma\to 2_1} Q^2_{\gamma 0} \langle 3_\gamma \to 2_1 \rangle^2 (1 - f_{32}z)^2, \\
\lambda_{\gamma,E2}(3_\gamma \to 4_1) &= cE^5_{3_\gamma\to 4_1} Q^2_{\gamma 0} \langle 3_\gamma \to 4_1 \rangle^2 (1 - f_{34}z)^2, \\
\lambda_{\gamma,E2}(2_\gamma \to 0_1) &= cE^5_{2_\gamma\to 0_1} Q^2_{\gamma 0} \langle 2_\gamma \to 0_1 \rangle^2 (1 - f_{20}z)^2 = \\
&= cE^5_{2_\gamma\to 0_1} B(E2,\ 2_\gamma^+ \to 0_1^+) = 1/5\, cE^5_{2_\gamma\to 0_1} B(E2;\ 0_1^+ \to 2_\gamma^+)
\end{aligned}\right\} \tag{1}$$

and then to pass from $\lambda_{\gamma,E2}(3_\gamma^+ \to 2_1^+)$ to $\lambda_{\gamma,E2}(3_\gamma^+ \to 2_\gamma^+)$, using the ratio of the intensity of conversion electrons or γ rays. In formulas (1), c is a constant, Q_{γ_0} is a quantity which is the same for all transitions between the γ band and the ground-state band; $\langle\ \rangle$ is an abbreviated notation for the Clebsch-Gordan coefficients; factors of the type of $(1-fz)^2$ were introduced by Mottelson [3] to allow for band shift; the numbers f are determined by the spins of the initial and final states; z is a constant for the given nucleus; the subscript γ relates to levels of the γ band, the subscript 1 to levels of the rotational ground-state band.

The reason for the similarity of the principal factors of these formulas is that the internal structure of all the initial and final states is the same; within each band, the states differ only in the rotation (the situation is the same as in the derivation of the "direct" and "reverse" Alaga rules. The corrections $(1-fz)^2$ take care of the slight variation introduced by the mixing of wave functions with K=0 and K=2.

In using these formulas, we calculate λ_γ,E2 for any transition between the γ band and the ground-state band for all 20 nuclei for which $B(E2;\ 0^+ \to 2_\gamma^+)$ has been measured (see Table 8). To pass from $\lambda_{\gamma,E2}(3_\gamma^+ \to 2_1^+)$, etc., to the probability of a $3_\gamma^+ \to 2_\gamma^+$ transition inside the γ band, it is necessary to know the relative probabilities of these discharge channels of the 3_γ^+ level (including the multipolarity of the transitions). The necessary information is available for only three nuclei out of the 20: for Er^{166}, Er^{168}, and Os^{190}. The calculations for these nuclei are given in [10]. However, since that date, the information on B(E2) for Er^{166} and Er^{168} has been refined in papers by Elbek et al. [11]. Currently, the result of the calculations of $\left|\frac{Q_\gamma}{Q_0}\right|$ has the form

$$\left|\frac{Q_\gamma}{Q_0}\right| = \begin{matrix} Er^{166} & Er^{168} & Os^{190} \\ 1.07 \pm 0.15 & 1.58 \pm 0.25 & 0.70 \pm 0.17 \end{matrix}\ .$$

Weighted average: $\left|\frac{Q_\gamma}{Q_0}\right| = 1.02 \pm 0.13.$

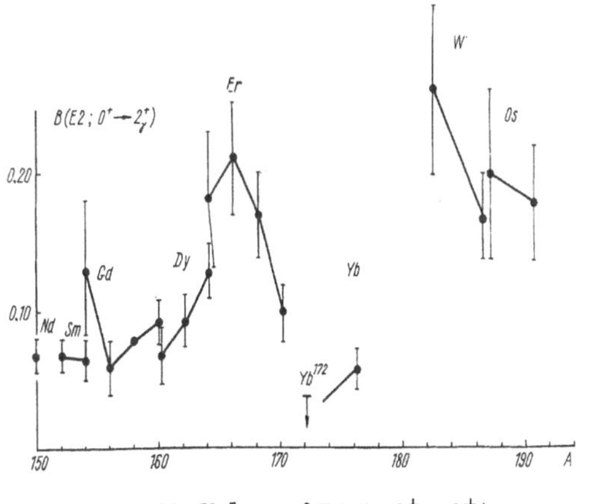

Fig. 15. Values of $B(E2;\ 0^+ \to 2_\gamma^+)$.

Thus, the quadrupole moments of the ground state and γ bands are close to each other. Correspondingly close are also the deformation parameters β. This conclusion is very important since it reveals considerable possibilities for different assessments. If it is assumed that $Q_\gamma = Q_0$, then λ_γ,E2 may be calculated for all the internal transitions in γ bands in all nuclei for which Q_0 is known. Further, if any transitions from these levels (for example, M1 intraband transitions or transitions to other levels) are observed experimentally, the values of λ may be calculated for all these transitions from the relative intensities.

It is quite likely that the quadrupole moments are similar not only in the 2_γ^+ level and the ground state, but in other excited states (at least in excitations to 2 MeV). As yet, however, the data are scanty: three 2_γ^+-levels and the 3^+, 1174 keV level of Yb^{172} for which $\left|\frac{Q_\gamma}{Q_0}\right| = 1.02 \pm 0.17$ [10]. Increase in experimental material is here extremely essential.

The value of $\left|\frac{Q_\gamma}{Q_0}\right|$ may be calculated in the Davydov-Filippov and Davydov-Chaban models. In the Davydov-Filippov model for $\gamma < 10°$, it is equal to $1-\gamma^2/2$, which for Er^{166} and Er^{168} is 0.97; for Os^{190} $\gamma = 21.4°$ and it is necessary to use the results in [12], which determine $\left|\frac{Q_\gamma}{Q_0}\right| =$ 0.896. None of these numbers contradicts the above data. In the Davydov-Chaban model, the ratio $\left|\frac{Q_\gamma}{Q_0}\right|$ ought to be a function of γ and μ. The corresponding calculations have not yet been made.

Values of g_K and g_R for the γ Band

The probabilities of type M1 transitions inside the γ band are proportional to $(g_K - g_R)_\gamma^2$. If $\lambda_{\gamma M1}$ is determined, it is possible to find $(g_K - g_R)_\gamma$. We determine the value of $\lambda_{\gamma M1}$ in the same way as $\lambda_{\gamma E2}$. If it were possible to observe a variation of the angular $\gamma-\gamma$ correlation in a magnetic field (the first γ arrives at the level of the γ band, the second leaves it), this would give the value of the magnetic moment μ, which depends on $(g_K - g_R)^2$ and on g_R. Since $\lambda_{\gamma M1}$ and μ depend differently on g_K and g_R, these values may be determined separately.

As yet, there is not an adequate set of values of $\lambda_{\gamma M1}$ and μ for any nucleus. The values of λ_{M1}, however, are known. It is unlikely that g_R for the γ band differs appreciably from g_R of the ground-state band ($\bar{g}_R = 0.3$), and therefore we may substitute this g_R and determine g_K. We then compare them with the value of g_K for the collective and two-particle levels. For the collective levels $g_K = 0$ and for the two-particle levels,

$$g_K = \frac{1}{K}\{g_{\Omega_1}\Omega_1 + g_{\Omega_2}\Omega_2\}.$$

So far, this is supposition. In support of it, one may say that the analogous formula in odd-even nuclei is well satisfied. E. P. Grigor'ev has carried out such an analysis for several states which are undoubtedly two-particle. It was found that additivity of $g_\Omega \Omega$ was well observed.

Intensities of Transitions between the γ Band and the Ground-State Band

The intensities of competing γ-transitions to levels of a lower band ought to be in a ratio determined by the Alaga rules. Thus, for E2 transitions, from 2_γ^+ levels:

$$B(E2,\ 2_\gamma^+ \to 0_1^+) : B(E2;\ 2_\gamma^+ \to 2_1^+) : B(E2;\ 2_\gamma^+ \to 4_1^+) = \langle_{2\to 0}\rangle^2 :$$
$$: \langle_{2\to 2}\rangle^2 : \langle_{2\to 4}\rangle^2 = 1/5 : 2/7 : 1/70 = 14 : 20 : 1 \ldots .$$

These ratios are satisfied only approximately. To explain this fact, Mottelson [3] advanced the hypothesis that some mixing of the wave functions of the bands occurs; in the 2_1^+ and 4_1^+ states, there is a component with K = 2, and in the 2_γ^+, 3_γ^+, ... states, a component with K = 0. The presence of these components gives rise to the possibility of the transitions (K = 0) → (K = 0) and (K = 2) → (K = 2). The additions are not large but they can affect the intensities; as a rule, the (K = 0) → (K = 0) transitions are accelerated by two orders. Formally, this amounts to the presence in the transition probability formulas of a correction factor of the type $(1 - fz)^2$, where $f = f(I_i, I_f)$, and z is a small constant, inherent in the given nucleus. Z may be determined from the ratios of the intensities of any pair of competing transitions. The values of z, determined from many pairs, ought to be the same. A corresponding analysis of experimental data was made by O. Nielsen [13] who found that the values of z, determined from different pairs, as a rule agreed; for different nuclei they were from 0.03 to 0.24; in the middle of the region of deformed nuclei, they were close to 0.05, increasing at the edges of the region. The position would be somewhat different if it were not for the exceptions:

W^{182}. From the ratio $\lambda_{\gamma E2}(2_\gamma^+ \to 2_1^+) : \lambda_{\gamma E2}(2_\gamma^+ \to 0_1^+)$, $Z = 0.046$. The observed intensity of the $2_\gamma^+ \to 4_1^+$ transition, however, is approximately one-fifth that resulting for this value of Z [14].

Os^{190}. From the ratio of the intensities in the three pairs

$$(2_\gamma^+ \to 2_1^+)/(2_\gamma^+ \to 0_1^+),\ (3_\gamma^+ \to 4_1^+)/(3_\gamma^+ \to 2_1^+) \text{ and } (4_\gamma^+ \to 4_1^+)/(4_\gamma^+ \to 2_1^+)$$

very divergent values of Z are obtained: 0.27; 0.41; 0.17 [10].

Multipolarity of Transitions Between the γ Band and Ground-State Band

The transition between these bands is mainly of E2 type. For $2^+ \to 0^+$ transitions, multipolarity is pure E2, according to the rules associated with the spins. In other transitions, the following additions are possible in principle:

in the transition	$2^+ \to 2^+$	$E0 + M1 + M3$
	$3^+ \to 2^+$	$M1 + M3$
	$3^+ \to 4^+$	$M1 + M3$
	$4^+ \to 2^+$	$M3$
	$4^+ \to 4^+$	$E0 + M1 + M3$

The additions E0 and M1 are forbidden according to the quantum number K. If, however, there is some mixing of the wave functions of the γ band and the ground-state band, the transitions may contain all the above-mentioned components.

The experimental data are quite indefinite. It would appear that there are additions of M1 transitions, but they are very small.

The existence of these additions recently gave rise to a keen argument. Lipas asserted that there ought not to be an M1 component for transitions between the γ band and the ground-state band, and if they were observed, it was due to the addition of two-particle states. A. S. Davydov and G. F. Filippov obtained a small addition of M1 (5-10%) also in their purely collective model. D. Grechukhin at the Dubno Conference (July, 1965) pointed out that the conclusion arrived at by Lipas was incorrect and that an addition of M1 component was also possible in transitions between purely collective levels of the γ band and the ground-state band.

LITERATURE CITED

1. Deformation of Atomic Nuclei [Russian translation], Moscow, Izd. Inostr. Lit., 1958.
2. Davydov, A. S., and Filippov, G. F., Zh. Éksperim. i Teor. Fiz., 35:440 (1958).
3. Hansen, P., et al., Nucl. Phys., 12:413 (1959).
4. Davydov, A. S., and Chaban, A. A., Nucl. Phys., 20:499 (1960).
5. Solov'ev, V. G., Preprint, Joint Institute for Nuclear Research, 1961.
6. Pyatov, N. I., and Solov'ev, V. G., Izv. Akad. Nauk SSSR, Ser. Fiz., 28:2 (1964).
7. Peker, L. K., Author's Abstract of Dissertation, 1964, p. 15.
8. Mottelson, B., and Nilsson, S., Kongl. Danske Vid. Selsk., Vol. 1, No. 8 (1959).

9. Nathan, O., and Nilsson, S., Alpha-, Beta- and Gamma-Ray Spectroscopy, Vol. 1, North Holland Publishing Company (1965).
10. Dzhelepov, B. S., Izv. Akad. Nauk SSSR, Ser. Fiz., 28:2 (1964).
11. Yoshizawa, Y., et al., Proceeding of Conference on Reactions Between Complex Nuclei (Apr., 1963), p. 289.
12. Davydov, A. S., and Rostovskii, V. M., Zh. Éksperim. i Teor. Fiz., 36:1788 (1959).
13. Nielsen, O., Proc. Rutherford Intern. Conf., Manchester (Sept., 1961).
14. Vitman, V. D., et al., Zh. Éksperim. i Teor. Fiz., 40:479 (1961).

β BANDS

Introduction

Alder et al. [1] remarked that in a deformed even–even nucleus there ought to be a band of excited 0_1^+, 2_1^+, 4_1^+, ... states with the quantum numbers $K=0$; $n_\beta=1$, $n_\gamma=0$ (or $\lambda=2$, $\nu=0$). They pointed out that a Coulomb field will excite strongly the 2_β^+ states; there ought to be observed transitions from them to a lower 0_1^+, 2_1^+, 4_1^+ band with the intensity ratios 1 : 10/7 : 18/7; radiation of M1 type in these transitions ought to be weak even if ΔI is equal to 0 or 1. When these predictions were made, only one type of 0^+ state with an energy of 1460 keV of Er^{166} was known in deformed nuclei of medium weight. At the present time, 17 0^+ states have been counted in nuclei of this group in excitation regions of up to 2 MeV (see Table 11). Probably some of them are 0_β^+ states, others are 0_2^+ states.

The Bohr-Mottelson model predicted only the occurrence of 0_β^+ states with an excitation energy of ~1 MeV. The Davydov-Filippov model disregarded vibrations of the nuclear surface and therefore no type-0_β^+ states appeared in it. The Davydov-Chaban model allows for vibrations and predicts the occurrence of type-0_β^+ states. They occur together with the softness parameter μ, and their energy depends substantially on μ more than on the experimental values. A simple and rapid method of determining values of $E(0_\beta^+)$ in the Stephens modification if $E(2_1^+)$ and $E(4_1^+)$ are known was found by N. S. Rabotnov and A. A. Seregin. A. S. Davydov and A. A. Chaban determined a formula for the energy of the 0_β^+ level:

TABLE 11. Properties of 0^+ Levels

Nucleus	$E(0^+)$, keV	$E(2_1^+)$, keV	$\frac{E(0^+)}{E(2_1^+)}$, exp	$\frac{E(0^+)}{E(2_1^+)}$, calc.	Δ, %
Nd^{150}	687	131	5.2	5.5	+5
Sm^{152}	685.0	121.79	5.62	6.0	+7
Gd^{154}	680.6	123.07	5.53	6.0	+8
Gd^{156}	1040	88.97	11.7	13.0	+10
Gd^{158}	1427	79.5	17.9	20.5	+13
Dy^{158}	991*				
Er^{164}	1248[2]				
Er^{166}	1460.3	80.57	18.12	19.5	+7
Yb^{168}	(1156)[1]				
Yb^{168}	(1197)[1]				
Hf^{174}	827[1]				
Hf^{174}	(1241)[1]				
Hf^{178}	1197	93.17	12.9	19.6	+34
Hf^{178}	1440	93.17	15.5	19.6	+21
W^{180}	(908)[2]				
Os^{188}	1086	155.03	7.01	7.2	+3
Os^{188}	1766	155.03	11.4	7.2	−58

1. Data of Élbek et al.
2. Data of Grigor'ev et al.
The remaining data are taken from [2].

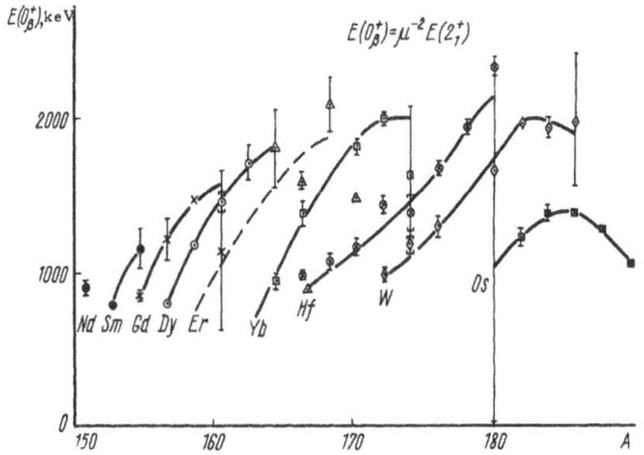

Fig. 16. Values of $E(0_\beta^+)$, calculated according to the μ formulas of Davydov and Chaban from the values of $E(2_1^+)$ and $E(4_1^+)$.

$$E(0_\beta^+) = \mu^{-2}E(2_1^+),$$

which ought to be quite good for small values of μ. N. A. Voinova has calculated $E(0_\beta^+)$ for this formula assuming values of μ based on energies of the 2_1^+ and 4_1^+ levels (Table 2). Fig. 5 of that lecture shows the values of μ from a J, μ set obtained from 2_1^+ and 4_1^+ levels. Many such values of μ may be calculated since 2_1^+ and 4_1^+ levels are known in many nuclei. The curves obtained are given in Fig. 16. The errors indicated are due to errors in the determination of $E(2_1^+)$ and $E(4_1^+)$. Points relating to the isotopes of one element agree satisfactorily with a parabolic curve; in some cases, there are not enough points and the curve has been drawn in such a way as not to disturb the general appearance of the family of curves. It becomes clear from the figure, for example, why the 0_β^+ band is not found in a nucleus such as W^{182}; the curves show that the 0_β^+ energy of W^{182} exceeds 2 MeV, while the energy of the $Ta^{182} \rightarrow W^{182}$ decay is only 1.83 MeV. It also becomes clear why 0_β^+ levels are observed predominantly near the boundary of a region of deformed nuclei since here the levels fall particularly low.

Since μ is not a smooth function of the atomic weight A, the energy of the 0^+ levels and the ratio $E(0_\beta^+)/E(2_1^+)$ also are evidently not a smooth function of A. Stephens et al. [2] determined μ from the energies of the 2_1^+ and 4_1^+ levels (assuming $\gamma = 0$), and then calculated the position of the 0_β^+ levels and compared them with the observed values. The average discrepancy between $E(0_\beta^+)$, calculated from μ and $E(0_\beta^+)_{exp}$ was ~17%.

LITERATURE CITED

1. Alder, K., et al., In the collection "Deformation of Atomic Nuclei." [Russian translation] Moscow, Izd. Inostr. Lit., 1953, p. 219.
2. Stephens, F., et al., Phys. Rev. Lett., 12:225 (1964).

RADIOACTIVE PROPERTIES OF THE NUCLEI OF THE HEAVIEST ELEMENTS

G. N. Flerov and V. A. Druin

Joint Institute for Nuclear Research

Introduction

The basic forms of radioactive disintegration of the transuranium elements are α decay, electron capture or β decay, and spontaneous fission, the part played by each of these forms of disintegration being different for different Z and A numbers of the nuclei concerned. The explanation of the relationships in the variation in properties of already known isotopes and the estimation of the disintegration energy and lifetime in relation to one form or another of radioactive transformation of uninvestigated nuclei are very important both for an understanding of the structure of the nucleus, and from the point of view of the synthesis of new elements, as well as for certain problems of astrophysics and cosmology.

The first part of this review considers the expected properties of isotopes which may be synthesized in the next few years, principal attention being devoted to questions of the spontaneous fission of nuclei, since in the first place, it is precisely this which plays an important part in the disintegration of the heaviest elements, and in the second place, this process is most complex in comparison with α and β decay and has been studied least of all.

The process of β decay has been quite well studied in the problem of weak interactions, and its basic relationships are satisfactorily described by the Fermi theory of electron–neutrino interactions.

The partial half-life of β decays is determined by the difference in energy between parent and daughter nuclei, and also by the selection rules according to spin and parity of the initial and final states. For an accurate prediction of the half-life of β decay, it is necessary to know the structure of the levels of the nuclei concerned. In the region of transuranium elements, however, the structure of the levels is such that there are very many low-lying states. It is therefore possible to assume for approximate estimates of the lifetime of a β-unstable nucleus that a daughter state is always to be found, the spin of which is equal to the spin of the parent state, and that the energy of the level is close to the energy of the ground state. In other words, in heavy nuclei, there is considerable probability of permitted first order transitions ($\Delta I = 0.1$; parity does not vary) or forbidden transitions ($\Delta I = 0$; parity varies). The transition energy may be assumed to be equal to the energy of a β transition between the ground states. With these assumptions regarding the transition energy as determined from the mass difference, it is easy to find $t_{\frac{1}{2}\beta}$ the accuracy of determining the half-life for β decay or for the capture of orbital electrons being sufficiently high, and the value of $t_{\frac{1}{2}\beta}$ for a large number of nuclei differing from the experimental value by a factor of not more than 5-10.

An analogous situation is found when α decay is examined. Matters are particularly favorable for determining the probabilities of α decay between the ground states of even–even nuclei. On the basis of a very simple theory of the penetration of the potential barrier, Gallagher and Rasmussen [1] obtained the semi-empirical formula

$$\log t_{\frac{1}{2}\alpha} = A_Z Q_{\text{eff}}^{-1/2} + B_Z + \log F,$$

where Q_{eff} is the effective energy of an α transition; A_Z and B_Z are constants dependent on Z; log F is the coefficient of forbiddenness for odd nuclei.

According to this formula, $t_{\frac{1}{2}\alpha}$ is determined with an accuracy of up to 30%, which is equivalent to an error of approximately 60 keV in the determination of Q_{eff}.

Thus, there are no great difficulties in predicting the lifetime of new isotopes and elements of the transuranium region with regard to β and α decay.

The situation in spontaneous fission calls for special consideration, to which we shall now pass.

Spontaneous Fission of Nuclei from the Ground State

It is known that many heavy nuclei are energy-unstable with regard to spontaneous fission into two fragments of approximately equal mass. This is because nuclei of average atomic weight are more closely "packed" than the nuclei of heavy elements. The mean binding energy per nucleon as a function of the mass number A increases slowly to ~ 8.5 MeV, then remains approximately constant and equal to this value in the region $50 < A < 150$, subsequently decreasing slowly with increase in A and attaining 7.4 MeV for uranium. As a result of the spontaneous fission for example of the nucleus of gold, an energy of about 130 MeV could be liberated.

However, despite the energy instability, the probability of spontaneous fission of the majority of even, heavy nuclei is not high, since this process is impeded by the presence of a considerable potential barrier. For fission of the nucleus of gold to take place, it would be necessary to overcome a potential barrier having a height of about 40 MeV. With increase in Z of the nucleus, the height of the barrier decreases, and in the region of the heaviest nuclei (Th–U), the ground state lies at approximately only 5 MeV below the top of the barrier. In this case, due to the quantum-mechanical tunnel effect, observation of spontaneous fission of a nucleus from the (unexcited) ground state becomes real.

Spontaneous fission was predicted theoretically by Bohr and Wheeler [2] on the basis of a model representation of the nucleus in the form of a liquid drop. In 1940, this phenomenon was discovered experimentally in the case of U^{238} [3]. In the course of the next 25 years, spontaneous fission was a subject of research at several physical laboratories. Due to the vigorous development of nuclear reactor engineering and charged particle accelerators a large number of transuranium elements have been synthesized under laboratory conditions. A wide range of nuclei has become available to experimenters engaged on fission research. Much experimental data on the process of spontaneous fission has been collected. In particular, it has been found that spontaneous fission of the nucleus into two fragments as in the case of induced fission, is accompanied by the emission of neutrons and γ quanta. At the same time, it has been shown that the average number of neutrons per fission event, on passing from U^{238} to Cf^{254}, increases from 2.4 ± 0.2 to 4.05 ± 0.19 and remains approximately constant for the different isotopes of one element.

A study of the γ quanta emitted in the spontaneous fission of Cf^{252} has shown that in one decay event an average of 10.3 photons with a total energy of 8.2 MeV are emitted; the γ-quanta spectrum was found to be quite similar to the spectrum observed in the fission of U^{235} by thermal neutrons.

For many nuclei, the energy spectrum and mass distribution of fission fragments have been investigated. The results of these experiments show that the mass distribution of the fragments has an asymmetric character, as in fission by thermal neutrons.

In some researches, the fission of nuclei into three parts has been studied [4, 5], it being found that the proportion of ternary fissions with emission of an energy α-particle forms about 1/300 of the total number of fission events. This result is also similar to the fission of U^{235} or Pu^{239} by the action of thermal neutrons.

It is thus clear that the emission of neutrons, α particles and γ quanta is not specific to spontaneous fission. In this sense, spontaneous fission may be regarded as a special case of the fission of nuclei at low excitation energies. The investigation of neutrons and γ rays in the spontaneous fission, for example, of Cf^{252}, is of interest rather from the method standpoint (due to the absence of background from the source of bombarding particles – reactor or accelerator) than from the theoretical standpoint.

With a view to explaining the mechanism of fission and the internal structure of the nucleus, it is our opinion that investigation of the probabilities of the spontaneous fission of different nuclei (or half-life with respect to spontaneous fission) offers the maximum interest. We shall therefore discuss the phenomena which occur in the first stage of the fission process up to the moment of rupture of the scission neck, if the terminology of the hydrodynamic model of the nucleus is used, and which are determined by the decay probability. The second state of the process evidently commences from the moment of rupture of the neck and proceeds like induced fission over the barrier.

Since spontaneous fission is regarded as a quantum-mechanical process of penetration through the potential barrier, a study of the half-life values of spontaneous fission may theoretically provide important information regarding the height of the potential barrier and the width of the sub-barrier region.

The fission process is associated with a radical rearrangement of the structure of the nucleus. A large number of nucleons participate in it, i.e., this process is essentially collective. Nevertheless, an important part is also played in the fission by single-particle effects, i.e., the features of the behavior of individual nucleons in the initiation of deformation of the nucleus as a whole. This is confirmed by the existence of high forbiddenness for the spontaneous fission of odd nuclei. The addition of only one nucleon to an even–even nucleus leads to a sharp reduction in the probability of fission.

In recent years, publications have appeared [6-8], in which an attempt has been made to connect the probability of spontaneous fission with the arrangement of the energy level of nucleons in various deformed nuclei. It was hoped to obtain in this way additional information on the single-particle and collective models of the nucleus from data on spontaneous fission half-life.

At the present time, extensive experimental material is available on the half-life of spontaneous fission, making it possible to establish some empirical formulas for the dependence of the probabilities of the process on the Z and A numbers of different nuclei.

a. Connection Between the Half-life of Spontaneous Fission T_{sf} and the Fissility Parameter Z^2/A

An analysis of the experimental results permits us to draw a conclusion regarding the strong dependence of the probability of decay on the fissility parameter Z^2/A, which as we know, in the liquid drop nuclear model characterizes the ratio of the Coulomb energy of repulsion of the nucleus and the stabilizing surface energy, and is a measure of the tendency of the nucleus to undergo fission. Figure 1 shows the dependence of T_{sf} on Z^2/A, based on recent data on the half-life of spontaneous fission. A general tendency is shown quite clearly towards an increase in fissility with increase in Z^2/A. However, in addition to this general monotonic dependence, there is clearly visible a dependence of the parabolic form of fission-probability on the mass number with fixed Z, and a considerably lower decay rate of the nuclei with odd Z or N. For given Z, the half-life values pass through a maximum with variation in A.

b. Forbiddenness Factors for Spontaneous Fission of Odd Nuclei

The first attempts to classify experimental data on the spontaneous fission of nuclei disclosed a characteristic feature. Nuclei containing an odd number of protons or neutrons could

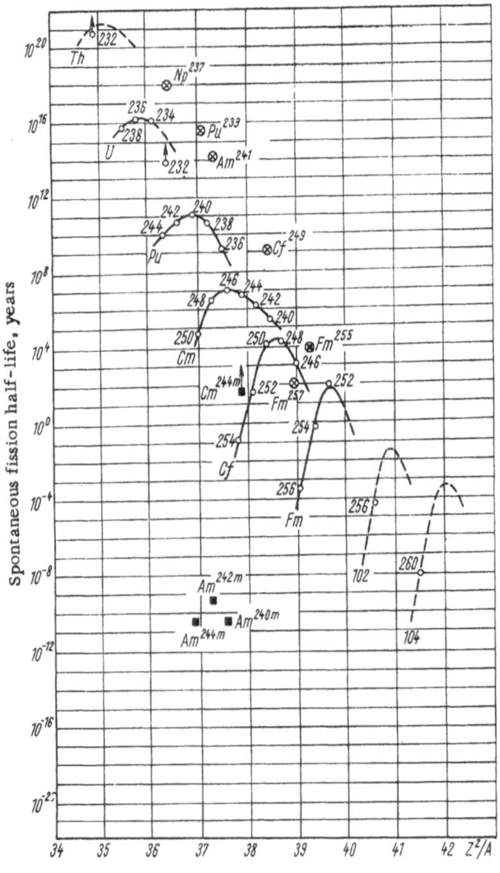

Fig. 1. Dependence of T_{sf} on Z^2/A of spontaneously fissile nuclei: ■) Spontaneously fissile excited states.

not be classified in any way. Figure 1 shows that the half-life values of odd nuclei are several orders (3-6) above those of the neighboring even–even isotopes. In other words for odd nuclei, there is a forbiddenness for spontaneous fission, but the factor of forbiddenness has no systematic form and lies in the range from $10^3 - 10^6$.

Theoretical Interpretation of the Relationships of Spontaneous Fission

Theory must explain three basic experimentally established facts, characterizing the relationships of the variation in half-life values of the spontaneous fission of different nuclei.

1. The smooth decrease in the half-life of nuclei with increase in the parameter Z^2/A.

2. The curves of parabolic form reflecting the variation in half-life with variation in A, with fixed Z for even–even isotopes.

3. Forbiddenness for spontaneous fission of nuclei having odd Z or N numbers.

Bohr and Wheeler [2] created the liquid drop or hydrodynamic model of the nucleus. According to this model, the nucleus is regarded as a drop of incompressible fluid possessing uniform volume density of the electric charge and approximately the same binding energy for each particle. Since a nucleon on the surface of the nucleus interacts with a smaller number of nucleons than a nucleon situated inside the nucleus, the total binding energy is decreased by a certain value, proportional to the surface of the nucleus. For this reason, a surface tension is produced. The stabilizing effect of the surface tension exceeds the destructive effect of the electrostatic forces of repulsion in nuclei of spherical shape. The dynamics of the nucleus are described by analogy with the motion of a liquid drop. In consequence of the deformation of the spherical drop, periodic vibrations of the surface are produced. These vibrations may be analyzed and their frequency ω_l determined. The frequency of vibrations of the nuclear surface taking Coulomb effects into account was determined by Ya. I. Frenkel' [9], who obtained the following expression for ω_l:

$$\omega_l = \left\{ l(l-1)\left[(l+2) - \frac{10\gamma}{2l+1}\right] \frac{U_s}{3r_0^2 MA} \right\}^{\frac{1}{2}}, \tag{1}$$

where γ is the ratio of the Coulomb energy $E_c = \frac{3}{5}\frac{(Ze)^2}{R}$ to the energy of the surface tension $E_S = 4\pi R^2\sigma = U_S \cdot A^{2/3}$ of the undeformed nucleus, i.e., $\gamma = \text{const}\ Z^2/A$; M is the mass of an individual nucleon; A is the number of nucleons in the nucleus.

For $l = 1$, $\omega_l = 0$, since deformations with $l = 1$ correspond to a displacement of the drop as a whole.

It follows from formula (1) that if $\gamma > \gamma_{sph}$, the vibrations become imaginary. This value is a minimum for $l = 2$, when $\gamma_{sph} = 2$.

If $\gamma > \gamma_{sph}$, $E_C > E_S$, and the nucleus is unstable to deformations. Thus, the condition of stability with respect to surface deformations will be $\frac{3}{5} \cdot \frac{l^2}{r_0}\frac{l}{U_s} \cdot Z^2/A < 2$, or $Z^2/A < (Z^2/A)_{sph} \approx 49$, or as is more often usually written $\frac{Z^2/A}{(Z^2/A)_{sph}} = x < 1$. For x = 1, even the spherical shape of the nucleus is unstable (the nucleus of "cosmium"). The experimental relationships of the variation of the half-life of spontaneous fission with increase in Z^2/A are in agreement with the predictions of the liquid drop model. If a straight line is drawn through the half-life values of even–even nuclei in Fig. 1, it intersects the axis of absissas ($T_{sf} \approx 10^{-21}$ sec) in the region of $Z^2/A \approx 48$, which is not in contradiction with the various estimates based on the hydrodynamic model.

The Bohr and Wheeler theory, in explaining the general aspects of fission, correctly describes the process in the first approximation, but in details the liquid drop model is not altogether satisfactory.

From the existence of forbiddenness for the spontaneous fission of odd nuclei of 10^3-10^6 relative to the corresponding even nuclei, it follows that odd nuclei have a higher fission barrier, or their shape must vary strongly during passage through the barrier. From the point of view of the liquid drop model, there should be no difference at all in the heights of the barrier of even and odd nuclei. According to this model, all nuclei in the ground state are spherically symmetrical, and the barrier is determined only by the value of x, which varies slightly on the addition of one nucleon. It is just as impossible to understand why the lifetime of heavier even–even isotopes of a given element decreases when the fissility parameter Z^2/A of those isotopes decreases.

In 1955 a paper by Swiatecki [10] was published, which although it does not resolve the problem, is one of the most successful attempts to bring the experimental data on the half-life of the spontaneous fission of even–even and odd nuclei into agreement with the deductions of the hydrodynamic model, and may be quite fruitful in predicting the properties of new elements. Swiatecki starts from two basic premises.

1. According to the liquid drop model, the logarithms of the half-life of spontaneous fission should lie on a straight line, decreasing with increase in Z^2/A. Actually, there is a parabolic curve and the experimental values of the logarithms of the half-life values deviate from the straight line by an amount δT.

2. The experimental masses of nuclei in the ground state differ from the value determined by the semi-empirical formula of Green [11] by the amount $\delta M = M_{exp} - M(A, Z)$.

Swiatecki made use of the hydrodynamic part of Green's formula, which did not take into account certain energy effects, such as the shell structure of nuclei, fluctuation in pairing energies, nonuniformity in the angular and radial distributions of the charges, and so forth, which contribute to the true value of the mass of a nucleus.

Quite a definite correlation was found between δT and δM, a variation in the lifetime by a factor of 10^5 corresponding to each millimass unit of mass deviation (~0.9 MeV). Swiatecki corrected the observed half-life values by introducing the empirical correction $K \cdot \delta M$ ($K \sim 5$, if δM is in millimass units).

The result was that a smooth dependence of $T_{sf} + K\delta M$ on Z^2/A was obtained for even–even nuclei. This work in fact shows that the liquid drop model is too simple for explaining all the fineness of the fission process, since it fails to take into account many details of nuclear structure.

Obviously, for producing an exact theory of spontaneous fission, it is essential to examine and understand the behavior of the nucleus as a whole, while taking into consideration the properties of the individual nucleons. Hydrodynamic concepts define the general character of the dependence of the initial fission energy on the atomic number and mass number, while the deviation from these mean values in different nuclei must be explained by the individual character of the nuclear states. The fact alone of the existence in some nuclei of high quadrupole-moments and low-lying rotational spectra shows that these nuclei possess marked nonsphericity. The generalized model successfully explains why nuclei acquire nonsphericity, in particular an axially symmetrical shape.

For explaining the peculiar character of the variation of the fission barriers of different nuclei, with which are associated the observed relationships in the half-life values of spontaneous fission, it is obviously necessary to take into account the internal structure of the deformed nucleus.

Such an approach to the classification of the half-life values of spontaneous fission was first proposed by Newton [12] and independently by Wheeler [6], but it was developed most fully and fruitfully by Johannsson [8]. Newton and Wheeler's ideas were based on deductions of the generalized model of the nucleus, while their subsequent development was directly associated with an analysis of the behavior of individual nucleons in deformed nuclei carried out by Nilsson [13].

The internal motion of the nucleons in axially symmetrical nuclei is characterized by the projection of the total angular momentum j on the axis of symmetry Ω_p. Single-particle levels for the spherical potential, characterized by the total momentum j, disintegrate in a spherical field to levels corresponding to different values of $|\Omega_p|$, the splitting of the levels, degenerated in the spherical nucleus, increasing with increase in deformation.

Thus, a state with momentum j splits into j + ½ states, each of which is characterized by the value Ω_p and definite parity. Parity may be positive or negative, depending on whether the orbital moment of a nucleon l will be even or odd on passing to zero deformation (in the spherical nucleus).

States differing only by the sign of Ω_p, i.e., by the direction of rotation relative to the axis of symmetry, are degenerate. Therefore, in nonspherical nuclei, every two neutrons and two protons may form a "shell."

The rotational motion of the deformed nucleus is characterized by the quantum numbers I, M, and K, i.e., by the total momentum and its projections on the spatially fixed axis and axis of symmetry of the nucleus.

For symmetrical deformations $K=\Omega=\sum_p \Omega_p$. All the even nucleons form filled shells, and therefore in the ground state of even–even nuclei the spin of the nucleus is $K = \Omega = 0$, and for the odd nucleus, it is determined by the Ω_p of the last odd nucleon: $K = \Omega_p$.

For the exact calculation of the position of the levels of individual nucleons in a strongly deformed nucleus, it is necessary to solve the Schrödinger equation in a nonspherical well with a diffuse boundary and to take spin-orbital interaction into account at the same time. The solution of this problem is accompanied by exceptional difficulties, even if it is confined to the spheroidal shape of well.

The single-particle eigenvalues and functions are calculated on an electronic computer and reduced to diagrams. Nilsson's diagrams are generally constructed up to $\delta = 0.3$; for some levels, the values of the energy of the levels is given up to $\delta = 0.5$.

The terms derived from different values of j for the spherical nucleus intersect for even very slight deformations. Subsequently, there is intersection of levels belonging to different shells of the spherical nucleus.

In the problem of spontaneous fission, the most important question connected with the Nilsson diagram is the behavior of the nucleons at the point of intersection of the levels. Johannsson suggested that nucleons, arranged in pairs (with antiparallel spins) on single-particle levels are readily able to pass to free levels, which intersect them in the deformation process, if this is energetically profitable. Such a transition is not forbidden, since the total spin of the nucleons is always equal to zero and parity is positive. In its turn, the result of this is that in the process of changing shape, the even–even nucleus is almost always in the lowest energy state.

The picture for odd nuclei is quite different. In such nuclei, there is a level on which there is only one nucleon with a momentum projection Ω_p, which at the same time is the spin of the nucleus K. With increase in spheroidal deformation, the state of the nucleus having the lowest energy may have a spin different from Ω_p. However, since the initial spin of the nucleus K must be conserved (integral of motion), the only levels available to the odd nucleon will be levels with $\Omega_p = K$. For conserving the momentum, a definite energy must be expended (provided the level of the last odd nucleon does not drop sharply with increase in deformation) on the excitation of the nucleon, and this results in an effective increase in the fission barrier. Johannsson attempted to explain the increase in the rate of fission of the heavier isotopes of a given element by estimating the variation magnitude of the barrier on the addition of two nucleons to a definite single-particle level. In the liquid-drop model, as the result of the addition of one or two nucleons, the surface energy and Coulomb energy of the nucleus vary. This variation is easily determined.

This was done in practice as follows.

The variation in barrier height according to the single-particle model was calculated by means of Nilsson diagrams. The variation in penetrability of the barrier due to a slight variation in its height was then estimated. The shape of the barrier must necessarily be known for calculating its penetrability. Johannsson used the shape of the barrier according to the hydrodynamic model.

The probability of the fission process was calculated according to the simplest Bohr-Wheeler formula

$$\lambda_f = \text{const} \cdot \exp\left[-K\int_{\delta_0}^{\delta_f} \sqrt{H}\, d\delta\right],$$

where δ_f is the deformation at the saddle point; δ_0 is the equilibrium deformation; K is the mass coefficient; H is the height of the barrier.

Taking a specific arrangement of the levels into account, the author found the corrections $\Delta \ln T_{sf}$ to the lifetimes of the nuclei, calculated according to the liquid drop model, and showed that with these corrections, the behavior of the half-life values of the fission became more regular. These values are grouped with a certain amount of scatter around a straight line.

For odd nuclei, with increase in the deformation, the energy of the odd nucleon increases much more rapidly than follows from hydrodynamic considerations. This leads to an effective increase in the fission potential-barrier, and consequently to an increase in lifetime. Such a qualitative explanation of the causes of forbiddenness for odd nuclei is quite convincing.

The deviation of the experimental data from the estimates according to the liquid-drop model, at least qualitatively, are quite well explained by single-particle effects. Considerable difficulty resides in the fact that for predicting the spontaneous fission half-life by the Johannsson method, it is necessary to estimate the deformation in the ground state and at the saddle point for every nucleus. In addition, for unknown nuclei with $Z > 100$ and $N > 156$, it is necessary to know the proton and neutron levels. It is therefore not surprising that Johannsson's attempt to calculate the half-life values of 102 and 104 elements resulted in a substantial discrepancy with the experimental data. Thus, for the isotope 102 of an element with a mass of 256, a spontaneous fission half-life of 0.02 sec was predicted theoretically, while experimentally it was found to be $1.5 \cdot 10^3$ sec; for 104^{260} calculation gave $5 \cdot 10^{-6}$ sec, and experimental data gave 0.3 sec.

However, the deduction, made on the basis of Johannsson's calculations and concerning the local character of the drop in spontaneous fission half-life values of nuclei close to the subshell $N = 152$, due to rarefaction of the 76 and 77 neutron levels, is deserving of attention.

These effects ought to diminish considerably for N > 158, and therefore the spontaneous fission half-life will increase with increase in N in agreement with the thesis of the liquid drop model, i.e., the existence of relatively long-life isotopes is possible, especially close to the shell N = 184.

Summing up, it should be pointed out that despite the definite successes of theory in explaining the existing relationships, the problem of estimating the properties of nuclei of the heaviest isotopes and new elements has not yet been finally solved, and only approximate concepts of the lifetime of new nuclei with regard to spontaneous fission can be obtained on the basis of extrapolation of empirical relationships. Figure 2 shows the dependence of T_{sf} on A of spontaneously fissile nuclei. It is clear that the presence of a neutron subshell N = 152 resulted in a sharp drop in half-life for isotopes with N > 152. Passing from a subshell to a still larger number of neutrons in the nucleus possibly leads again to an increase in stability of the nuclei. It is at present impossible to determine where this region is situated. Attempts are being made to answer this question in laboratories engaged in the synthesis of transuranium elements.

Viola and Wilkins [14] recently attempted to obtain a semi-empirical formula for estimating the lifetime of heavy elements. This work is a further development of the Swiatecki, Dorn, and Johannsson calculations.

It is assumed that the fission barriers vary smoothly as a function of the fissility parameter Z^2/A. The dependence of the fission barrier E_f on Z^2/A may be written in the form

$$E_f = f\left(\frac{Z^2}{A}\right) = a \log T_{sf} + b.$$

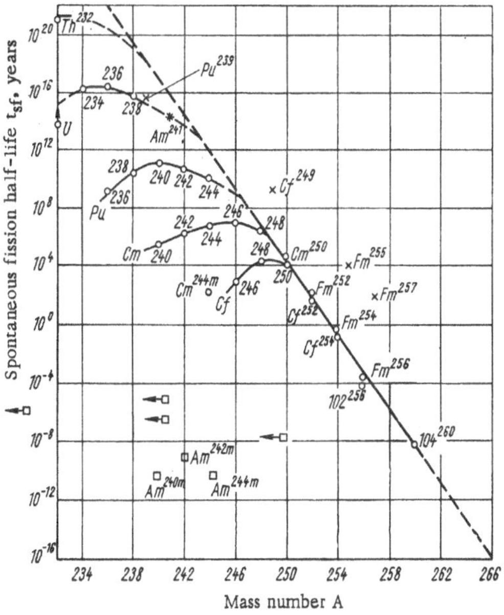

Fig. 2. Dependence of T_{sf} on A of spontaneously fissile nuclei: ☐) Spontaneously-fissile excited states.

Experimental results on fission barriers and half-life values of spontaneous fission of even−even nuclei are used as starting data for obtaining the coefficients a and b. It is found that for even−even nuclei, the best agreement is obtained for the following coefficients:

$$\log T_{sf} = 6.884 E_f - 21.5.$$

Here T_{sf} is the time in years, and E_f is the fission barrier in MeV.

It follows from this that with variation in the height of the barrier by 1 MeV, the half-life varies by $10^{6.884}$.

When deformation of a nucleus occurs, the structure of the ground state is evidently distorted at the saddle point, for which additional energy is required, this being equivalent to an effective increase in the hydrodynamic barrier. An experimental measure of the variation in height of the barrier δE_f, due to the different nuclear effects, is the deviation of the experimental mass (M_{exp}) from the smooth mass hydrodynamic surface M(A, Z), i.e.,

$$\delta E_f = \delta m = M(A, Z) - M_{exp}.$$

Then

$$\log T_{sf} = -6.884 E_f + 6.884 \delta m - 21.5, \text{ or}$$
$$\log T_{sf} = -3.344 Z^2/A + 133.86 + 6.884 \delta m.$$

Unlike Swiatecki and Dorn [10, 15], who used the simplest expression for the mass M(A, Z), including volume, surface, Coulomb and symmetry terms, Viola and Wilkins [14] introduce corrections for the additional stability of the nucleus, due to the deformation in the ground state. Thus, the variation in structure of each nucleus is taken care of by the term δm representing the difference between the experimental mass and the total hydrodynamic mass for the spherical nucleus and a correction term taking into account the initial deformation, i.e.,

$$\delta m = M(A, Z) - M_{exp} = M_{hydr} + M_{def} - M_{exp}.$$

The "experimental masses" are calculated on the basis of the closed disintegration energy cycles, for M_{hydr} Seeger's mass formula [16], without shell and deformation corrections, is used, and M_{def} is obtained from the agreement of $\log T_{sf}$ with the experimental data on the half-life of known even−even isotopes.

Figure 3 shows the dependence of $\log T_{sf}$ on N for different elements. For isotopes between U and Fm, the error in T_{sf} does not exceed a factor 3.9, corresponding to an inaccuracy of ~86 keV in the determination of the mass.

Good agreement is also obtained for the new isotopes Cf^{256} (T_{exp} from 8 to 30 days; T_{calc} = 2.4 days). Fm^{257} (experimental 120 years, calculated 65 years), 102^{256} (experimental 25 min, calculated 1.9 h), 104^{260} (experimental 0.3 sec, calculated 0.21 sec for α-decay and 1.2 sec for T_{sf}).

For Fm^{258} the calculated value of T_{sf} = 5 h. The value of T_{sf} = 11 days, to be found in the literature, has recently been found to be incorrect. The same applies to 102^{254} (T_{sf} = 15 days, T_{exp} = 6 sec, are incorrect data).

An essential conclusion arrived at by the authors was the prediction of the existence of an islet of relative stability of nuclei above the neutron number 160. This concerns particularly the closed neutron shell N = 184, where stability to spontaneous fission ought to be very high.

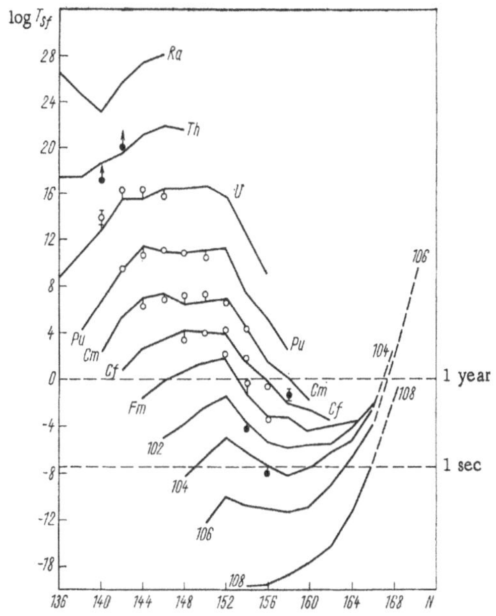

Fig. 3. Dependence of log T_{sf} on N of spontaneously fissile nuclei.

Nuclei with N = 184 will evidently have a very short life as regards both spontaneous fission and α-decay.

The nucleus 126 of an element with mass number 310, containing 126 protons and 184 neutrons, is of particular interest. If a closed shell Z = 126 and N = 184 exists, it is quite probable that the doubly magic nucleus 126^{310} ought to be closely packed.

Swiatecki made estimates of the size of the fission barrier, obtaining the value of 10 MeV.

The probability of tunnel penetration through such a high barrier is very slight, and the half-life of spontaneous fission ought to be not less than the half-life of spontaneous fission of uranium (barrier of about 5 MeV). The energy of α decay of the nucleus 126^{310}, according to estimates, ought to be 12 MeV. Nevertheless, the half-life may attain quite high values (hours or more).

Spontaneous Fission from Anomalous Excited States

As follows from the first part of the review, the probability of spontaneous fission of nuclei from the ground state depends on the deformation of the nuclei, configuration of the nucleon levels, spin, parity, and so forth. It is therefore natural to expect that for excited states, which are characterized not only by energy variation but also by variation of the other parameters of the nucleus, the probability of spontaneous fission ought to vary substantially. The experimental study of processes of this type is greatly impeded by the considerable competition of other forms of disintegration. For excited states, the principal competing process is obviously γ radiation occurring in times of 10^{-12}–10^{-14} sec in the absence of forbiddenness factors. The observation of spontaneous fission, therefore, is possible only when there is a high forbiddenness for γ radiation owing to one factor or another.

In cases where the fission of nuclei is studied at excitation energies somewhat below the activation energy (subbarrier fission), the probability of fission is comparable with the probability of the emission of γ quanta.

Thus, for example, in the capture of thermal neutrons by Th^{232} [17], the excitation energy is lower than the barrier by approximately 1 MeV, the fission cross section being $(0.06 \pm 0.02) \cdot 10^{-27}$ cm^2 for the complete thermal-neutron capture cross-section of the order of 10 barn. Although the excitation energy of the compound nucleus Th^{233} is clearly defined and the spin of the initial state can assume only the two values $\pm {}^1/_2 \hbar$, it is nevertheless extremely difficult to carry out quantitative calculations of the probability of fission of Th^{233}, for example according to the Johannsson model, because of the considerable uncertainties in the behavior of the nucleon levels and the impossibility of allowing accurately for the fission barrier.

The second possibility, spontaneous fission from excited states at an excitation energy of several MeV and high forbiddenness for γ radiation, has been found to be extremely unlikely.

However, in 1961, S. M. Polikanov and collaborators [18], using the high-frequency method of indicating spontaneous fission events, and high intensities of accelerated multiple-charged ions (C, N, O^{16}, Ne) in the cyclotron of the Joint Institute for Nuclear Research, observed the spontaneous fission of Am^{242} from the excited state.

Subsequently, work was conducted on these lines both in the Soviet Union (Joint Institute for Nuclear Research), and other countries. The experimental results obtained to date amount mainly to the following:

1. Spontaneous fission from excited states is evidently a common phenomenon. Already seven isomers are known [19] (Am^{242}, Am^{240}, Am^{244}, Np^{228}, see Fig. 2), although the method used in the experiments permitted recording of the spontaneous fission of nuclei only in a comparatively narrow time range (from 10^{-3} to 10^4 sec) and the investigation of a limited number of isotopes.

2. In every case, spontaneous fission occurs much more rapidly from excited states than from the ground state. For individual isotopes, the acceleration factor attains $10^{20}-10^{24}$.

3. The isomeric ratio in all the reactions investigated is a very low value (from 10^{-3} to 10^{-5}). This is due to the low value of the probability of formation of a given state, but not to competition with α-decay and γ-quanta emission. Experiments to discover α-decay branches and γ radiation, carried out at the Joint Institute for Nuclear Research and at Berkeley [22] showed that it was not possible to increase the isomeric ratio by a factor of more than 20–40, allowance being made for competing disintegration processes.

4. Spontaneously fissile isomers are obtained not only in nuclear reactions with heavy ions, but also in reactions with α particles, deuterons and neutrons with an energy of 14 MeV.

5. The energy of the excited state from which spontaneous fission occurs is evidently not high. In particular, for the isomer Am^{242}, which has been studied most, it has been shown in experiments with deuterons that the excitation energy did not exceed 2 MeV [20].

6. The energy spectrum of the fission fragments in the limits of the experimental accuracy does not differ from a similar spectrum of fragments in the fission of Am^{241} (by thermal neutrons [21]).

In the experimental study of the spontaneous fission of nuclei from excited states, various hypotheses have been advanced for their explanation. After the first investigations of S. M. Polikanov et al. [18], L. A. Sliv and L. K. Peker suggested a hypothesis, according to which the energy of the isomeric level is 3–4 MeV, this being necessary to explain the increase in the probability of fission by a factor of about 10^{20}. The spin of this level, according to their assumption, must be very high (20−24 ℏ) to explain the low probability of the emission of γ quanta.

However, subsequent experiments did not confirm these hypotheses; the excitation energy of the isomeric level of Am^{242} was found to be not more than 2 MeV, and the spin, estimated from experiments with neutrons, could scarcely attain such high values.

A somewhat different approach to the isomer problem was made in a paper by D. F. Zaretskii and M. G. Urin [23] who assumed that the excitation energy of the level was not high and that the probability of fission increased owing to a reduction in the effective mass assisting tunneling through the potential barrier. This hypothesis, however, is in contradiction with experimental results of Vandenbosch et al. [24].

These authors attempted to discover spontaneous fission from the known 34 msec isomeric state of Cm^{244} with an energy of 1042 keV, formed in the β-decay of Am^{244}, corresponding to the rupture of a pair of neutrons coupled in the ground state. On rupture of the nucleon pair, the mass coefficient in the isomeric state ought to be less than in the ground state. Owing to this, there ought to be a reduction in the lifetime for spontaneous fission of $10^{15}-10^{20}$ times, i.e., it was to be expected that the half-life of spontaneous fission of the isomer Cm^{244} would be several seconds, while experiment gave a lower limit of $\geq 1.4 \cdot 10^2$ years.

Instead, as follows from the first part, an appreciable acceleration of spontaneous fission may be caused by a variation in the levels or by a variation in deformation. For example, the acceleration factor in the transition from Cf^{250} to Cf^{254} is $\sim 10^7$.

Therefore, as working hypothesis, one may put forward the supposition that in addition to the ground state with β_0 deformation and a definite arrangement of the levels, the nucleus may be situated in excited states, in which either the initial deformation differs substantially from β_0 or the arrangement of the nucleon levels is such that it assists fission. This hypothesis, which was considered in a private discussion with O. Bohr in 1964, is not very readily amenable to quantitative calculations, since the energies of all the nucleons for different deformations of the nucleus must be taken into account, as well as the dependence of the penetrability of the barrier on deformation.

If this state differs substantially according to the deformation or configuration of the levels, the so-called isomeric ratio will depend strongly on the method of obtaining the isomer. In particular, it seems to us to be not impossible that the unusualness (abnormality) is connected with the abnormality of the isomeric state, which is formed in the first stage of a nuclear reaction, i.e., not all the channels of a nuclear reaction in the initial and subsequent stages lead to the same probability of the occurrence of an isomeric anomalous state.

This hypothesis is still less amenable to theoretical analysis. However, a number of qualitative conclusions may be made on the most general considerations.

In particular, from this point of view, the isomeric ratio in the capture of thermal neutrons in the reaction Am^{241} (n, γ) Am^{242} ought to be very small, since the spin of the nucleus varies by $\pm ^1/_2$ ℏ, and there are no direct reactions. Experimentally, it has only been possible to estimate the limit value of the isomeric ratio, which is 10^{-7}.

With increase in the energy of the incident particles, the number of initial reaction channels and the probability of direct processes ought to increase. This ought to result in a sharp increase in the isomeric ratio in nuclear reactions, producing protons with an energy of several hundred MeV.

In the case of the formation of compound nuclei, high excitations of the initial nucleus ought also to result in a high isomeric ratio. In reactions with heavy ions passing through a compound nucleus, a very large number of channels appear and the isomeric ratio ought to increase.

If this hypothesis is found to be correct, excited states of this type may also occur in other nuclei in various parts of the table of elements. For the lighter nuclei, these states will not disintegrate by spontaneous fission, but instead there may be unusual forms of α-decay, anomalous γ transitions, etc.

LITERATURE CITED

1. Gallagher, C., et al., Nucl. Chem., 3:333 (1957).
2. Bohr, N., and Wheeler, J., Phys. Rev., 56:426 (1939).
3. Petrzhak, K. A., and Flerov, G. N., Zh. Éksperim. i Teor. Fiz., 10:1013 (1940).
4. Titterton, E. W., and Brinkley, T. A., Nature, 187:228 (1960).
5. Perfilov, N. A., et al., in the book "Physics of Atomic Nuclear Fission," Moscow, Gosatomizdat, 1962.
6. Wheeler, J., In the collection "Niels Bohr and the Development of Physics," ed. W. Pauli, Pergamon. [Russian translation Moscow, Izd. Inostr. Lit., 1958, p. 214.]
7. Sasakawa, T., and Yacuno, M., Progr. Theor. Phys., 20:315 (1958).
8. Johannsson, S. A., Nucl. Phys., 12:449 (1959).
9. Frenkel('), Ya., Phys. Rev., 55:987 (1939).
10. Swiatecki, W. J., Phys. Rev., 100:937 (1955).
11. Green, A. E., Phys. Rev., 95:1006 (1954).
12. Newton, J. O., Progr. in Nucl. Phys., 4:234 (1955).
13. Nilsson, S. V., In the collection "Deformation of Atomic Nuclei" [Russian translation], Moscow, Izd. Inostr. Lit., 1958.
14. Viola, V. E., and Wilkins, B., Preprint ANL, "Fission Barriers and Lifetimes in the Trans-Radium Elements," 1965.
15. Dorn, D. W., Phys. Rev., 121:1740 (1961).
16. Seeger, P. A., Nucl. Phys., 25:1 (1960).
17. Korneev, E. I., et al., Zh. Éksperim. i Teor. Fiz., 37:41 (1959).
18. Polikanov, S. M., et al., Zh. Éksperim. i Teor. Fiz., 42:1454 (1962).
19. Polikanov, S. M., et al., Zh. Éksperim. i Teor. Fiz., 44:804 (1963); Druin, V. A., et al., Zh. Éksperim. i Teor. Fiz., 40:1261 (1961); Flerov, G. N., et al., Zh. Éksperim. i Teor. Fiz., 45:1396 (1963); Linev, A. F., et al., Preprint Joint Institute for Nuclear Research, D-1693, Dubna, 1964; Dobanov, Yu. V., et al., Preprint Joint Institute for Nuclear Research, D-1801, Dubna, 1964; Hyde, E. K., Private Communication.
20. Flerov, G. N., et al., Rev. Roum. Phys., 10:217 (1965).
21. Polikanov, S. M., et al., in press.
22. Stephens, Diamond, Private communication.
23. Urin, M. G., and Zaretskii, D. F., Paper C 382a. Congress on Nuclear Physics, Paris, 1964; Zaretskii, D. F., and Urin, M. G., Zh. Éksperim. i Teor. Fiz., 43:1021 (1962).
24. Vandenbosch, R., et al., J. Inorg. Nucl. Chem., 26:219 (1964).

MAGNETIC PROPERTIES OF STRONGLY DEFORMED ATOMIC NUCLEI

A. Z. Hrynkiewicz and S. Ogaza

Polish People's Republic

Introduction

The structure of low-lying energy levels of nuclei has been relatively well studied for two limit groups, i.e., spherical nuclei close to filled shells and strongly deformed nuclei. In the first case, short-range pairing forces predominate, to some extent stabilizing the shell structure; in the second case, long-range quadrupole forces orient the orbits of the outer nucleons in space in such a manner as to lead to a stable deformation of the nucleus.

In this lecture, a comparison is made between the theoretical descriptions of the magnetic properties of strongly deformed nuclei with experimental results. We shall confine ourselves to a consideration of nuclei with the atomic weight 150 < A < 190, since in the last few years a large amount of experimental data on magnetic properties has been accumulated for these nuclei, making it possible to attempt an analysis of these data.

Magnetic Dipole Moment of the Nucleus and Reduced Probabilities of Magnetic Dipole Transitions

Experimental data on the magnetic properties of nuclei are obtained from the experimentally determined dipole moments of the nuclear states, and the probabilities of magnetic dipole transitions between two nuclear states. The operator of the multipole magnetic moment of order L is defined as

$$M_\mu(ML) = \mu_N \sum_{k=1}^{A} \left[\nabla r_k^L Y_{L\mu}(r)\right]\left[g_s^{(k)} \mathbf{s}_k + \frac{2}{L+1} g_l^{(k)} \mathbf{l}_k\right], \tag{1}$$

where

$$\mu_N = \frac{e\hbar}{2Mc} = 5.054 \cdot 10^{-24} \text{ erg/gauss} = 3.15 \cdot 10^{-12} \text{ eV/gauss}$$

is the nuclear magneton; $\mathbf{s}_k$ and $\mathbf{l}_k$ are the operators of spin and orbital moment of the k-th nucleon; and $g_s^{(k)}$ and $g_l^{(k)}$ are the spin and orbital nuclear factors of the k-th nucleon.

The operator M_μ is an irreducible tensor operator of rank L. Its matrix elements may be represented by means of the Wigner-Eckardt theorems in the form

$$\langle I'm' | M_\mu(ML) | Im \rangle = (-)^{I'-m'} \begin{pmatrix} I' & L & I \\ -m' & \mu & m \end{pmatrix} \langle I \| M(ML) \| I' \rangle. \tag{2}$$

The parity of the operator $M_\mu(ML)$ is determined by the formula $\pi(ML) = (-)^{L+1}$. Since only operators with positive parity have average values differing from zero in states with a definite parity, the nucleus has only odd magnetic moments (dipole, octupole, etc.). The dipole magnetic moment plays the most important part in describing the magnetic properties of nuclei. (The octupole magnetic moment has so far been measured for 13 nuclei only.) The dipole magnetic moment μ of a nuclear state having the total angular momentum I is determined (arbitrarily) as the mean value of the operator M_0 (M1) in the state m = 1.

$$\mu = \sqrt{\frac{4\pi}{3}} < II|M_0(M1)|II> = \sqrt{\frac{4\pi}{3}} \begin{pmatrix} I & 1 & I \\ -I & 0 & I \end{pmatrix} < I \| M(M1) \| I > =$$

$$= \frac{\sqrt{4\pi} I}{[3(2I+1)(I+1)I]^{1/2}} < I \| M(M1) \| I >. \tag{3}$$

It follows from the properties of the 3j symbol that the magnetic dipole moment is equal to zero for I = 0. The vector operator of the dipole magnetic moment in μ_N units may be represented in the form

$$\mu = g\mathbf{I} = \sum_{k=1}^{A} (g_s^{(k)} \mathbf{s}_k + g_l^{(k)} \mathbf{l}_k), \tag{4}$$

where **I** is the operator of the angular momentum of the nucleus, and g is the nuclear factor. The probability of the transition of M1 from the I state to the I' state is determined by the formula

$$T(M1;\ I \uparrow I') = \frac{16\pi}{9} \cdot \frac{k^3}{\hbar} B(M1;\ I \uparrow I'), \tag{5}$$

where $k = 2\pi/\lambda$ is the wave number of the emitted photon, and B(M1) is the reduced probability of transition:

$$B(M1;\ I \to I') = \sum_{mm'} |< I'm' | M_{m-m'}(M1) | Im > |^2 = \frac{1}{2I+1} |< I' \| M(M1) \| I > |^2. \tag{6}$$

Theoretical calculations of the magnetic moments of nuclei and probabilities of magnetic transitions are complicated by the fact that it is necessary to know not only the wave functions of the nuclear states, but also the structure of the operator of the magnetic moment. In formulas (1) and (4), additivity of the magnetic moments of the individual nucleons is assumed. This assumption is not exactly satisfied for two reasons.

1. The virtual meson currents between nucleons in the nucleus not only result in a variation in the values of the g-factors compared with their values for free nucleons, but also in the occurrence of a number of exchange terms [1]. This effect may have substantial significance for light nuclei (for example, it may be the cause of the anomalous values of μ for H^3, He^3), but for heavy nuclei, it may be assumed to be small ($\sim A^{-1}$). This is confirmed by the good agreement between the magnetic moments of nuclei with a doubly magic core ±1 nucleon, and Schmidt's values.

2. Velocity-dependent interactions, such as for example spin-orbital interactions, also lead to corrections which do not merely amount to renormalizing the g-factors. Bohr and Mottelson [2] allowed for spin-orbital interaction between two nucleons and obtained the following correction for the dipole moment:

$$\delta\mu = \pm 0.15 \tau_3 \frac{2j+1}{2j+2},$$

where

$$j = l \pm s, \quad \tau_3 = \begin{cases} 1 & \text{for } p \\ -1 & \text{for } n. \end{cases}$$

If the enumerated effects are ignored and it is assumed that the operator of the magnetic moment is composed additively of the operators of the magnetic moments of all the nucleons [see formula (4)], it should be borne in mind that the values of the nucleon g-factors included in it will differ from the values of the g_s- and g_l-factors for free nucleons:

$$g_s = \begin{cases} +5.587, \\ -3.826, \end{cases} \quad g_l = \begin{cases} 1 & \text{for protons} \\ 0 & \text{for neutrons}. \end{cases}$$

We shall indicate two very important causes of the renormalization of the g-factors.

1. Variation in the electromagnetic structure of the nucleons, with which are associated the anomalous values of the g_s factors for free nucleons, caused by interaction of nucleons in the nucleus. Drell and Walecka [3] found that this effect diminished g_s by approximately 7%.

2. The relativistic corrections for nucleons close to the Fermi surface of heavy nuclei lead to a diminution in the orbital magnetic moment of the proton by approximately 5% (for average values of A) [3].

All the corrections enumerated lead to a variation in the values of magnetic moments by 0.1-0.2 μ_N. If we confined ourselves to such accuracy, agreement of theory with experiment could be expected if we knew the exact wave functions describing the states of the nucleus. This assumption, however, will obviously never be satisfied, since the states of the nucleus are always described approximately by means of different models.

In the case of odd nuclei, we use the operator of the magnetic moment in the form

$$\mu = \sum_i \mu_i + \mu_{ext} \qquad (7)$$

in agreement with formula (4). The first term represents the operator of the magnetic moment of paired particles of the core, the second the operator of the magnetic moment of the external unpaired nucleon. In zero approximation, $\Sigma\mu_i = 0$. The inclusion of interaction may alter both terms of formula (7). This is due to the fact that usually we endeavour to express the mean value of the operator for the excited states in terms of the mean value for the modified operator for the states of zero approximation.

First of all, we shall examine the magnetic properties of nuclei on the basis of the purely phenomenological model, then we shall study the experimental methods and results, of measurements of the magnetic properties of nuclei in the region $150 < A < 190$. The last part of the lecture gives a discussion and analysis of these results with the help of a more exact theoretical approach, taking into account effects such as residual pair interaction, spin–spin interaction and also the coupling of the motion of an external nucleon with the motion of the core through Coriolis forces. This microscopic approach leads to a deeper understanding of the magnetic properties and connects them with other nuclear characteristics.

Phenomenological Description of the Magnetic Properties of Strongly Deformed Nuclei

The states of a deformed odd nucleus are described by the phenomenological model as states of one nucleon in a field formed by an even-even core with stable deformation. It is usually assumed that the field of the core has axial symmetry. The angular momentum of the core $\mathbf{R}$ is directed perpendicularly to the axis of symmetry. The total angular momentum of the nucleus $\mathbf{I} = \mathbf{R} + \mathbf{j}$ where $\mathbf{j}$ is the angular momentum of the unpaired nucleon, which is not a good quantum number, but its projection on the axis of symmetry $j_3 = K$ is the integral of motion. The total angular momentum of the nucleus and its projection on the axis of symmetry ($I_3 = K$) and on an arbitrarily defined direction in space ($I_2 = M$) are also conserved.

Such a system is described in a particularly simple manner in what is called adiabatic approximation. The physical significance of this approximation is that the particle will move sufficiently rapidly compared with the rotation of the core. Mathematically, this is expressed by the possibility of representing the wave function of the nucleus in the form of derivatives of two functions: χ_K, describing the motion of the particle and D^I_{MK}, describing the motion of the core. The functions D^I_{MK} are wave functions of a symmetrical gyroscope which are dependent on the Euler angles. If the deformation is symmetrical with respect to reflection at the center of the system, the wave function must be antisymmetrized as follows:

$$\Psi_{IMK} = \frac{\sqrt{2I+1}}{4\pi}\left[D^I_{MK}\chi_K + (-)^{I-K} D^I_{M-K}\chi_{-K}\right]. \tag{8}$$

The energy levels of the system described by function (8) form a rotational band with the definite quantum number K. The levels of this band depend on two phenomenological parameters: the moment of inertia J and (in the case of $K = 1/2$), the decoupling parameter α.

We represent the operator of the magnetic moment in the form

$$\mu = g_K \mathbf{j} + g_R \mathbf{R}, \tag{9}$$

where the first term is connected with the internal structure, and the second with the collective rotation. The mean value of μ in the state ψ_{IMK} for $M = I$ has the form

$$\mu = \langle IKI|\mu|IKI\rangle = g_R I + (g_K - g_R)\frac{K^2}{I+1}\left[1 + (-)^{I+1/2}(2I+1)\, b_0 \delta_{K^{1/2}}\right]. \tag{10}$$

The additional term with b_0, differing from zero for $K = 1/2$, is connected with the symmetry property of the wave function. It leads to decoupling of the internal and collective motions.

For even–even nuclei, the magnetic moments of the states of the ground rotational band are determined only by the rotational q_R factor

$$\mu = g_R I. \tag{11}$$

The reduced probabilities of M1-transitions between two adjacent levels of the same rotational band in odd nuclei has the form

$$B(M1;\, I+1 \to I) = \frac{3}{4\pi}\mu_N^2 (g_K - g_R)^2 K^2 |\langle I+1K10|IK\rangle|^2 \left[1 + (-)^{I-1/2} b_0 \delta_{K^{1/2}}\right]^2. \tag{12}$$

It will be noted that so far we have used only the properties of symmetry of the wave function and the assumption regarding adiabaticity without entering into the details of the structure of the nucleus, the magnetic properties having been described by the three parameters: g_R, g_K, and b_0 (for $K = 1/2$). The correctness of formulas (10) and (12) has not yet been confirmed with the same accuracy as that of the formulas for the energies of rotational states or for the probability of E2-transitions between the states of the rotational band.

For $K \neq 1/2$ in formulas (10) (12) the two parameters g_R and g_K are separated. For their determination, two independent experiments are thus necessary, at least one of which should be the measurement of the magnetic moment. For the band $K = 1/2$, three independent experiments (one being the measurement of μ) are necessary.

Review of Experimental Methods of Determining the Magnetic Properties of Nuclei

All experimental methods of determining the magnetic moments of nuclei make use of the interaction of the latter with an external magnetic field or with the intraatomic magnetic field. In a magnetic field H, the spin of a nucleus precesses with the Larmor frequency

$$\omega = g \frac{\mu_N}{\hbar} H. \tag{13}$$

This frequency corresponds to the Zeeman splitting of the levels

$$\Delta E = \hbar\omega. \tag{14}$$

Methods of measuring nuclear g-factors consist in determining the precession frequency or splitting of the levels in a known magnetic field H. The sign of the g factor is determined by the direction of the angular velocity of precession with respect to the direction of the field H.

The magnetic moments of the ground states of stable nuclei and long-lived states are determined mainly by means of the methods of magnetic nuclear resonance and resonance in atomic beams, and also by methods of atomic spectroscopy. In resonance methods, the frequency of the Larmor precession in an external magnetic field is measured; in the optical method, the splitting of the hyperfine structure, caused by interaction of the magnetic moment of the nucleus with the intraatomic magnetic field, is measured. Most (about 80%) of the measurements carried out have been made by means of these three basic methods. Resonance methods give the highest accuracy. Their errors are due mainly to inaccuracy in the measurement of the magnetic field and inaccuracy in calculating the diamagnetic corrections. In optical methods, the accuracy is much worse, due to the difficulties in calculating the intraatomic magnetic fields.

For radioactive nuclei, the possibility of using the methods enumerated is limited by the amount of substance available for study and the lifetime of the investigated nuclei. These two factors are interrelated. As a rule, the shorter the lifetime of nuclei, the smaller is the amount of them available for experimental use.

The method of magnetic nuclear resonance requires macroscopic amounts of substance. The nucleus having the shortest lifetime, for which it has been possible to use this method, is the nucleus H^3 ($T_{1/2}$ = 12 years).

The optical method may be used for making measurements on a much smaller quantity of substance. The lower limit of this amount is determined by the attainable intensity of light, on which the exposure time depends. The best achievement of the optical method in the region of short-lived nuclides is the measurement of the magnetic moment of Tl^{199} having a half-life of 7.4 h.

In the atomic beam method, it is possible to use the radiation of the nuclei investigated for detecting the effect sought; for this reason, it is possible to work with microscopic quantities of substance and to make measurements for nuclear states with a lifetime of the order of minutes. For example, the magnetic moment of N^{13} having a half-life of 10 min has been measured.

Measurements of the g factors of short-lived excited states are based on the following effects.

1. Zeeman splitting of γ lines in Mössbauer spectra.

2. Perturbation of the angular correlation of cascade γ-transition of an extranuclear magnetic field.

3. Perturbation of the angular distribution of γ radiation, emitted after Coulomb excitation, under the action of an external magnetic field.

4. Perturbation by an external magnetic field of the angular distribution in the resonance scattering of γ radiation.

The method using the Mössbauer effect is similar to the optical method. It consists in determining the splitting of the hyperfine structure of γ lines caused by the interaction of the nucleus with the intraatomic magnetic field. If the magnetic moment of the nucleus is known in one of its states between which a γ transition is occurring, it is possible to calculate the magnetic moment of the second state and the value of the magnetic field. Recent work on the Mössbauer effect in Coulomb excitation [4] permits a considerable extension of the range of application of this method, which previously it has been possible to use in a few cases only. The limitation in the application of the Mössbauer method from the point of view of short lifetimes of the investigated states is due to the increase in the natural width of the γ line ($\Gamma = \hbar/\tau$). It is possible to measure the splitting exactly if the experimental line width 2Γ is less than the value of the splitting itself, this being expressed by the condition

$$2\Gamma \leqslant \frac{\mu H}{I}, \text{ whence } \tau \geqslant \frac{2I\hbar}{\mu H}. \tag{15}$$

The lower limit attainable in practice for the lifetime τ is 10^{-9}–10^{-10} sec. The upper limit for the lifetime of states for which it is still possible to make measurements is determined by the technical difficulties of observing very narrow lines.

In the observation of angular correlation and angular distribution of γ radiation, as well as in the above-mentioned resonance methods, the frequency of the precession of the nucleus in an external magnetic field of given value is determined.

The fundamental difference is that in resonance methods, the nucleus makes a large number of precessions, and the frequency of a high-frequency magnetic field producing transitions between the levels of Zeemann splitting is measured, while in studying the influence of a magnetic field on the angular distribution of γ radiation, in many cases the nucleus in its lifetime makes only a small part of a complete revolution, and the frequency of precession is determined according to the angle and the time of a revolution.

The accuracy of the measurement of g factors of short-lived excited states is in most cases 10–15%, and attains 2–3% only under particularly favorable conditions.

With regard to short lifetimes, the applicability of these methods is limited by the attainable values of the magnetic fields, since the angle of rotation of the nucleus is proportional to the lifetime and to the value of the magnetic field. For magnetic fields produced under laboratory conditions, the lower limit for τ is 10^{-10} sec. This limit may be lowered by using intraatomic fields, which in the rare-earth region attain values of several megaoersted. This is possible, however, only in cases where the time of electron spin relaxation τ_r exceeds the period of Larmor precession:

$$\tau_r > \frac{I\hbar}{\mu H}. \tag{16}$$

This means that the width of the levels of the hyperfine structure must be less than the splitting of the levels, otherwise the magnetic field becomes indefinite. Consequently, the measurements may be made only for nuclei situated in the lattices of magnetic substances or in paramagnetic substances having a high value of electron relaxation time.

The upper limit of the lifetimes of excited states for which it is possible to measure magnetic moments is $\sim 10^{-5}$ sec. This limit is due to the increase in the background of random coincidences, unavoidable in the measuring technique employed.

The second experimental source of information on the magnetic properties of deformed odd nuclei is that of the reduced probabilities of magnetic dipole transitions between levels of the rotational band. Knowing B(M1), the parameter $(g_K - g_R)$ may be calculated by means of formula (12).

The most direct method of measuring the reduced probability of an M1 transition is the measurement of the mean lifetime τ of the rotational state by the method of delayed coincidences. For the calculation of B(M1), however, it is necessary to know, in addition to τ, also the ratio of the intensities of type E2 and M1 radiations, δ^2, and sometimes the ratio of the intensity of the investigated transition to the total intensity of the other transitions with given level $I/\Sigma I_r$. Then

$$B(M1) = \frac{5.69 \cdot 10^{-5} E^{-3}}{\tau(1+\alpha)(1+\delta^2)\left(1 + \frac{\sum_r I_r}{I}\right)}, \tag{17}$$

In this formula B(M1) is expressed in the units μ_N^2, the transition energy E is in keV and τ is in sec.

An indirect method of measuring B(M1) for the ground-state band is to determine the value of B(E2) from the effective cross section of Coulomb excitation of the rotational state of this band. From the experimental value of B(E2), it is possible to calculate the mean lifetime of the given state and use it in formula (17).

The methods of determining B(M1) described require several independent and complicated measurements. In rare cases only, therefore, has the value of $(g_K - g_R)^2$ an error of less than 10%.

Experimental Data on the Magnetic Properties of Deformed Nuclei in the Region 150 < A < 190

Table 1 presents the experimental values of the g factors of excited nuclear states with $K \neq 1/2$ in the region $150 < A < 190$. It gives only the most accurate and reliable results of measurements.

For even–even nuclei, the values of g_R of the first excited rotational 2^+ states were measured in 16 cases. For the nuclei Er^{166} and Hf^{180}, the values of g_R are known for the two excited 2^+ and 4^+ states of the ground rotational band. According to the rotational model, these values ought to agree. Due to the very large errors, it is difficult to judge the experimental values for Hf^{180}, but for Er^{166} both values agree in the limits of the experimental accuracy.

Table 2 presents data on the magnetic properties of the ground rotational bands with $K \neq 1/2$ for odd nuclei in the region $150 < A < 190$. Table 2 gives the values of the magnetic moments of the ground states and the values of $(g_K - g_R)^2$ calculated from the probabilities of M1 transitions between the levels of the rotational band. The last two columns give the values of g_R and g_K, calculated on the basis of measured values of μ and the mean values of $(g_K - g_R)^2$.

TABLE 1. Experimental Values of g Factors of Excited Nuclear States with K ≠ 1/2 for Nuclei in the Region 150 < A < 190

Nucleus	Energy of the state, keV	$T_{1/2}$, nsec	I_π ($[Nn_z \Lambda]K$)	Nuclear g-factor
Sm^{152}	122	1.4	2+	+0.350 (30)
Eu^{153}	103	3.3	3/2+ [411] 3/2	+1.34 (6) [5]
Sm^{154}	82	2.7	2+	+0.30 (3)
Gd^{154}	123	1.2	2+	+0.367 (30)
Gd^{155}	87	6.7	3/2+, 5/2+	+0.376 (26)
Gd^{156}	89	2.2	2+	+0.320 (30)
Dy^{160}	87	1.8	2+	+0.282 (34)
Dy^{161}	26	27	5/2−[523] 5/2	+0.52 (7)
Er^{166}	80	1.82	2+	+0.32 (2)
	265	0.12	4+	+0.276 (24)
Er^{168}	80	1.92	2+	+0.27 (3)
Yb^{170}	84	1.6	2+	+0.334 (5)
Tu^{171}	129	0.5	7/2+	+0.27 (5)
Yb^{172}	79	1.57	2+	+0.304 (34)
Lu^{175}	114	0.097	9/2+ [404] 7/2	+0.49 (5)
Hf^{177}	113	0.52	9/2− [514] 7/2	+0.232 (13)
Hf^{178}	93	1.50	2+	+0.356 (35)
				+0.29 (2)
Hf^{180}	93	1.51	2+	+0.371 (32), 0.29 (4) [6]
	309	0.08	4+	+0.5 (1)
Ta^{181}	482	10.9	3/2+ [402] 5/2	+1.29 (2)
W^{182}	100	1.37	2+	+0.239 (20)
W^{184}	111	1.28	2+	+0.279 (20)
W^{186}	122	1.01	2+	+0.368 (42)
Re^{187}	206	570	9/2− [514] 9/2	+1.11 (1)
Os^{186}	137	0.84	2+	+0.316 (28)
Os^{188}	155	0.73	2+	+0.23 (3)

TABLE 2. Experimental Data on the Magnetic Properties of Ground Rotation Bands with K ≠ 1/2 of Odd Nuclei in the Region 150 < A < 190

Nucleus	Quantum numbers of ground state	Magnetic moment of ground state	$(g_K - g_R)_1^2$	$(g_K - g_R)_2^2$	g_R	g_K
$_{63}Eu^{153}$	[413] 5/2+	+1.507 (3)	0.031 (6)	0.033 (5)	0.476 (8)	+0.654 (3)
Gd_{91}^{155}	[521] 3/2−	−0.242 (2)	—	0.62 (9)	0.311 (36)	−0.476 (24)
Gd_{93}^{157}	[521] 3/2−	−0.3225 (10)	—	0.66 (8)	0.271 (30)	−0.539 (20)
$_{65}Tb^{159}$	[411] 3/2+	+1.90 (5)	1.67 (30)	1.72 (30)	0.486 (58)	+1.788 (43)
Dy_{95}^{161}	[642] 5/2+	−0.455 (10)	0.32 (16)		0.255 (100)	−0.345 (41)
Dy_{97}^{163}	[523] 5/2−	+0.635 (14)	—	0.0024 (8)	0.289 (8)	+0.240 (6)
$_{67}Ho^{165}$	[523] 7/2−	+4.01 (8)		0.68 (10)	0.504 (52)	+1.329 (27)
Er_{99}^{167}	[633] 7/2+	−0.564 (7)	—	0.157 (16)	0.149 (30)	−0.249 (9)
Yb_{103}^{173}	[512] 5/2−	−0.6778 (25)	0.66 (11)	0.46 (10)	0.257 (50)	−0.483 (20)
$_{71}Lu^{175}$	[404] 7/2+	+2.230 (11)	0.138 (10)	0.110 (13)	0.360 (16)	+0.716 (6)
$_{71}Lu^{177}$	[404] 7/2+	+2.236 (10)	—	0.141 (19)	0.347 (28)	+0.722 (6)
Hf_{105}^{177}	[514] 7/2−*	+1.044 (59)*	0.0018 (2)	0.0022 (12)	0.253 (13)	+0.211 (13)
Hf_{107}^{179}	[624] 9/2+	−0.47 (3)	0.221 (18)	0.141 (28)	0.263 (42)	−0.186 (11)
$_{73}Ta^{181}$	[404] 7/2+	+2.35 (1)	0.203 (10)	0.208 (25)	0.320 (8)	+0.771 (4)
$_{75}Re^{185}$	[402] 5/2+	+3.1718	—	1.41 (14)	0.419 (43)	+1.609 (17)
$_{75}Re^{187}$	[402] 5/2+	+3.2043	1.82 (43)	1.47 (15)	0.404 (44)	+1.630 (17)

The values of $(g_R - g_K)^2$, found from the probabilities of M1 transitions both between the first excited and ground levels and between the second and first excited levels of the ground-state rotational band, are known with relative good accuracy in eight cases. It follows from the rotational model that

$$\frac{(g_K - g_R)_1^2}{(g_K - g_R)_2^2} = 1.$$

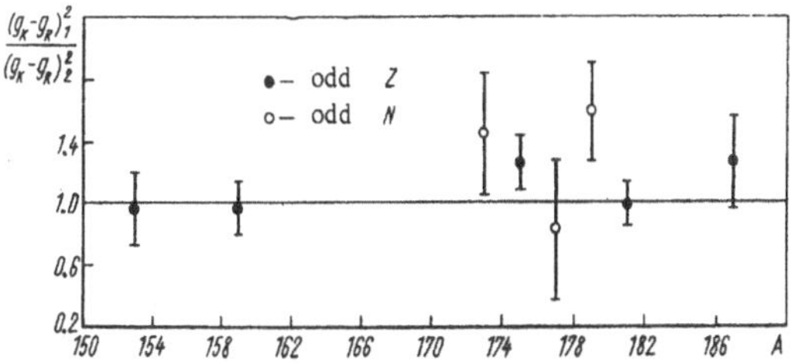

Fig. 1. Ratio of the parameters $(g_K - g_R)^2$ for M1 transitions between states of the same rotational band for nuclei in the region 150 < A < 190.

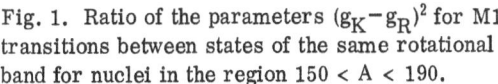

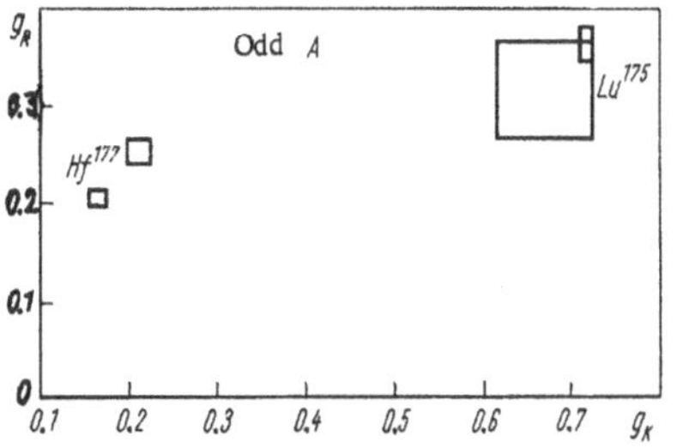

Fig. 2. Values of the parameters g_R and g_K for Lu^{175} and Hf^{177}.

TABLE 3. Experimental Information on the Magnetic Properties of the Principal Rotational Band of Tu^{169} [8]

	Experiment	Calculations*
$\mu_{1/2}$	−0.231 (1)	(−0.231)
$\mu_{3/2}/\mu_{1/2}$	−2.33 (4)	(−2.33)
$\mu_{5/2}$	+0.61 (12)	+0.74
$\mu_{7/2}$	+1.31 (18)	+1.45
$\tau_{3/2}$, нсек	5.93 (17)	5.38
$\tau_{5/2}$, нсек	0.090 (4)	(0.090)
$\tau_{7/2}$, нсек	0.43 (3)	0.41
$\delta^2_{3/2\to1/2}$, 10^{-3}	~1	1.03
$\delta^2_{5/2\to3/2}$, 10^{-2}	~2.4	2.17
$\delta^2_{7/2\to5/2}$, 10^{-4}	~5	7
$\lambda_{5/2}$	0.104 (5)	0.108
$\lambda_{7/2}$	54 (12)	61
$\lambda_{9/2}$	0.440 (53)	0.495
$\lambda_{13/2}$	1.13 (17)	1.14
$\Gamma_{\gamma 3/2\to1/2}$, 10^5 sec^{-1}	5.3 (5)	5.72

* Parameters used in calculations: $g_K = -1.57$; $g_R = 0.406$; $b_0 = -0.16$, obtained from the values of $\mu_{1/2}$, $\mu_{3/2}/\mu_{1/2}$, and $\tau_{5/2}$.

Figure 1 shows the experimental values of this ratio. There is no great deviation from unity.

De Boer and Symons [7] measured the probabilities of M1 transitions between higher states of rotational bands of odd nuclei in the region 150 < A < 190 and calculated the corresponding values of $|g_K - g_R|$. They supposed that despite high experimental errors, a systematic increase in these values with increase in the spin of the rotational state, amounting to ~0.1-0.2 per unit spin was noticeable. This effect in the author's view was due to an admixture of adjacent Nilsson states caused by Coriolis interaction.

For two nuclei Lu^{175} and Hf^{177} we know the magnetic moments of two states and two probabilities of M1 transitions for the same rotational bands. It is thus possible to calculate two values of g_R and g_K for these nuclei. The values obtained are compared in Fig. 2. For Lu^{175} agreement is satisfactory. The discrepancy observed for Hf^{177} is due in all probability to an error in the determination of the magnetic moment of the ground state by the optical method.

Measurements for bands with $K = 1/2$ have been made only for Tu^{169}, Tu^{171}, and W^{183}. A particularly large amount of experimental material has been obtained for the ground-state rotational band of Tu^{169} [8]. For this band, the magnetic moments of the first four states, the lifetimes of three excited states (by the delayed coincidence method), four intensity ratios

$$\lambda_I = \frac{J(E2;\ I \to I-2)}{J(M1;\ I \to I-1) + J(E2;\ I \to I-1)},$$

three intensity ratios of E2 and M1 transitions $\delta^2 = J(E_2)/J(M_1)$ and the width of the γ line for $3/2 \to 1/2$ transitions (from the Mossbauer spectrum) have been measured.

All these data are given in Table 3. The most accurate experimental values of $\mu_{1/2}$, $\mu_{3/2}/\mu_{1/2}$ and $\tau_{3/2}$ were used as basis for calculating g_R, g_K, and b_0, which were then used for calculating the other values given in Table 3. Comparison of the values obtained with the experimental values serves as check of the internal agreement of the phenomenological model.

Analysis of the Experimental Data on Magnetic Properties of Deformed Nuclei in the Region $150 < A < 190$

The parameters g_R, g_K, and b_0 (for $K = 1/2$), used in the phenomenological model and obtained experimentally, may be calculated theoretically on the basis of a more exact model of the nucleus. Such a model is the superfluid model, which takes into account the residual pair interactions in the approximation of free quasiparticles. Allowing for pair correlations leads to a substantial variation in the magnetic moment of even–even nuclei. In the case of odd nuclei, the internal g-factors do not vary in this approximation compared with the single-particle picture, since the magnetic moments of a particle and a quasiparticle are the same. This follows from the fact that the spins of a particle and a hole make the same contribution to the orbital magnetic moment.

In the calculations, use is made of formula (7) determining the operator of the magnetic moment.

In calculating g_R, allowance for pair correlations results in a variation of the values of the moments of inertia compared with the rigid body values. In addition, allowance is made for the perturbation caused by Coriolis force, which leads to a polarization of the core due to the virtual rupture of the nucleon pairs and, as a result of this, to an additional contribution of the core to the magnetic moment.

In calculations of g_K, the following perturbations of the single-particle states of the unpaired nucleon are additionally taken into account.

1. Spin–spin interaction producing spin polarization of the core. By allowing for this effect, it is possible to express the internal magnetic moment of the nucleus as a magnetic moment of the external nucleon with corresponding renormalization of the spin factor g_s.

2. Interaction with the core, due to Coriolis forces, which mixes the single-particle states. In the first approximation of the theory of perturbations, this results in a renormalization of the factor g_K.

Rotational g Factors for Even–Even Nuclei

In calculating g_R, a start is usually made from the Inglis cranking model. In this model, it is assumed that the nucleon moves in a deformed field rotating under the influence of external forces. With this approach, it is possible to calculate the moment of inertia of the rotating nu-

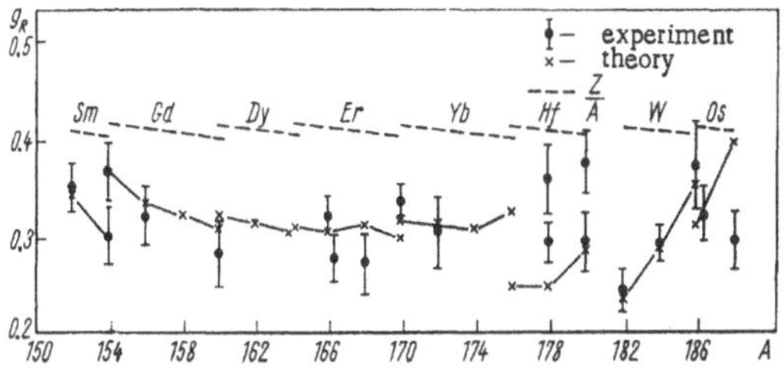

Fig. 3. Rotational g factors for even–even nuclei in the region 150 < A < 190.

cleus from the value of the additional energy necessary for the nucleon to follow the rotating field adiabatically.

Nilsson and Prior [9] by allowing for Coriolis forces in the cranking model obtained the following expression for the g factor of an even–even nucleus:

$$g_R = \frac{J_p}{J_p + J_n} + (g_s^p - 1)\frac{W_p}{J} + g_s^n \frac{W_n}{J}, \tag{18}$$

where J is the total moment of inertia of the nucleus, and J_p and J_n are the contributions of the protons and neutrons to the moment of inertia. The last two terms in formula (18) are corrections due to spin polarization of the core, and they may be ignored, since they are small compared with the first term and compensate each other to a considerable extent, the coefficients $(g_s^p - 1)$ and g_s^n being close to each other in their absolute values and opposite in sign.

In addition, Nilsson and Prior in their calculations allowed for residual pair interactions, which results in a variation of the total moment of inertia compared with the rigid body moment of inertia, and varied the neutron and proton contributions. This is due to the different constants of pair interaction and the different distributions of single-particle levels close to the Fermi surfaces for neutrons and protons. Since the pairing energy is higher for protons than for neutrons, the contribution of the protons to the moment of inertia is less than that of the neutrons, and therefore g_R is less than the rigid body value:

$$g_R < \frac{Z}{A}. \tag{19}$$

Nilsson and Prior calculated the moments of inertia and rotational g factors for deformed even–even nuclei, determining the values of the pair interaction constants from the difference in mass of even–even and odd nuclei. The best agreement with the experimental differences in mass for the region 150 < A < 190 was given by the values

$$G_n = \frac{18}{A}\ \text{MeV and}\ G_p = \frac{25}{A}\ \text{MeV}.$$

It was found that despite the sensitivity of the moments of inertia to variations in the deformation parameter, the values of g_R depended little on this parameter, since J_p and J_n vary similarly with deformation.

In Fig. 3, the experimental values are compared with Nilsson and Prior's data and with the rigid body values of Z/A. Nilsson and Prior's data are in good agreement with the experimental values.

Rotational g Factors for Odd Nuclei

In odd nuclei, the external nucleon makes a substantial contribution to the moment of inertia of the nucleus. From formula (18) it is clear that

$$g_R\ (\text{even--even} + p) > g_R\ (\text{even--even}),$$
$$g_R\ (\text{even--even} + n) < g_R\ (\text{even--even}).$$

The variation in the moment of inertia caused by coupling of the motion of the external nucleon with the motion of the nucleons of the rotating core may be described by a Coriolis force. In the first approximation of the perturbation theory, allowance for this interaction leads to renormalizing of the g_R factor,

$$g_R = g_R^0 + \delta g_R + \delta' g_R, \tag{20}$$

where g_R^0 is the rotational factor of the neighboring even–even nucleus (A−1); δg_R is the contribution of the variation in the moment of inertia, and $\delta' g_R$ is the contribution of spin polarization of core nucleons. If the blocking effect (in which a level occupied by a particle is inaccessible to pairs) is ignored, then in first approximation of the perturbation theory

$$\delta g_R = \frac{J_{\text{odd}} - J_{\text{even}}}{J_{\text{odd}}} \left(g_l - g_R^0\right) \tag{21}$$

and for nuclei having an odd number of protons

$$g_R = 1 - \left(1 - g_R^0\right) \frac{J_{\text{even}}}{J_{\text{odd}}} + \delta' g_R, \tag{22}$$

TABLE 4. Rotation g Factors of Odd Nuclei in the Region 150 < A < 190

Nucleus	$[Nn_z\Lambda]K$	g_R^0	$J_{\text{even}}/J_{\text{odd}}$	$\delta' g_R$	g_R Experiment	g_R Theory	
			Odd Z				
Eu^{153}	[413] 5/2	0.341	0.557	−0.14	0.476 (8)	0.49	0.454
Tb^{159}	[411] 3/2	0.319	0.870	−0.07	0.486 (58)	0.34	0.455
Ho^{165}	[523] 7/2	0.304	0.837	+0.11	0.504 (52)	0.53	0.450
Lu^{175}	[404] 7/2	0.305	1.007	−0.03	0.360 (16)	0.27	0.305
Ta^{181}	[404] 7/2	0.280	0.990	−0.08	0.320 (8)	0.21	0.338
Re^{185}	[402] 5/2	0.284	0.980	−0.02	0.419 (43)	0.28	0.283
Re^{187}	[402] 5/2	0.352	0.979	−0.03	0.404 (44)	0.34	0.292
			Odd N				
Gd^{155}	[521] 3/2	0.367	0.562	+0.05	0.311 (36)	0.26	0.258
Gd^{157}	[521] 3/2	0.333	0.726	+0.04	0.271 (30)	0.28	0.312
Dy^{161}	[642] 5/2	0.318	0.433	−0.05	0.225 (100)	0.09	0.055
Er^{167}	[633] 7/2	0.303	0.651	−0.08	0.149 (30)	0.12	0.238
Yb^{173}	[512] 5/2	0.308	0.858	+0.01	0.257 (50)	0.27	0.296
Hf^{177}	[514] 7/2	0.245	0.859	+0.04	0.253 (13)	0.25	0.259
Hf^{179}	[624] 9/2	0.245	0.717	−0.08	0.263 (42)	0.10	0.132

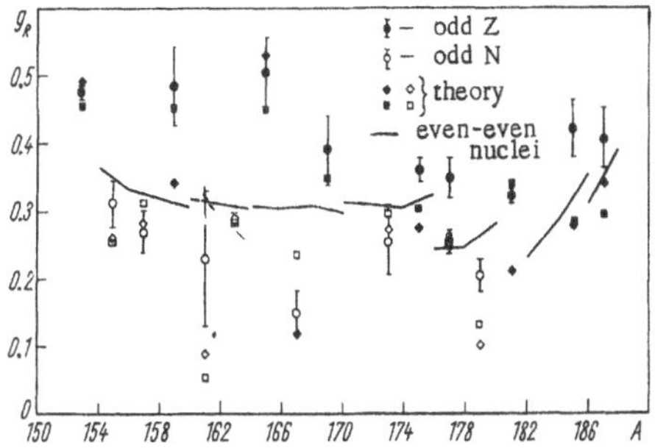

Fig. 4. Rotational g-factors for odd nuclei in the region 150 < A < 190.

and for nuclei with an odd number of neutrons

$$g_R = g_R^0 \frac{J_{\text{even}}}{J_{\text{odd}}} + \delta' g_R. \tag{23}$$

Grin' and Pavlichenkov [10] calculated the spin contribution $\delta' g_R$. For calculating the theoretical values of the rotational g-factors for odd nuclei, they used values of J_{odd} and J_{even} found from the experimental data of the energies of the first rotational levels of a given nucleus of mass A and a nucleus of mass A−1. Table 4 gives the experimental and theoretical values of g_R for odd nuclei in the region 150 < A < 190. Grin' and Pavlichenkov's data are corrected in the sense that in calculating J_{odd} and J_{even} they used the energies of all the known states of the rotational bands. The values of g_R^0 of neighboring even–even nuclei are taken from the paper by Nilsson and Prior. The last column gives the results of new unpublished calculations by Prior. A comparison between experimental and theoretical results is given in Fig. 4. The agreement of theory with experiment is not so good as in the case of even–even nuclei but the main features of the experimental data reflect the theory. It is possible that this agreement could be improved by taking the blocking effect into account and by increasing the accuracy of measurement.

Internal Nuclear Factors of Odd Nuclei

Taking pair correlations into account in the independent quasiparticle approximation does not affect the internal nuclear magnetism, and therefore the internal g-factors may be calculated by means of the Nilsson wave functions. The expression for g_K obtained by Nilsson for a particle in the field of the deformed core (for $K \neq 1/2$) has the form

$$g_K K = g_l K + (g_s - g_l) \langle \nu | s_3 | \nu \rangle. \tag{24}$$

The mean value of the operator of the projection of spin on the axis of symmetry of the nucleus in the Nilsson state $|\nu\rangle$ is expressed by the formula

$$\langle \nu | s_3 | \nu \rangle = \frac{1}{2} \sum_l \left(a_{lK-1/2}^2 - a_{lK+1/2}^2\right), \tag{25}$$

where a_{lK} are the expansion coefficients of $|\nu\rangle$ according to states $|Nl\Lambda\Sigma\rangle$. The g_K factors calculated in this way deviate systematically from the experimental values, if the values of g_l and g_s for free nucleons are used in the calculations.

Mottelson drew attention to the fact that these deviations could be due to spin polarization of the core by the external nucleon. Spin interaction of the outer nucleon with core nucleons tends to render the spins of identical nucleons antiparallel and the spins of a neutron and proton parallel. Due to the opposite signs of the g_s factors of proton and neutron, spin polarization ought to diminish quite considerably the contribution made by spin to the magnetic moment, since the contributions of many nucleons will be added together coherently. This effect may lead to renormalizing of the value of g_s of the external nucleon, which is equivalent to the introduction of an effective operator of the magnetic moment acting on an invariant single-particle state.

de Boer and Rodgers [11] made an analysis of the experimental values of g_K and showed that formula (24) would give correct results if instead of g_s an effective g factor

$$g_s^{\text{eff}} \approx 0.6 g_s$$

was introduced.

In the case of a rotational band with K = 1/2, the value of $(g_K - g_R)b_0$ like g_K, may be expressed by g factors of the outer nucleon. For Tm^{169} it is found that agreement with experiment is obtained only by introducing two different values of the effective g_s factor: for g_K, $g_s^{\text{eff}} \approx 0.80\, g_s$ must be used, and for $(g_K - g_R)b_0$, $g_s^{\text{eff}} \approx 0.55\, g_s$.

The main contribution to the renormalizing of g_s is made by the part of residual spin–spin interaction which has spherical symmetry in spatial, variable interacting nucleons. It is to be expected that a sufficiently exact estimate of the effect will be obtained by means of an additional interaction of the simple form

$$V = \frac{1}{2} V_0 \sum_{i \neq y} \sigma_i \sigma_j, \tag{26}$$

where V_0 is a constant depending on isospin.

Bochnacki and Ogaza [12] made calculations with this interaction and obtained good agreement of the g_K factors with experimental data.

It was found that in first approximation of the perturbation theory, the term with $\langle \nu | s_3 | \nu \rangle$ in the expression for the magnetic moment is added to the single-particle state $\alpha_\nu^+ |0\rangle$ of three-particle state $\alpha_{\nu'}^+ \beta_{\nu''}^+ \alpha_\nu^+ |0\rangle$, produced as the result of virtual rupture of core pairs. The effect of spin polarization of the core by the external nucleon may be reduced to the occurrence of a certain field

$$U = V_0 \langle \nu | s_3 | \nu \rangle \sum_i \sigma_3^{(i)}, \tag{27}$$

where summation is over all the nucleons of the core. If we take into account the secondary effect in which the polarized nucleons of the core induce an additional field, again acting on the core, we obtain the following final formula for the mean value of the projection of spin:

$$\langle S_3 \rangle = \langle \nu | s_3 | \nu \rangle \frac{1}{1-\alpha}, \tag{28}$$

where α determines the polarizability of the core:

$$\alpha = -8V_0 \sum_{\nu'\nu'' \neq \nu} \frac{|\langle \nu' | s_3 | \nu'' \rangle|^2 (u_{\nu'} v_{\nu''} - v_{\nu'} u_{\nu''})^2}{E_{\nu'} + E_{\nu''}}, \tag{29}$$

where u_ν, v_ν are the Bogolyubov transformation factors; $E_{\nu'}$, $E_{\nu''}$ are the energies of the quasiparticles.

The resulting renormalization of the mean values of spin projection is equivalent to renormalization of the g_s factor in the operator of the magnetic moment

$$g_s^{\text{eff}} - g_l = (g_s - g_l) \frac{1}{1-\alpha}. \tag{30}$$

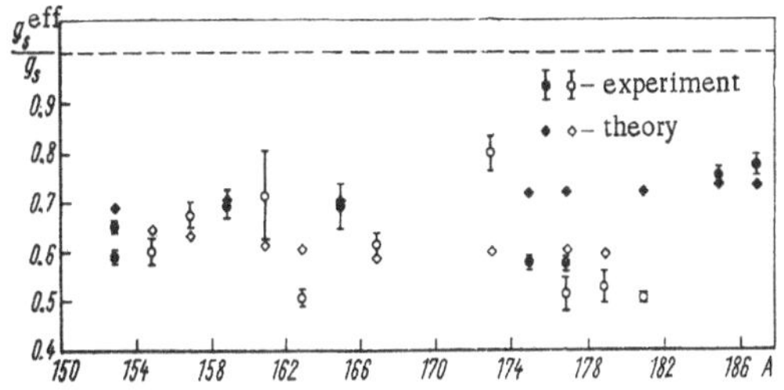

Fig. 5. Effective g_s factors for odd nuclei in the region 150 < A < 190.

Bochnacki and Ogaza calculated g_s^{eff} for deformed nuclei in the region 150 < A < 190, using the quasiparticle energies and parameters λ and Δ of Pyatov and Solov'ev [13]. The values of the interaction constants were determined from the splitting of the two-quasiparticle doublet states $K_1 \pm K_2$ of even–even nuclei. The average value of V_0 obtained from all known doublet splittings is

$$V_0 = \frac{59}{A} \text{ MeV.}$$

The results of the calculations of g_s^{eff}/g_s are compared with experiment in Table 5 and Fig. 5. Spin polarization of the core satisfactorily explains the observed values. There still remains, however, some difference between experimental and theoretical values, possibly due to the fact that the quadrupole term in spin–spin interaction and the effect of Coriolis forces have not been taken into account. Coriolis interaction in the form

$$H_c = -\frac{\hbar^2}{2J}(I_+j_- + I_-j_+) \tag{31}$$

TABLE 5. Magnetic Parameters Related to Internal Structure of Odd Nuclei in the Region 150 < A < 190.

Nucleus	Experiment		Theory
	g_K	g_s^{eff}/g_s	g_s^{eff}/g_s
$_{63}Eu^{153}$	+0.654 (3)	0.591 (4) 0.635 (4)*	0.690
Gd_{91}^{155}	−0.476 (24)	0.599 (29)	0.647
Gd_{93}^{157}	−0.539 (20)	0.678 (24)	0.635
$_{65}Tb^{159}$	+1.788 (43)	0.697 (28)	0.708
Dy_{95}^{161}	−0.345 (41)	0.714 (85)	0.613
Dy_{97}^{163}	+0.240 (6)	0.507 (13)	0.607
$_{67}Ho^{165}$	+1.329 (27)	0.691 (42)	0.703
Er_{99}^{167}	−0.249 (9)	0.615 (22)	0.586
Yb_{103}^{173}	−0.483 (20)	0.796 (33)	0.597
$_{71}Lu^{175}$	−0.716 (6)	0.578 (8)	0.720
$_{71}Lu^{177}$	+0.722 (6)	0.569 (9)	0.717
Hf_{105}^{177}	+0.211 (13)	0.512 (32)	0.602
Hf_{107}^{179}	−0.186 (11)	0.528 (31)	0.589
$_{73}Ta^{181}$	+0.771 (4)	0.503 (6)	0.719
$_{75}Re^{185}$	+1.609 (17)	0.748 (16)	0.726
$_{75}Re^{187}$	+1.630 (17)	0.767 (16)	0.727

in first approximation of the perturbation theory leads to additions of K ± 1 states to the K (K ≠ $^1/_2$) state. This alters the internal factor g_K by the amount

$$\delta g_K = \frac{1}{K}(A_{K+1} + A_{K-1}), \quad (32)$$

where

$$A_{K\pm 1} = \mp \frac{\hbar}{2J} \sum_{K\pm 1} \frac{<K|j\mp|K\mp 1><K\mp 1|\mu\pm|K>}{E_{K+1}-E_K}(u_{K\pm 1}u_K + v_{K\pm 1}v_K)^2. \quad (33)$$

Here

$$\mu_\pm = (g_l - g_R^0)j_\pm + (g_s - g_l)s_\pm,$$

and J and g_R^0 are the moment of inertia and rotational g factor of the neighboring even−even nucleus (A−1).

Bochnacki and Ogaza [14] calculated the corrections resulting from Coriolis interaction, for the nucleus Eu^{153}. The experimental value of g_s^{eff}/g_s thus corrected is shown in Table 5 and Fig. 5. It will be seen that agreement between theory and experiment is better. Estimates of similar corrections for other nuclei (Dy^{161}, Dy^{167}, Lu^{175}, Lu^{177}, and Ta^{181}) show that the corrected values of g_s^{eff} also approach the theoretical values.

Conclusion

For the more detailed and profound study of the magnetic properties of deformed nuclei, experimenters will have to increase the accuracy of measurements and widen their ranges. It appears to be particularly essential to acquire more abundant experimental material on rotational bands with K = ½ and to make experiments for excited internal states and also for vibrational β- and γ-states. It is furthermore necessary to increase the accuracy of determining the M1-transition probabilities, and obtain information on these transitions not only between the states of one rotational band but also between different internal states. It is desirable also to carry out measurements for nuclei with A > 220 and for new deformation regions.

The application of the Mössbauer effect in Coulomb excitation opens up fresh avenues for accurate measurements of the magnetic moments of low-lying excited states of many nuclei. By using the internal magnetic fields acting on nuclei in magnetic lattices, it is possible to measure the magnetic moments of states having a lifetime of 10^{-12} sec and perhaps even less. The use of germanium γ-ray detectors having a high energy-resolving power in combination with modern β-spectrometers will undoubtedly increase the accuracy of determining the M1-transition probabilities.

LITERATURE CITED

1. Blin-Stoyle, R. J., Theories of Nuclear Moments, Oxford Univ. Press, London, 1957.
2. Bohr, A., and Mottelson, B. R., Nuclear Structure and Energy Levels (in print).
3. Drell, S. D., and Walecka, J. D., Phys. Rev., 120:1069 (1960).
4. Seyboth, D., et al., Phys. Rev. Lett., 14:954 (1965); Lee, Y. K., et al., Phys. Rev. Lett., 14:957 (1965).
5. Atzmony, U., et al., Phys. Rev., 136:1237 (1964).
6. Deutsch, M., and Hrynkiewicz, A. Z. (1960) (unpublished).
7. de Boer, J., and Symons, G. D., Congr. International de Physique Nucléaire, Paris, 1964, p. 541.
8. Bowman, J. D., et al., Nucl. Phys., 61:682 (1965).

9. Nilsson, S. G., and Prior, O., Mat. Fys. Medd. Dan. Vid. Selsk., Vol. 29, No. 16 (1961).
10. Grin', Yu. I., and Pavlichenkov, I. M., Zh. Éksperim. i Teor. Fiz., 41:954 (1961).
11. de Boer, J., and Rogers, J. D., Phys. Rev. Lett., 3:304 (1963).
12. Bochnacki, Z., and Ogaza, S., Nucl. Phys., 69:186 (1965).
13. Pyatov, N.I., and Solov'ev, V. G., Izv. Akad. Nauk SSSR, Ser. Fiz., 28:1617 (1964).
14. Bochnacki, Z., and Ogaza, S., Acta Phys. Polonica, 27:649 (1965).

References to experimental work are given only where they are not considered in the following review articles.

Alder, B. K., and Steffen, R. M., Electromagnetic moments of excited nuclear states, Ann. Rev. Nucl. Sci., 14:403 (1964).

Bodenstedt, A. E., and Rogers, J. D., Magnetic Moments of Excited Nuclear States, Perturbed Angular Correlations, Amsterdam, 1964.

Nathan, O., and Nilsson, S. G., Collective Nuclear Motion and the Unified Model, Alpha, Beta-, and Gamma-Ray Spectroscopy, Amsterdam, 1965.

Nilsson, S. G., Nuclear Magnetic Dipole Moments and Electrical Quadrupole Moments, Perturbed Angular Correlations, Amsterdam, 1964.

Ogaza, S., Properties of magnetically strongly deformed atomic nuclei, Report IFJ, No. 362 (1964).

SOME PROPERTIES OF DEFORMED ODD-A NUCLEI IN THE RARE-EARTH ELEMENT REGION (REVIEW OF EXPERIMENTAL DATA)

K. Ya. Gromov

Joint Institute for Nuclear Research

Introduction

In the review by Mottelson and Nilsson [1] experimental data on deformed nuclei with odd A were analyzed for the first time. It was shown that all the fundamental properties of such nuclei could be well explained on the basis of the concepts of the generalized model concerning single-particle states (Nilsson's systems of levels) and the rotational levels associated with them.

After publication of the review [1], very many experimental data appeared concerning this group of nuclei. New results were obtained in theoretical studies.

We shall discuss the following questions:

1. Classification of matrix elements (log ft) for β decay of nuclei with odd A in the region $150 < A < 190$.

2. Collective states of nuclei of this group.

3. Three-quasiparticle states.

ANALYSIS OF EXPERIMENTAL DATA ON β-TRANSITION PROBABILITIES IN DEFORMED ODD NUCLEI OF THE RARE-EARTH REGION

In the analysis of β-transition probabilities in deformed odd nuclei, in addition to the usual selection rules:

$$\Delta I = 0.1 \text{ and } \Delta\pi = \text{no for allowed transitions}$$

and

$$\Delta I = 0.1 \text{ and } \Delta\pi = \text{yes for first order transitions,}$$

it is necessary to take into account the selection rules, according to the asymptotic quantum numbers, first formulated by Alaga. These selection rules for allowed and first-order forbiddenness transitions are given in Table 1.

TABLE 1. Selection Rules for β Transition According to Asymptotic Quantum Numbers

Transition	ΔK	$\Delta\Lambda$	Δn_z	ΔN
Allowed (a)	0 1	0 0	0 0	0 0
First-order forbiddenness (I)	0	0	$\begin{cases}+1\\-1\end{cases}$	+1 −1
	0	1	0	±1
	1	1	0	±1
	1	0	$\begin{cases}1\\-1\end{cases}$	1 −1

We use the customary notations: a − allowed β-transitions; 1 − β-transitions of first-order forbiddenness; u − β transitions which do not violate the selection rules according to asymptotic numbers (unhindered); h − β transitions which violate the selection rules according to asymptotic quantum numbers (hindered).

V. G. Solov'ev [2] proposed an additional classification of β transitions, dividing them into three groups I, II, and III. According to this classification:

Group I includes:

a) β disintegrations in which the initial and final states are ground states of the (proton or neutron) system;

b) particle transitions (i.e., transitions in which one particle appears or disappears in the proton or neutron system) in the case of invariable number of particle pairs;

c) hole transitions with unit variation in the number of pairs.

Group II includes:

a) hole transitions in the case of invariable number of particle pairs;

b) particle transitions when the number of particle pairs varies by one unit.

Group III includes:

a) transitions with variation in the number of quasiparticles of the proton (neutron) system by more than one unit;

b) transitions in which, when the number of quasiparticles varies by one, the position of other quasiparticles is varied.

Such a classification is introduced to emphasize the function of pair correlations in the analysis of β-transition probabilities. Indeed, in the independent particle model, which does not take into account pair correlations, the pairs of particles occupy all the low levels down to a level with the number N/2. (N is the number of particles), while the higher levels are unoccupied (see Fig. 1). Thus, group IIa and IIb transitions are transitions with variation in the position of two particles. Such transitions are forbidden in the independent particle model. Pair correlations result in there being a probability of finding pairs of quasiparticles in states lying above the Fermi surface. In this case, group IIa and IIb transitions occur with variation in the position of only one quasiparticle. This type of transition is not forbidden in the superfluid model; their probability may be calculated.

Group III transitions are forbidden in the independent particle model and the superfluid model. This type of transition has not been found in the β decay of deformed odd nuclei.

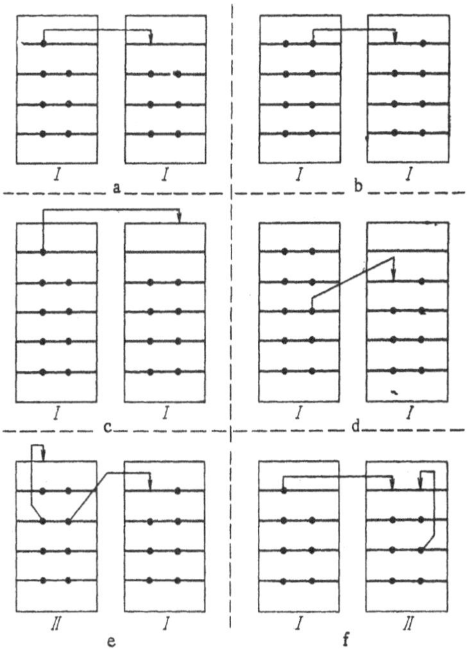

Fig. 1. β transitions of groups I and II in nuclei with odd A. The right-hand and left-hand columns show diagrammatically the proton (or neutron) and neutron (or proton) systems of the nucleus, respectively. β transitions are shown by arrows from the left-hand to the right-hand system; a-d) group I transitions; e and f) group II transitions.

TABLE 2. Allowed Unhindered β_f Transitions in Odd Nuclei

β_f transition	Additional classification	R	$\log ft_{exp}$	$\log ft_{exp}R\eta$
	7/2−[523]⇄5/2−[523]			
$Ho^{165} \leftarrow Er^{165}$	I I	0.42	4.64 ± 0.02	4.26 ± 0.02
$Tu^{165} \leftarrow Yb^{167}$	I I	0.58	$4.74^{+0.07}_{-0.03}$	$4.50^{+0.07}_{-0.02}$
$Ho^{161} \rightarrow Dy^{161}$	I I	0.31	4.8 ± 0.2	4.16 ± 0.2
$Ho^{163} \leftarrow Er^{163}$	I I	0.36	4.81 ± 0.01	4.42 ± 0.01
$Ho^{159} \rightarrow Dy^{159}$	I I	0.39	5	4.5
$Ho^{167} \rightarrow Er^{167}$	I I	0.52	4.8	4.4
$Tb^{161} \leftarrow Gd^{161}$	I I	0.26	4.80 ± 0.05	4.21 ± 0.05
	9/2− [514] ⇄7/2− [514]			
$Lu^{175} \leftarrow Yb^{175}$	I I	0.32	4.82 ± 0.05	4.22 ± 0.05
$Ta^{179} \leftarrow W^{179}$	I I	0.44	4.6	4.2

TABLE 3. Allowed Hindered β Transitions in Odd Nuclei

βf transition	Additional classification	R	$\log ft_{exp}$	$\log ft_{exp} R\eta$
		$3/2+ [411] \rightleftarrows 3/2+ [651]$		
$Eu^{153} \leftarrow Gd^{153}$	II I	0.2	7.0	6.3
		$3/2- [541] \rightleftarrows 3/2- [521]$		
$Eu^{155} \leftarrow Sm^{155}$	II I	0.55	6.9	6.4
		$5/2- [532] \rightleftarrows 3/2- [521]$		
$Eu^{155} \leftarrow Sm^{155}$	II I	0.22	5.6 ± 0.1	4.95 ± 0.1
$Ho^{161} \leftarrow Er^{161}$	I I	0.53	5.5 ± 0.3	5.2 ± 0.3
$Tb^{159} \leftarrow Gd^{159}$	II I	0.07	6.6 ± 0.1	5.5 ± 0.1
		$5/2+ [413] \rightleftarrows 5/2+ [642]$		
$Eu^{155} \rightarrow Gd^{155}$	I II	0.15	7.3	6.3
$Eu^{157} \rightarrow Gd^{157}$	I II	0.22	7.5	6.7
		$3/2+ [411] \rightleftarrows 5/2+ [642]$		
$Tb^{161} \rightarrow Dy^{161}$	I I	0.15	7.8 ± 0.1	7.0 ± 0.1
		$5/2+ [413] \rightleftarrows 7/2+ [633]$		
$Ho^{165} \leftarrow Dy^{165}$	II I	0.21	5.7	4.9
		$1/2+ [411] \rightleftarrows 3/2+ [651]$		
$Tu^{165} \rightarrow Er^{165}$	I II	0.2	6.6	4.9
		$1/2- [523] \rightleftarrows 5/2- [512]$		
$Tu^{171} \rightarrow Er^{171}$	I I	0.15	6.3 ± 0.1	5.5
		$7/2+ [404] \rightleftarrows 7/2+ [633]$		
$Ta^{175} \rightarrow Hf^{175}$	I II	0.12	6.7	5.8
$Ta^{177} \rightarrow Hf^{177}$	I II	0.06	7.9	6.7
		$7/2+ [404] \rightleftarrows 9,2+ [624]$		
$Ta^{181} \leftarrow W^{181}$	I I	0.34	6.7 ± 0.2 -0.1	$6.3 + 0.2$ -0.1
$Lu^{177} \leftarrow Yb^{177}$	I I	0.41	6.2	5.7
$Lu^{177} \rightarrow Hf^{177}$	I II	0.15	6.3 ± 0.1	5.5 ± 0.1
$Ta^{177} \rightarrow Hf^{177}$	I I	0.33	8.2	7.7
$Ta^{179} \rightarrow Hf^{179}$	I I	0.25	~ 6.7	~ 6.1
$Re^{183} \leftarrow Os^{183}$	I I	0.46	7.2 ± 0.2	6.9 ± 0.2

Tables 2-4 show all the currently known experimental values of $\log ft_{exp}$ for allowed unhindered (au), allowed hindered (ah) and first-order forbiddenness unhindered (1u) β-transitions in deformed odd nuclei of the rare-earth group.

All the β transitions are distributed in individual groups and neutron states. These states are shown at the head of each group; first the proton state, then the neutron state. The first column shows the nuclei between which the β transition occurs. The second column shows the additional classification. The first index (I or II) relates to the proton system of the nucleus, the second to the neutron system.

The fourth column gives the experimental values of the matrix elements for β transition [$\log ft_{exp}$]. Where it is possible to determine an error in the experimental determination of $\log ft_{exp}$ this is indicated. In most cases, where the error in the determination is not shown and there is no approximately, less or equal sign, the error in our opinion does not exceed 0.3. The third column gives the values of the superfluid correction R to the experimental values of $\log ft_{exp}$. These corrections were calculated by V. G. Solov'ev [2]. The correction R is the product of the correction for the neutron system R_N and for the proton system R_Z of the nucleus.

TABLE 4. β Transitions of First-Order Forbiddenness, Unhindered in Odd Nuclei

β transition	Additional classification	R	$\log ft_{exp}$	$\log ft_{exp}R\eta$
	3/2+ [411] ⇄ 3/2− [521]			
$Tb^{159} \leftarrow Gd^{159}$	I I	0.24	6.7	5.9
$Tb^{159} \leftarrow Dy^{159}$	I I	0.30	7.0	6.3
$Tb^{161} \rightarrow Dy^{161}$	II I	0.38	6.8±0.1	6.2±0.1
$Eu^{155} \leftarrow Sm^{155}$	I I	0.38	6.6±0.1	6.0±0.1
$Tb^{157} \rightarrow Dy^{157}$	I I	0.24	6.8±0.1	6.0±0.1
$Tb^{157} \rightarrow Gd^{157}$	I I	0.27	7.2	6.5
	1/2−[523] ⇄ 1/2+ [633]			
$Ho^{165} \leftarrow Dy^{165}$	I I	0.33	6.2±0.1	5.6±0.1
	1/2+ [411] ⇄ 3/2− [521]			
$Ho^{161} \leftarrow Er^{161}$	II I	0.18	7.4	6.4
$Eu^{155} \leftarrow Sm^{155}$	I I	0.5	6.8	6.2
$Tu^{165} \rightarrow Er^{165}$	I II	0.12	7.4±0.1	6.5±0.1
$Tu^{163} \rightarrow Er^{163}$	I II	0.24	6.4±0.1	5.8±0.1
	1/2+ [411] ⇄ 1/2− [521]			
$Tu^{169} \leftarrow Er^{169}$	I I	0.33	6.4±0.1	5.7±0.1
$Tu^{171} \rightarrow Yb^{171}$	II I	0.39	6.2±0.1	5.6±0.1
$Tu^{167} \rightarrow Er^{167}$	I I	0.48	6.3±0.1	5.8±0.1
	1/2+ [411] ⇄ 1/2− [521]			
$Tu^{165} \rightarrow Er^{165}$	I I	0.54	6.6±0.1	6.1±0.1
$Tu^{163} \rightarrow Er^{163}$	I I	0.54	6.5	6.0
$Tu^{173} \rightarrow Yb^{173}$	II I	0.24	6.3±0.1	5.5±0.1
	1/2+ [411] ⇄ 1/2− [510]			
$Re^{183} \leftarrow Os^{183}$	I II	0.27	6.3	5.7
$Ta^{181} \leftarrow Hf^{181}$	II I	0.1	7.2	6.1
	1/2+ [411] ⇄ 3/2− [512]			
$Re^{187} \leftarrow W^{187}$	II I	0.02	7.4	5.7
	9/2− [514] ⇄ 9,2+ [624]			
$Re^{183} \leftarrow Os^{183}$	I I	0.46	6.3	6.0
$Ta^{181} \leftarrow W^{181}$	I I	0.37	6.9±0.3	$6.5^{+0.3}_{-0.1}$
	7/2+ [404] ⇄ 1/2− [514]			
$Ta^{175} \rightarrow Hf^{175}$	I I	0.31	6.5	6.0
$Lu^{175} \leftarrow Yb^{175}$	I I	0.32	6.25±0.1	5.15±0.1
$Lu^{177} \rightarrow Hf^{177}$	I I	0.27	6.6±0.1	6.0±0.1
$Ta^{177} \rightarrow Hf^{177}$	I I	0.21	6.6±0.1	5.9±0.1
	7/2+ [404] ⇄ 7/2− [503]			
$Ta^{175} \rightarrow Hf^{175}$	I I	0.41	6.2	5.8
$Ta^{177} \rightarrow Hf^{177}$	I I	0.40	6.4	6.0
$Ta^{183} \rightarrow W^{183}$	I II	0.11	6.9	5.9
$Ta^{185} \rightarrow W^{185}$	I II	0.19	6.5	5.8
	7/2+ [404] ⇄ 9/2− [505]			
$Ta^{175} \rightarrow Hf^{175}$	I I	0.41	6.4	6.0
$Ta^{185} \rightarrow W^{185}$	I II	0.16	∼7.3	∼6.5
	5/2+ [402] ⇄ 3/2− [512]			
$Re^{187} \rightarrow W^{187}$	I I	0.30	7.9	7.4
$Re^{185} \rightarrow W^{185}$	I I	0.38	7.5	7.1
$Re^{183} \rightarrow W^{183}$	I I	0.46	7.7	7.4

The fifth column gives the values of the matrix elements for β transitions, taking into account the superfluid corrections $\log ft_{exp}R\eta$. The statistical factor

$$\eta = \langle I_i K_i \lambda_f - K_i | I_f K_f \rangle$$

is also taken into account.

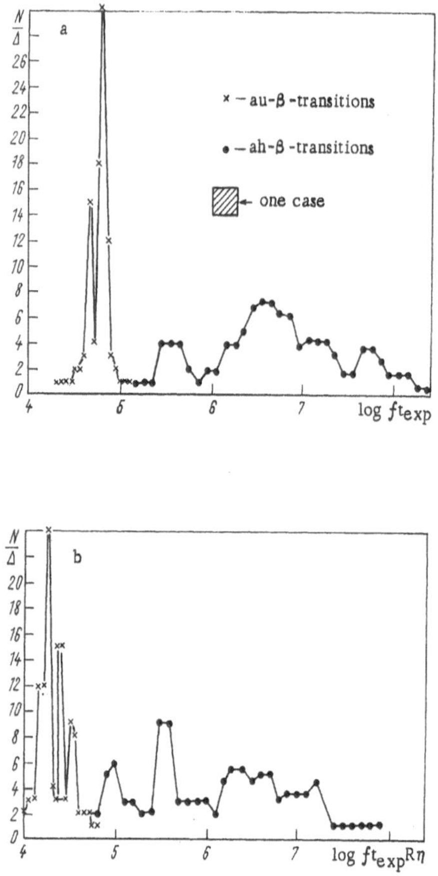

Fig. 2. Distribution of the values of log ft_{exp} (a) and log $ft_{exp}R\eta$ (b) for au and ah transitions. To one value of log ft_{exp} or log $ft_{exp}R\eta$ there corresponds a rectangle, the area of which is equal to the area of the shaded rectangle. The base of the rectangle is equal to twice the error of measurement of log ft_{exp}.

Figures 2 and 3 show the distributions of the values of log ft_{exp} and log $ft_{exp}R\eta$ for the β-transition groups considered. Errors in the determination of the values of log ft were taken into account in constructing these distributions. To each value of log ft there corresponds a rectangle, the area of which is equal to the area of the shaded rectangle in Fig. 2. The base of the rectangle is equal to double the error in the determination of log ft_{exp}.

The analysis made shows that all the values of log ft_{exp} measured to date may be arranged in the following limits:

au transitions $4.6 < \log ft_{exp} < 4.8$;
ah transitions $5.5 < \log ft_{exp} < 8.0$;
lu transitions $6.2 < \log ft_{exp} < 7.9$.

These results may be compared with the results of the analysis of the values of log ft_{exp} by Mottelson and Nilsson [1]:

au transitions $4.5 < \log ft_{exp} < 5.0$;
ah transitions $6.0 < \log ft_{exp} < 7.5$;
lu transitions $5.5 < \log ft_{exp} < 7.5$.

We obtained the following limits for the values of log $ft_{exp}R\eta$:

au transitions $4.2 < \log ft_{exp}R\eta < 4.5$;
ah transitions $4.9 < \log ft_{exp}R\eta < 7.6$;
lu transitions $5.5 < \log ft_{exp}R\eta < 7.4$.

They may be compared with the results of the analysis by V. G. Solov'ev [2]:

au transitions $4.0 < \log ft_{exp}R\eta < 4.7$;
ah transitions $5.5 < \log ft_{exp}R\eta < 6.5$;
lu transitions $5.5 < \log ft_{exp}R\eta < 6.5$.

The following conclusions may be drawn from the results of classifying the values of the matrix elements for the β decay of deformed nuclei with odd A.

1. Pair correlations substantially influence the probability of β transitions. The experimental values of log ft_{exp} for group II β-transitions do not differ substantially from the log ft_{exp} values for group I β-transitions. This is in agreement with calculations according to the superfluid model; the R corrections for group II transitions differ substantially from the corrections for group I only when β decay occurs to high energy levels. The independent particle model requires division of the log ft_{exp} values for I and II β-transitions into two groups.

2. If the superfluid model corrections are taken into account, the scatter of the values of log $ft_{exp}R\eta$ compared with log ft_{exp} does not alter. Evidently, this is because the superfluid

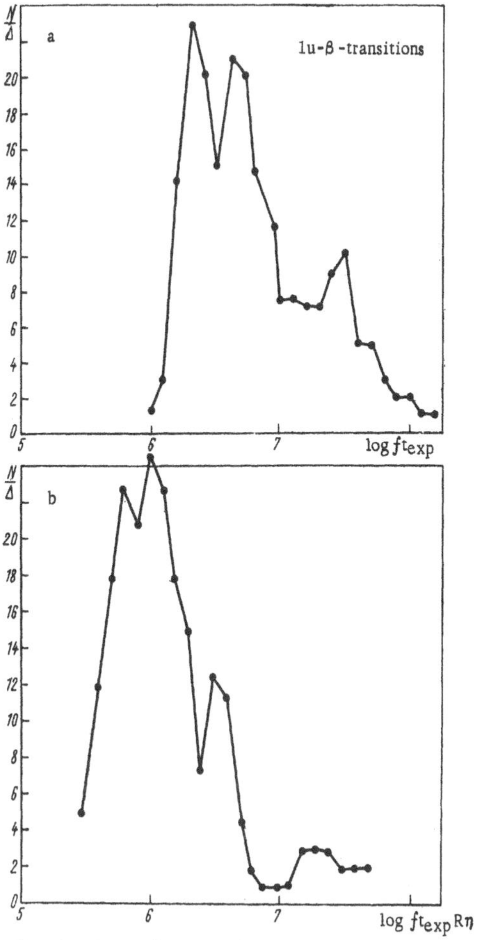

Fig. 3. Distribution of log ft_{exp} and log $ft_{exp}R\eta$ values for 1u β-transitions. The method of construction is the same as for Fig. 2.

model uses averaged parameters for describing the entire group of deformed rare-earth nuclei. If the limits of log ft_{exp} and log $ft_{exp}R\eta$ are examined in each subgroup of transitions between definite states of the Nilsson scheme, a diminution of the range between these limits is observed more often for log $ft_{exp}R\eta$ than for log ft_{exp}.

3. The log ft_{exp} values for allowed unhindered β-transitions are clearly segregated in a group with the limits $4.6 < \log ft_{exp} < 4.8$. This shows that essentially the generalized model describes these transitions correctly. This conclusion is also useful for the analysis of experimental data. Observation of log ft_{exp} in the range 4.6-4.8 confirms that we are dealing with the au β-transition. In the rare-earth region, only two β-transitions are possible:

$$^{7}/_{2}^{-}\ [523] \rightleftharpoons {}^{5}/_{2}^{-}\ [523]$$

and

$$^{9}/_{2}^{-}\ [514] \rightleftharpoons {}^{7}/_{2}^{-}\ [514]$$

LITERATURE CITED

1. Mottelson, B., and Nilsson, S. G., Mat. Fys. Skr. Dan. Vid. Selsk., Vol. 1, No. 8 (1959).
2. Soloviev (Solov'ev), V. G., Mat. Fys. Skr. Dan. Vid. Selsk., Vol. 1, No. 1 (1961).

COLLECTIVE EXCITED STATES OF DEFORMED NUCLEI WITH ODD A IN THE RARE-EARTH REGION

In the generalized model of the nucleus, we imagine a deformed nucleus with odd A to be an even−even core, having an odd particle (neutron or proton) coupled to it. Thus, the properties of nuclei with odd A ought, it appears, to be determined by the properties of the even−even core and the properties of the odd particle, moving in the field of this core. An even−even core represents essentially a neighboring even−even nucleus, and it is possible to expect that nuclei with odd-A levels similar in nature to the levels of even−even nuclei will occur. In particular, it is possible for levels to occur in nuclei with odd A which are similar in properties to the collective γ- and β-vibrational levels in even−even nuclei. It is, however, necessary to explain in what degree an odd nucleon influences the motion of an even−even core; in what degree the motions of the odd nucleon and core are coupled; and whether it is possible to distinguish experimentally, for example, single-particle levels and collective levels.

The first experimental indications that levels of a collective nature are excited in deformed odd nuclei were obtained by Nathan and Popov [1] and Gallagher et al. [2]. In the last two or three years, a number of collective levels in nuclei with odd A have been identified. Available data on collective levels in nuclei with odd A have been gathered and analyzed [3-5].

We shall examine the properties of collective levels in nuclei with odd A, which may be expected on the basis of the generalized model; the indications by which collective levels in nuclei with odd A are identified experimentally.

1. Single-particle motion in a deformed nucleus with odd A is characterized by the projection of angular momentum on the deformation axis $K_0 = \Omega$. The vibrational motion is the projection of the angular momentum of the vibrations on the deformation axis ν. Thus, the projection of the angular momentum on the symmetry axis of the vibrational state will be

$$K = |K_0 \pm \nu|.$$

In contrast to even−even nuclei, the projection of the angular momentum on the symmetry axis in odd nuclei in the ground state is not equal to zero. In odd nuclei, therefore, it is possible for two γ-vibrational levels to occur of the type

$$K_1 = |K_0 - 2| \text{ and } K_2 = K_0 + 2.$$

The occurrence of β-vibrational levels with $K = K_0$ is also possible. The parity of the β- and γ-vibrational levels agrees with the parity of the state with which this level is associated.

2. The energy of the levels of rotational bands of vibrational states in odd nuclei ought to be described by the usual formula

$$E_I = E_0 + A[I + (I+1) + a(-1)^{I+1/2}(I+1/2)\,\delta_K^{1/2}] - A[K(K+1) + a(-1)^{K+1/2}(K+1/2)\,\delta_{K1/2}].$$

We know that in even−even nuclei the moment of inertia of the vibrational state (parameter A in the formula for E_I) differs from the moment of inertia of the ground state by no more than 10-20%. Obviously, the moments of inertia in the ground state and collective state in odd nuclei are similar. The moments of inertia of single-particle states in odd nuclei often differ appreciably from one another and this fact is used as additional argument for the identification of the states. The assumption of equality of the moments of inertia of the ground and vibrational states may also be used for the identification of vibrational states.

For γ-vibrational levels with $K = |K_0 - 2| = 1/2$, the decoupling parameter in the formula for the energy of the rotational levels is equal to zero. This fact may be used as essential indication of γ-vibrational states, since for all single-particle states of deformed nuclei in the rare-earth region, the decoupling parameter a differs appreciably from zero. For β-vibrational levels with $K = K_0 = 1/2$, the decoupling parameter a is evidently close in value to the decoupling parameter a of the ground state.

3. The Coulomb excitation probability B(E2) of γ-vibrational levels in even–even nuclei is equal to from four to six single-particle units. It is to be expected that the excitation probability of γ-vibrational levels in odd nuclei is close to B(E2) for even–even nuclei:

$$B_{\gamma}(E2)_{e\text{-}e} \cong B_{K_0-2}(E2) + B_{K_0+2}(E2).$$

4. Identification of collective levels in odd nuclei is facilitated if it can be shown that characteristics of single-particle states cannot be assigned to the level under consideration. This is due, firstly, to the fact that in this case it is easier to distinguish a single-particle level from a collective level; secondly, if the single-particle and collective levels with similar characteristics lie close to each other, theory predicts intense mixing of these levels, and it becomes fundamentally impossible to identify one of them as a collective level and the other as a single-particle level.

Table 5 presents all the currently available information on γ-vibrational levels in nuclei with odd A in the region of the rare-earth elements.

As will be gathered from the table, most of the available experimental data concern γ-vibrational levels of the type $K_1 = K_0 - 2$. In most cases, the spins and parities of these levels have been clearly determined (12 out of 17 cases). It is difficult to interpret levels with such spins and parities as single-particle levels, since in the Nilsson schemes there are no appro-

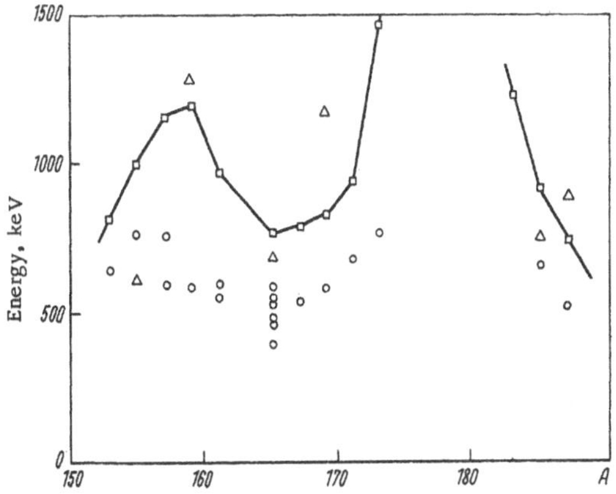

Fig. 4. Dependence of the energies K_0-2 and K_0+2 on A: γ-vibrational levels in nuclei with odd A; O) type $K_1 = K_0 - 2$ levels; Δ) type $K_2 = K_0 + 2$ levels; □) $2+-\gamma$ = vibrational levels in even–even nuclei with mass number (A−1).

TABLE 5. Experimental Data on γ-Vibrational States in Odd Nuclei

Nucleus	K_π γ-vibrational state	Type of γ-vibration	$K_\pi[Nn_zΛ]$ ground state	E, keV γ-vibrational state	E, keV γ-vibrational state	A, keV γ-vibrational field	a, keV γ-vibrational field	A, keV ground field	B(E2)	Reference
Eu^{153}	(1/2+)	K_0-2	5/2+ [413]	634.6	811	—	—	11.78	—	[1]
Tb^{155}	(1/2+)	K_0-2	3/2+ [411]	761 ± 2	997.3	15.1 ± 1.2	$+(0.10\pm0.08)$	13.25 ± 0.03	—	[2]
	(7/2+)	K_0+2	3/2+ [411]	616 ± 2 706 ± 2	997.3	—	—	13.25 ± 0.03	—	
Tb^{157}	1/2+	K_0-2	3/2+ [411]	597 ± 1	1155	$12.9\pm0,6$	$+(0.03\pm0.05)$	12.42 ± 0.03	—	[3,4]
	(1/2−)	K_0-2	5/2− [532]	770 (1100)	1155	—	—	—	—	
Tb^{159}	1/2+	K_0-2	3/2+ [411]	580 ± 1	1185	11.8 ± 0.9	$+(0.017\pm0.076)$	11.74 ± 0.01	1.5	[5,6]
	7/2	K_0+2	3/2+ [411]	1270	1185	—	—	11.74 ± 0.01	2.0	
Dy^{161}	1/2+	K_0-2	5/2+ [642]	545	966.0	—	—	5.8	1.2 0.4	[7, 8]
Ho^{161}	3/2−	K_0-2	7/2− [523]	593	966.0	—	—	—	—	[7]
Ho^{165}	3/2−	K_0-2	7/2− [523]	514.2 ± 1.7	761.8	10.3 ± 2.7	—	10.65 ± 0.08	1.9	[9, 10]
	11/2−	K_0+2	7/2− [523]	687	761.8	10.2 ± 1.0	—	10.65 ± 0.08	1.7	
Dy^{165}	3/2+	K_0-2	7/2+ [633]	538.62	761.8	8.854 ± 0.009	—	9.1262+ +0.0014	—	[11—13]
	3/2−	K_0-2	1/2− [521]	465.40 (573.56)	761.8	11.05	—	10.671± ±0.002	—	
	1/2−	K_0-2	5/2− [512]	386.00 (570.25)	761.8	—	—	11.2136± ±0.0011	—	
Er^{165}	(1/2−)	K_0-2	5/2− [523]	590	861	—	—	—	—	[14]
	1/2+	K_0-2	5/2+ [642]	460.5 (507.6)	861	—	—	—	—	
Er^{167}	3/2+	K_0-2	7/2+ [633]	531.8	787	—	—	8.21+0.19	1.5	[15—17]
Tm^{169}	3/2+	K_0-2	1/2+ [411]	570	822.4	13.1 ± 4.3	—	12.467 ± 0.007	1.0	[18]
	5/2+	K_0+2	1/2+ [411]	1170	822.4	—	—	12.467 ± 0.007	1.5	
Tm^{171}	3/2+	K_0-2	1/2+ [411]	675	930	12.4	—	12.127 ± 0.003	—	[19—20]
	5/2+	K_0+2	1/2+ [411]	912	930	12.1	—	12.127 ± 0.003	—	
Lu^{173}	3/2−	K_0-2	1/2− [541]	≥759.5(≥887.7)	1467	—	—	8.52 ± 0.05	—	[21]
Re^{185}	1/2+	K_0-2	5/2+ [402]	645.8	904	24	—	18.25 ± 0.25	1.0	[22—23]
	9/2+	K_0+2	5/2+ [402]	750 ± 25	904	—	—	18.25 ± 0.25	1.0	
Re^{187}	1/2+	K_0-2	5/2+ [402]	511.6	730	—	—	20.35 ± 0.26	1.3	[24—26]
	9/2+	K_0+2	5/2+ [402]	880 ± 20	730	—	—	20.35 ± 0.26	1.7	

Remarks. 1. Energy of γ-vibrational states in nuclei with odd A. The excitation energy of the single-particle state on which is based the given γ-vibration state has been deducted. The values of the total energies are given in brackets.

2. The energy of the 2^+ γ-vibrational state in the neighboring even – even nucleus with mass number (A – 1).

3. Experimental values of the constants in the formulas

$$B_{\text{single-part.}} = \quad E_I = E_0 + AI(I+1) + B(I+1)^2 I^2 \quad K \neq 1/2$$
$$E_I = E_0 + A[I(I+1) + a(-)^{I+1/2}(I+1/2)] \quad K = 1/2$$

4. Value in single-particle units. $3 \cdot 10^{-5} A^{4/3}$ $e^2 \cdot 10^{-48}$

5. Literature references for this table are given at the end of this section.

priate single-particle states at these excitation energies. All the γ-vibrational type (K_0-2) levels in stabilized nuclei given in Table 5 were Coulomb excited. The values of B(E2), as was to be expected for γ-vibrational states, were nearly two single-particle units. Such values are difficult to understand if the levels considered are interpreted as single-particle levels. In cases where rotational bands were found for γ-vibrational type (K−2) states, the values of the parameter A for these bands were close to the value of A for rotational bands of the ground state. In the isotopes Tb^{155}, Tb^{157}, and Tb^{159} levels $K_1 = K_0-2 = {}^1/_2$ and the associated rotational bands were observed. The decoupling parameter for bands based on these states is equal to zero in the limits of the experimental accuracy. The nearest single-particle state in this region (${}^1/_2{}^+$ [411]) has a parameter $a \simeq -0.8$.

There are fewer experimental data on γ-vibrational levels of type $K_2 = K_0 + 2$. These data are less definite. In all the cases given in the table, except Tb^{155}, levels of type $K_2 = K_0+2$ were Coulomb excited. The experimental values of B(E2) for the excitation of these levels lie in the limits of one or two single-particle units. It may be assumed that the position of levels of type $K_2 = K_0 + 2$ has been determined, even if approximately, in Tb^{159}, Ho^{165}, Tm^{169}, Re^{185}, and Re^{187}.

There are still fewer data on β-vibrational levels of type $K = K_0$ in odd nuclei. From some sources [6, 7] experimental data have been obtained, showing that in the nuclei Tb^{157} and Tb^{155}, there are levels for which $K = K_0$, and γ transitions from which there is an appreciable type-E0 addition. This may indicate that these levels are similar to 0^+ β-vibrational levels in even–even nuclei.

Figure 4 shows the energies of γ-vibrational levels in odd nuclei (difference in the energy levels of type $K_0 \pm 2$ and K_0) plotted against the mass numbers of the nuclei in which they have been found. For comparison, values are given of the energy of 2^+ γ-vibrational levels in the neighboring (A−1) even–even nucleus (even–even core). It is difficult as yet to find any regularities in the behavior of γ-vibrational levels in odd nuclei. It is only possible to make some general assumptions.

1. Levels of type $K_1 = K_0-2$ in all nuclei lie below levels of type $K_2 = \Omega + 2$. Tb^{155} is an exception. However, experimental data in this case are not sufficiently reliable and need refinement.

2. The difference in the energies of levels of type $K_1 = K_0-2$ and K_0 in an odd nucleus is always less than the energy of the 2^+ γ-vibrational level in the neighboring even–even nucleus.

These conditions are in good agreement with the conclusions of V. G. Solov'ev (see this Collection, p. 23) regarding the fact that on splitting the γ-vibrational level in an odd nucleus, the state with smaller K drops more than the state with larger K.

LITERATURE CITED

1. Nathan, O., and Popov, V. J., Nucl. Phys., 21:631 (1960).
2. Gallagher, C. J., et al., Nucl. Phys., 19:18 (1960).
3. Peker, L. K., et al., Izv. Akad. Nauk SSSR, Ser. Fiz., 28:287 (1964).
4. Sheline, R. K., Shelton, W. N., Moks, H. F., and Carter, R. F., Phys. Rev., 136B:351 (1964).
5. Ghatovich, V., and Gromov, K., Preprint Joint Institute for Nuclear Research P-2086 (1965).
6. Person, L., and Ryde, H., Ark. Fys., B.25, N.29, (1964); Person L., et al., Ark. Fys., B.24, N.34:451 (1962).
7. Werner, L., et al., Preprint ZEKI Possendorf, June (1965). Material for Working Conference on Nuclear Spectroscopy and Nuclear Theory, Dubna, July, 1965. Preprint Joint Institute for Nuclear Research.

LITERATURE CITED FOR TABLE 5

1. Suter, T., et al., Nucl. Phys., 29:33 (1962).
2. Person, L., and Ryde, H., Ark. Fys., B.25(29):397 (1964).
3. Person, L., et al., Ark. Fys., B.24(34):451 (1962).
4. Werner, L., Paper presented at the Dubna Conference, July, 1965.
5. Person, L., Ark. Fys., B.25:307 (1963).
6. Diamond, R. M., et al., Nucl. Phys., 43:560 (1963).
7. Funke, L., et al., Nucl. Phys., 55:41 (1964).
8. Epokhina, K. I., et al., Program and Theses of Papers Presented at the 15th Annual Conference on Nuclear Spectroscopy, Minsk, 1965. Moscow-Leningrad, "Nauka," 1965.
9. Person, L., et al., Ark. Fys., 23:1 (1963).
10. Diamond, R. M., et al., Nucl. Phys., 43:560 (1963).
11. Sheline, R. K., et al., Phys. Rev., 136:B.351 (1964).
12. Schult, O. W., et al., Leit. f. Phys., 182:171 (1961).
13. Bondarenko, V. A., et al., Program and Theses of Papers Presented at the 15th Annual Conference on Nuclear Spectroscopy, Minsk, 1965, Moscow-Leningrad, "Nauka," 1965.
14. Zvol'ska, V., et al., Paper presented at the Dubna Conference, July, 1965.
15. Zvol'ska, V., Dissertation, Joint Institute for Nuclear Research, 1964.
16. Gromov, K. Ya., et al., Izv. Akad. Nauk SSSR, Ser. Fiz., 26:1023 (1962).
17. Gangrskii, Yu., and Lemberg, I., Izv. Akad. Nauk SSSR, Ser. Fiz., 26:1027 (1962).
18. Diamond, R. M., et al., Nucl. Phys., 43:560 (1963).
19. Artna, A., and Johns, M. W., Canad. J. Phys., 39:1817 (1961).
20. Peker, L. K., Izv. Akad. Nauk SSSR, Ser. Fiz., 28:289 (1964).
21. Valentin, J., et al., Nucl. Phys., 31:351 (1962).
22. Johns, M. W., et al., Canad. J. Phys., 35:1159 (1957).
23. Nathan, O., and Popov, V. J., Nucl. Phys., 21:631 (1960).
24. Gallagher, C. J., et al., Nucl. Phys., 19:18 (1960).
25. Nathan, O., and Popov, V. J., Nucl. Phys., 21:631 (1960).
26. Widemann, T. W., et al., Compt. Rend., 260:3926 (1965).

THREE-QUASIPARTICLE STATES IN DEFORMED NUCLEI WITH ODD A

If, in a deformed nucleus with odd A, a neutron or proton pair is ruptured, a three-quasiparticle state is formed. The projection of the angular momentum on the symmetry axis for the three quasiparticle state is obtained as the sum of the projections of the angular momenta of the three quasiparticles:

$$K = |\Omega_1 \pm \Omega_2 \pm \Omega_3|.$$

We obtain a multiplet consisting of four levels. The parity of the three-quasiparticle state is determined as usual. Since the three-quasiparticle state is associated with the rupture of a pair of particles, its energy is greater than or close to the energy of the gap 2Δ. This means that three-quasiparticle states have an energy of about 1 MeV or above. The average energies of some three-quasiparticle states were calculated by V. G. Solov'ev [1]. The superfluid model of the nucleus does not take account of the interaction between quasiparticles forming a three-quasiparticle state. Therefore, only the mean energies of three-quasiparticle states are given in [1]. N. I. Pyatov and A. S. Chernyshev [2] studied the splitting of three-quasiparticle states, while taking account the interaction between the quasiparticles. The highest state was found to be that in which the asymptotic spins Σ of the nucleons of the ruptured pair were parallel and the spin of the odd nucleon was antiparallel to them ($\uparrow\uparrow\downarrow$). States with parallel spins ($\uparrow\uparrow\uparrow$) and two other states ($\uparrow\downarrow\uparrow$) and ($\uparrow\downarrow\downarrow$) are lower. The parameters for calculating the splitting energies of multiplets were taken from the splitting energies of two-quasiparticle states. On the basis of the results of N. I. Pyatov and A. S. Chernyshev, it is not difficult to calculate the splitting energy for any three-quasiparticle state in the rare-earth region. The total splitting energy may attain 1 MeV. Experimental data on splitting energies of three-quasiparticle multiplets are practically nonexistant, and it is therefore difficult to compare the calculations made in [2] with experiment.

Three-quasiparticle states ought to have an energy of 1 MeV or more. This determines the experimental difficulties of detecting and studying such states. Two cases may be mentioned in which the problem of detecting states of a three-quasiparticle nature is made easier. V.G.Solov'ev [1] pointed out the possibility of experimentally detecting three-quasiparticle states in β decay. Three-quasiparticle states may be divided into two types: A type (3a) and (3p) state, when all three quasiparticles are identical, and type (2n, p) and (2p, n) states. Three-quasiparticle states of type (3p) and (3n) will not be well excited in β decay, since then more than one quasiparticle will be produced in the proton or neutron system of the nucleus (group III transition, see p. 184).

The situation is different with β decay on a type (2p, n) or (2n, p) level. Such β transitions may be represented diagrammatically as

$$p_1 \to p_2 p_3 n_4 \quad \text{(state } 2p,\ n)$$

or

$$n_1 \to n_2 n_3 p_4 \quad \text{(state } 2n,\ p)$$

where the subscripts 1, 2, 3, 4 are a set of asymptotic quantum-numbers characterizing the single-quasiparticle states. If state 1 is identical with state 2 (or 3) in both the proton and neutron systems of the nucleus, one particle is produced in each case. Such β transitions belong to group I or II; their probability is comparable with the probability of β decay to single-particle states. If states 1, 2, and 3 are different, more than one quasiparticle is produced in one of the systems; the β transition belongs to group III and its probability will be low. Thus, if state 1 is identical with state 2, β decay to three-quasiparticle states may be observed. Particularly outstanding cases occur when states 3 and 4 are single-quasiparticle neutron and proton states, between which allowed, unhindered β-transitions occur:

$$p_{3,4} 7/2^- [523] \rightleftarrows n_{4,3} 5/2^- [523],$$
$$p_{3,4} 9/2^- [514] \rightleftarrows n_{4,3} 1/2^- [514].$$

We have seen that particularly fast β transitions ($\log ft_{\rm exp}$ in the limits 4.6-4.8) are associated with transitions between these states. β transitions to three-quasiparticle levels of the type mentioned will also be allowed, unhindered transitions; it may be expected that the values of $\log ft_{\rm exp}$ for β transitions of such type will also be in the limits of 4.6-4.8. Such three-quasiparticle levels are relatively easy to detect experimentally. At the present time, several cases of the excitation of three-quasiparticle states of this type are known. In Er^{165} on the disintegration of Tm^{165} [3-5], in Lu^{177} on the disintegration of Yb^{177} [6]; in Ho^{161} on the disintegration of Er^{161} [7] and in Dy^{163} on the disintegration of Tb^{163} [8]. We shall examine two of them: the 1428 keV level in Er^{165} and the 885 keV level in Dy^{163}. N. A. Bonch-Osmolovskaya et al. [4], and T. Kudarova and V. Zvol'ska [5] measured the multipolarity of transitions from the 1428 keV level of Er^{165} (Fig. 5) and found that this level had $3/2^+$ spin and parity. The level is populated in 12% of the cases of the decay of Tm^{165}. Preibisz et al. [3] measured the energy of the decay $Tm^{165} \to Er^{165}$ and determined $\log ft$ for K capture on the 1428 keV level; it was equal to 5.1 + 0.2. These ex-

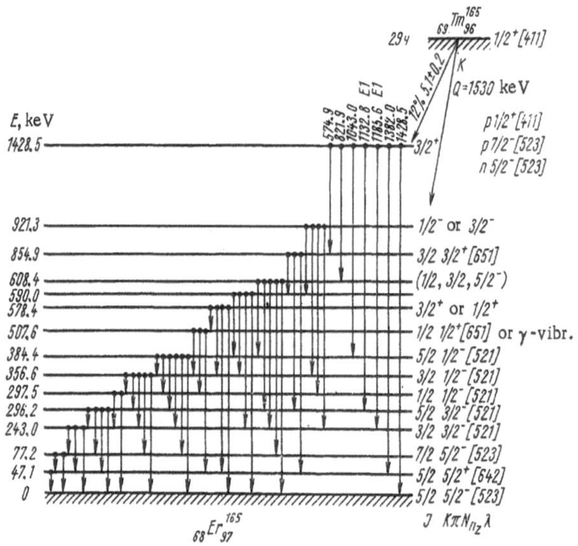

Fig. 5. Decay diagram of Tm^{165}.

perimental factors leave the only possibility of interpreting the 1428 keV level as a three-quasiparticle state of type $p_1\ {}^1/_2{}^+$ [411] $-p_2\ {}^7/_2{}^-$ [523] + n ${}^5/_2{}^-$ [523].

This state cannot be interpreted as a single-quasiparticle state, since there are no single-quasiparticle states to which allowed, unhindered β-transition is possible, while β transitions of another type ought to have a higher value of log ft.

Werner et al. [8] measured the value of log ft for the β transition from Tb^{163} to the 885 keV level of Dy^{163}. It was equal to 4.8 ± 0.1. The ground state of Tb^{163} evidently has the characteristics ${}^3/_2{}^+$ [411]. There is obviously only one possible interpretation of the 885 keV level, i.e., as a three-quasiparticle level of type

$$p3/2^+ [411],\quad p7/2^- [523],\quad n5/2^- [523]$$

with spin ${}^1/_2$ and positive parity.

Another possibility of experimentally detecting three-quasiparticle levels is the three-quasiparticle isomeric states in odd nuclei. It was in just this way that three-quasiparticle states in odd nuclei were first detected. O. B. Nilsson et al. [9] discovered the isomeric state in Lu^{177} with $T_{1/2}$ = 155 days. The energy of the isomeric state was 969 keV. This state was found experimentally to have ${}^{23}/_2{}^-$ spin and parity. It is impossible to explain this level as a single-particle level, since there are no single-particle states having such a high spin; in Nilsson's schemes, there are no levels having a spin higher than ${}^{13}/_2$. The only possibility is to interpret this level as a three-quasiparticle level of the type

$$n_1 9/2^+ [624] + n_2 7/2^- [514] + p_3 7/2^+ [404].$$

In the same investigation, a three-quasiparticle state of type ${}^{23}/_2{}^+$ was observed in Hf^{177}; p_1 ${}^7/_2{}^+$ [404], $p_2\ {}^9/_2{}^-$ [514], and n ${}^7/_2{}^-$ [514] with an energy of 1315 keV were produced on β decay of the isotopic state in Lu^{177}. Experimental data on three-quasiparticle states in Lu^{177} and Hf^{177} are confirmed and defined in [10]. L. K. Peker [11] pointed out that three-quasiparticle isomers ought to be sought in odd nuclei where two-quasiparticle isomers are found in neighboring even–even nuclei. Indeed, in Hf^{178}, a neighbor nucleus of Lu^{177}, an isomeric state ($T_{1/2}$–4.8 sec) of type 8^- is observed at an energy of 1148 keV. Gallagher and Solov'ev [12] interpreted this as a two-quasiparticle state of type $n_1\ {}^9/_2{}^+$ [624] + $n_2\ {}^7/_2{}^-$ [514]. It is natural to expect that in Lu^{177}, the three-quasiparticle state which is lowest in energy will be obtained by adding to this two-quasiparticle state a proton in the same state as the ground state of Lu^{177}, i.e., ${}^7/_2{}^+$ [404]. The two-quasiparticle analog of the three-quasiparticle state in Hf^{177} is a state of type 8^-, p ${}^7/_2{}^+$ [404], +p ${}^9/_2{}^-$ [514] with an energy of 1480 keV in Hf^{178}.

Thus, the quasiparticle states in Ta^{179}, ${}^{23}/_2{}^-$; Re^{181}, ${}^{21}/_2{}^-$; Re^{189}, ${}^{25}/_2{}^-$; and Ir^{191}, ${}^{23}/_2{}^-$ have been predicted. They have not yet been found but evidently they ought to exist.

The detection and study of the properties of three-quasiparticle states is of considerable value in confirming the predictions of the superfluid model of the nucleus. It is very important to obtain information on the complete set of levels of three-quasiparticle multiplets, and on the population and discharge processes of these levels, etc.

LITERATURE CITED

1. Solov'ev, V. G., Zh. Éksperim. i Teor. Fiz., 43:246 (1962).
2. Pyatov, N. I., and Chernyshev, A. S., Izv. Akad. Nauk SSSR, Ser. Fiz., 28:1173 (1964).
3. Preibisz, Z., et al., Phys. Lett., Vol. 14, No. 3 (1965).
4. Bonch-Osmolovskaya, N. A., Wang Ch'eng-ju, and Gromov, K. Ya., Preprint Joint Institute for Nuclear Research P-2239 (1965).
5. Kudarova, T., Zvol'ska, V., and Veis, M., Paper presented at the Dubna Conference, July, 1965.
6. Johansen, H. S., et al., Phys. Lett., No. 8, p. 61 (1964); Ewan, G. T., Bull. Amer. Phys. Soc., 9:18 (1964).
7. Gromov, K. Ya., et al., Preprint Joint Institute for Nuclear Research, P-1852 (1964).
8. Werner, L., et al., Paper presented at the Dubna Conference, July, 1965.
9. Jorgensen, et al., Phys. Lett., 1:321 (1962).
10. Kristensen, L., et al., Phys. Lett., 8:57 (1964).
11. Peker, L. K., Izv. Akad. Nauk SSSR, Ser. Fiz., 28:306 (1964).
12. Gallagher, C. J., and Soloviev (Solov'ev), V. G., Mat. Fys. Skr. Dan. Vid. Selsk., Vol. 2, No. 2 (1962).

SOME PROBLEMS IN THE STUDY OF SPHERICAL ODD–ODD NUCLEI

L. K. Peker

USSR

ODD–ODD NUCLEI

At the present time, odd–odd nuclei are the object of quite intensive researches. This concentration of the attention of theorists and experimenters is due to the fact that, for various reasons, nuclei of this category have been studied least of all, and that by the study of odd–odd nuclei it is possible to obtain unique information on the residual p–n interaction which is difficult to obtain in any other way.

We shall examine some results of these researches. We know that the interaction of nucleons in a nucleus may be divided approximately into two parts. One of them results in the formation of a self-consistent field in which quasiparticles are moving.

The effects due to this field are described by the shell model (for spherical and deformed nuclei) and many of them are known from investigations of nuclei having an odd mass number A.

More particularly, it may be assumed that nucleons, in the same way as in odd nuclei, successively fill the proton and neutron levels of the self-consistent field (of the shell model) with definite values of the angular momenta $j_{p(n)}$, and that the angular momentum $I_{p(n)}$ of an odd nucleon group on a level being filled is the approximate integral of motion.*

In this case, the quantum characteristics of the states of odd–odd nuclei may be represented in the form of combinations of quantum characteristics of odd proton and neutron groups. For example, the spin of a state with the configuration

$$\{j_n^{\nu_n} I_n;\ \ j_p^{\nu_p} I_p\},$$

where ν_n; ν_p are the number of nucleons in the level being filled, is determined by the expression $\mathbf{I} = \mathbf{I}_p + \mathbf{I}_n$. Consequently, a multiplet of levels with $|I_p - I_n| \leq I \leq |I_p + I_n|$, corresponds to the given configuration.

The magnetic moments of levels in the approximation are determined by the expression

$$\mu = \frac{I(I+1)(g_{j_n} + g_{j_n}) + [j_n(j_n+1) - j_p(j_p+1)](gj_n - g_{j_p})}{2(I+1)}, \qquad (1)$$

*For this the p–n interaction should not be very strong.

where $g_{j_{p(n)}}$ is the gyromagnetic ratio of the proton or neutron on a level with momentum $j_{p(n)}$.

However, the motion of quasiparticles in a self-consistent field does not really occur freely, but between them act short-range forces, usually termed residual forces. In particular, in odd–odd nuclei, the residual interaction of an odd proton and odd neutron has an unusually high value.

We shall examine the effects caused by this interaction. In the most general form, the state of two interacting particles depends on their spatial $\mathbf{r}_1$, $\mathbf{r}_2$, spin $\boldsymbol{\sigma}_1$, $\boldsymbol{\sigma}_2$, and isospin τ_1, τ_2 coordinates, i.e., the operator of the pair interaction is the function

$$V(1,2) = V(\mathbf{r}_1, \sigma_1, \tau_1; \mathbf{r}_2, \sigma_2, \tau_2). \tag{2}$$

Since the interaction energy is scalar, the operator V (1, 2) ought to contain only those combinations of coordinates which satisfy the requirements of invariancy relative to displacements, rotations and inversion of the corresponding coordinate systems.

At the same time, V (1, 2) may comprise an interaction dependent only on $r_{12} = |\mathbf{r}_1 - \mathbf{r}_2|$, as well as a noncentral tensor interaction dependent on the angles between the vectors $\mathbf{r}_{12} = \mathbf{r}_1 - \mathbf{r}_2$ and $\boldsymbol{\sigma}_1$, $\boldsymbol{\sigma}_2$. Usually, the tensor potential $f(r_{12})\, S_{12}$ is determined so that it becomes zero on spatial averaging. From the vectors $\mathbf{r}_{12}, \boldsymbol{\sigma}_1$, and σ_2, it is possible to construct only one scalar value of S_{12}, satisfying the enumerated conditions:

$$S_{12} = \frac{3(\sigma_1 \cdot r_{12})(\sigma_2 \cdot r_{12})}{r_{12}^2} - (\sigma_1 \cdot \sigma_2). \tag{3}$$

Consequently, the operator V (1, 2), satisfying the principles of invariance, may be written in the form

$$V(1,2) = V_0 f_0(r_{12}) + V_1 f_1(r_{12})(\sigma_1 \cdot \sigma_2) + V_2 f_2(r_{12}) S_{12} + (\tau_1 \cdot \tau_2)\left[V_0' f_0'(r_{12}) + V_1' f_1'(r_{12})(\sigma_1 \cdot \sigma_2) + V_2' f_2(r_{12}) S_{12}\right]. \tag{4}$$

The operator V (1, 2) contains six unknown functions and six parameters. It cannot be used at the present time. The problem may be simplified by assuming that all six functions $f(r_{12})$ are identical, their explicit form being selected on the basis of physically reasonable assumptions. In particular, it is possible to use a function $f(r_{12})$ having the form of a Gaussian distribution

$$f(r_{12}) = e^{-\frac{|r_1 - r_2|^2}{\rho^2}}, \tag{5}$$

where ρ is the radius of action of the residual forces.

In consequence, the expression V (1, 2) contains seven parameters subject to determination and probably they are all important. Thus, even in such a greatly simplified form, the problem still remains insoluble.

Evidently, at the present time, it would be more rational to conduct the investigation of the problem of p–n interaction not by simplifying the potential V (1, 2) but by complicating it, commencing with the simplest of the possible potentials, and complicating it only when the experimental data compel this.

We shall revert to the examination of the properties of the multiplet of the odd–odd nucleus and of the configurations $\{j_p^{\nu_p} I_p;\ j_n^{\nu_n} I_n\}$. The removal of degeneration in such a multiplet is one of the most important effects of the residual p–n interaction. By its action, the multiplet splits. The character of the splitting (i.e. the dependence of the excitation energy of the level on its spin I) is mainly determined by just this interaction. The investigation of such multiplets of odd–odd nuclei, therefore, also enables important information to be obtained on the characteristics of $V_{pn}(1,2)$.

In accordance with our program, we select the simplest of its permissible expressions [1, 2]:

$$V_{p,n} = V_0 + V_1(\sigma_p \cdot \sigma_n). \tag{6}$$

Here $V_{0,1}$ corresponds to $V_{0,1} \cdot f_{0,1}(r_{12})$ in formula (4); V_0 to Wigner forces; V_1 to central forces depending on the spins of the nucleons.

In this approximation, the character of the splitting of the multiplet depends on the signs of V_0 and V_1 (corresponding to the forces of attraction and repulsion) and on their magnitudes.

On the basis of general considerations, some conclusions may be drawn concerning the part played by V_0 and V_1 in the splitting of the multiplets.

In particular, it has been shown [1, 2] that in the case where the number of protons or neutrons in a level with $j_{p/n}$ equal to $^1/_2$; $^5/_2$; $^9/_2$ $\nu_{p(n)} = (2j+1)/2$ (i.e., when the level is half-filled), splitting of the multiplet is due to spin forces $V_1(\sigma_p \cdot \sigma_n)$ only, and does not depend at all on V_0. Therefore, a study of the multiplets of such configurations is of very great interest, as it enables information on the second term of equation (6) to be obtained.

It has also been shown in [1, 2] that in the case of other types of configurations, important conclusions may be drawn regarding the spins of the lower levels of a multiplet, if the respective signs, but not the values, of V_0 and V_1 are known.

At the present time, the signs and values of V_0 and V_1 are determined only from an analysis of experimental results. Fuller information may be obtained if the energies and spins of all the levels of the multiplet are known from experiment. In such a case, the multiplet is analyzed in detail for different assumptions concerning the value and signs of V_0 and V_1. Such complete experimental data, however, are available for only a very small number of nuclei, and owing to difficulties of various kinds, calculations have so far only been made for particularly simple odd–odd nuclei having one particle (or hole) per level (the nuclei ${}_9F^{18}_9$, ${}_{19}K^{40}_{20}$, ${}_{21}Sc^{42}_{22}$, ${}_{39}Y^{90}_{51}$, ${}_{41}Nb^{92}_{51}$, ${}_{81}Tl^{206}_{125}$, ${}_{81}Tl^{208}_{127}$, ${}_{83}Bi^{208}_{125}$, ${}_{83}Bi^{210}_{127}$ and a few others).

Comparing the results of calculations with experimental data on multiplets in these nuclei, L. A. Sliv et al. came to the conclusion that the Wigner forces in both heavy and light nuclei were forces of attraction, i.e., $V_0 < 0$, while the sign of the spin-dependent part of V_1 varied on passing from light to heavy nuclei. In F^{18}, Y^{90}, Nb^{92}, the signs of V_0 and V_1 are the same and less than 0 (forces of attraction), in heavy nuclei Bi^{210}, the signs of V_0 and V_1 are opposite; $V_0 < 0$ and $V_1 > 0$. It has also been found that in ${}_{83}Bi^{210}_{157}$, $V_0 \approx -5\,V_1$.

From data on the binding energies of other odd–odd nuclei, it also follows that V_0 is substantially greater than V_1 [1]. It is possible on the basis of these results, to formulate rules for determining the spins of the lower states of multiplets.

For configurations of particle–particle or hole–hole types, the spin of a lower level of a multiplet is determined by the parity of the Nordheim number $N = j_p + l_p + j_n + l_n$, and consequently by the relationships

1. $$j_p = l_p \pm {}^1/_2, \quad j_n = l_n \mp {}^1/_2,$$

$$\begin{cases} I = |I_p - I_n|, & A \leqslant 92 \quad \text{(A1)} \\ I = |I_p \pm I_n|, & A \sim 210 \quad \text{(B1)} \\ \text{levels } I = I_p + I_n \text{ and } I = I_p - I_n \text{ are close} \end{cases}$$

2.

$$j_p = l_p \pm {}^1/_2,\ j_n = l_n \pm {}^1/_2,$$

$$\begin{cases} I = |I_p \pm I_n|, \quad A \leqslant 92 & \text{(A2)} \\ \text{levels } I = I_p + I_n \text{ and } I = I_p - I_n \text{ are close} & \\ I = |I_p - I_n|, \quad A \sim 210. & \text{(B2)} \end{cases}$$

3. For a configuration of particle–hole type, independently of j_p and j_n, in all nuclei

$$I = I_p + I_n - 1. \qquad \text{(A3; B3)}$$

4. When one of the levels with j = $^1/_2$; $^5/_2$; $^9/_2$ ... is half-filled, the parity π and spin of the lower state of the multiplet 1 are connected by the relationships:

$$\pi = (-)^{I+1} \text{ in nuclei with } A \leqslant 92 \qquad \text{(A4)}$$

$$\pi = (-)^{I} \text{ in heavy nuclei with } A \sim 210 \qquad \text{(B4)}$$

Since the amount of experimental data on ground-state spins is much greater than that of complete data on multiplets, their analysis and comparison with the enumerated rules in principle permits verification of the conclusions made with regard to the signs of V_0 and V_1.

In 1960, Brennan and Bernstein [4] even before a theoretical treatment had been made, analyzed experimental data on the spins of odd–odd nuclei with A ≤ 116, and found empirical rules agreeing with the rules A1; A2; A3 for light nuclei. They came to the conclusion that V_0 and V_1 in nuclei with A ≤ 116 corresponded to forces of attraction (< 0).

A re-analysis was recently made [8] of experimental data on the properties of lower states of spherical odd–odd and odd nuclei in the region A = 1-214, obtained up to the end of 1964 [5-7]. As a result, in our view, the most probable values of spins and configuration of protons and neutrons in ground states of odd–odd nuclei were selected.*

If the analysis made is correct, the following conclusions may be drawn [8].

1. Rule A1 is confirmed or does not contradict experiment in 37 nuclei, and is not confirmed in two nuclei $_{55}S^{138}_{83}$ and $_{55}I^{136}_{83}$ in the region A = 1-150.

Rule B1 is confirmed in $_{83}Bi^{210}_{127}$, and possibly in $_{83}Bi^{212}_{129}$.†

2. Rule A2 is confirmed or does not contradict experiment in 33 nuclei, and is not confirmed in three nuclei in the range A = 1-150 ($_{29}Cu^{58}_{29}$, in which lowering of the level with I = 1 below the level with I = 0^+, expected in the ground state, may be due to pair correlations, as in $_{83}Bi^{210}_{127}$).

No configuration of type B2 has yet been found in any of the heavy nuclei examined.

3. Rule A3, B3 is confirmed in 28 nuclei and is not confirmed in 10.

* The ground states of odd–odd nuclei represent a more or less complex mixture of configurations, and we indicate only the principal component of this mixture. For example, in the nuclei $^{47}Ag^{108}_{61}$, Ag^{110}_{63} the ground states contain a mixture of the configurations $p(g_{9/2})^{-3}_{9/2}+;\ n(g_{7/2})^{1}_{7/2}+$ $p(g_{9/2})^{-3}_{7/2}+;\ n(d_{5/2})^{-1}_{5/2}+$. We assume that the first component is the principal component, since in that case, the high probability of β decay (log ft = 4.7) can be better explained. Analogous considerations determined the choice of principal components of mixtures of configurations in other nuclei.

† Lowering of the level with I = 1 below the level with I = 0 may be due to the influence of pair correlations or tensor forces.

4. Rule A4 is confirmed or does not contradict experiment in the ground states of 74 nuclei with A = 1-214, and is not confirmed in one nucleus $_{29}Cu^{68}_{39}$. If it is empirically extended to the case $I_{p(n)} \neq j_{p(n)}$ it has the form:

$$\pi = (-)^{l+1} \cdot (-)^{I_p - j_p} \cdot (-)^{I_n - j_n}. \quad (A4') \tag{8}$$

The spin of the ground state was determined in 71 nuclei as a function of parity of the Nordheim number N:

$$I = |I_p - (-)^N I_n|. \tag{9}$$

In the case where $j_{p(n)} = 1/2$, identification of the configurations was usually more reliable than in the case where $j \neq 1/2$.

Rule A4 is confirmed in 50 cases of such a type and is not satisfied in one (Cu^{68}). If the known excited states with $j_{p(n)} = 1/2$ are also taken into account, rule A4 is confirmed in more than 80 cases and is not satisfied in one (Cu^{68}).

Rule B4 is not confirmed in a single case.

The following conclusions may be drawn from the analysis made and the results of calculations by Sliv et al.:

1. Rules A1-A4 permit prediction of the spins of lower states of multiplets of spherical odd-odd nuclei with a high degree of accuracy, if the configuration of the multiplet can be established.

2. The region of odd-odd nuclei in which $V_0 < 0$ and $V_1 < 0$ extends to values of A = 150 [8].

Consequently:

1. The shell model is suitable for describing the spins of odd-odd nuclei.*

2. The values of I_p and I_n are approximate integrals of motion, and consequently, the interaction of protons and neutrons is not very strong.

3. The simplest expression $V_{pn} = V_0 + V_1 (\sigma_p \sigma_n)$ in most cases (except $j_{p(n)} = 1/2$) is sufficient to describe the principal features of multiplet splitting and as yet does not require complications, in particular it does not require the introduction of tensor forces.

In the case of the configuration with $j_{p(n)} = 1/2$, the situation is more complicated.

It was pointed out above that the spins of the lower states of these multiplets (doublets), satisfying rule A4 and not satisfying rule B4 contradict Sliv's conclusions regarding the positive sign of V_1 in the region A = 196-214, if this interaction alone is responsible for splitting of the multiplet.

In addition, as already mentioned, V_1 is small ($V_1 \approx 0.1$-$0.2\ V_2$), and splitting of the multiplet caused by it is small, and judging by estimates ought not to exceed several tens of keV.

During the experiments, doublet splitting was found in $_{81}Tl^{206}_{125}$: $s_{1/2}\, p_{1/2}$ – 305 keV; $s_{1/2}\, f_{5/2}$ – 387 keV; $s_{1/2}\, p_{3/2}$ – 167 keV (dp reaction); in $_{81}Tl^{198}_{117}$: $s_{1/2}\, f_{5/2}$ 283 keV; $s_{1/2}\, p_{1/2}$ – 117 keV [5, 9], and so forth.

It has recently been shown by the groups of Rasmussen [3] and Sliv [2] that the inclusion of tensor forces in V_{pn} makes it possible to take these characteristics into account, at least

*This also applies to magnetic moments.

qualitatively, in particular rule A4 may be established for nuclei with A = 196-214. It should be noted that we are here obviously dealing with the first real appearance of tensor forces in $V_{p,n}$.* This imposes a complication of the selected potential which is now written in the form

$$V_{p,n} = V_0 + V_1(\sigma_p \cdot \sigma_n) + V_2 S_{12}. \tag{10}$$

The effect of tensor forces in order of magnitude is the same as the effect of the usual central spin forces, and in multiplet splitting, where Wigner forces play a principal role, it is quite weak. In particular, according to Rasmussen [3], it never violates the rules A1, A2, A3.

In conclusion, it should be mentioned that all the calculations so far made of the spectra of odd–odd nuclei can lay claim only to a qualitative description of the characteristics of multiplet splitting, mainly a descriptive of the order of sequence of several low levels, since there is a large number of factors, difficult to take into account at the present time, but nevertheless important.

The forces dependent on isospin, pair correlations, interaction with the core, and the mixing of the configurations which they cause, are indubitably of great importance, and by taking them into account, we shall be assisted in understanding the fine details of behavior of the levels in a multiplet.

LITERATURE CITED

1. de Shalit, A., and Walecka, J. D., Phys. Rev., 120:1790 (1960); de Shalit, A., and Walecka, J. D., Nucl. Phys., 22:184 (1961); de Shalit, A., Nucl. Phys., 22:677 (1961).
2. Sliv, L. A., et al., Zh. Éksperim. i Teor. Fiz., 40:946 (1960); Kharitonov, Yu. I., Dissertation LFTI (1965), Izv. Akad. Nauk SSSR, Ser. Fiz., 28:315 (1964).
3. Kim, Y. E., and Rasmussen, J. O., Nucl. Phys., 47:184 (1963); Kim, Y. E., Phys., 131:1712 (1963); Rasmussen, J.O., and Kin, Y. E., Izv. Akad. Nauk SSSR, Ser. Fiz., 29:94 (1965); Phys. Rev., Vol. 135, B44 (1964).
4. Brennan, M. N., and Bernstein, A. M., Phys. Rev., 120:927 (1960).
5. Dzhelepov, B. S., et al., Decay Schemes of Radioactive Nuclei with A ≥ 100, Moscow, Izd. Akad. Nauk SSSR, 1963.
6. Dzhelepov, B. S., and Peker, L. K., Decay Schemes of Radioactive Nuclei with A < 100, Moscow, Nauka, 1965.
7. Nuclear Data Sheets, 1963-1964.
8. Peker, L. K., and Novikov, Yu. N., Yad. Fiz., in press.
9. Erskine, J. R., Phys. Rev. Lett., 14(12):A3 (1965); Mukherje, P., Phys. Rev. Lett., 13:238 (1964).

NUCLEAR ISOMERISM

Essentially, any known low excited states may be called isomeric states of nuclei, since they are all metastable. Their lifetime is 10^{-15} sec, i.e., it is many orders longer than typical nuclear time, equal to $\sim 10^{-21}$–10^{-22} sec. It is, however, customary to include in isomeric states only long-lived states. From the point of view of the concepts of the shell model, it is convenient also to introduce a multipolarity limitation of γ transitions, with reference to isomeric transitions of the type M2, M3, M4; E3, E4, and so forth.

*Some authors [3] previously introduced tensor forces, although this was not necessary, since the effects they were considering could be explained by other causes: pair correlations, interaction of particles with the core and consequently mixing of the configurations, etc.

In the subsequent discussion, therefore, by isomeric states we shall understand either states with $T_{1/2} \geq 10^{-6}$ sec, or states emitting on disintegration γ quanta of high multipolarity ($\geq$ M2).

The relatively long lifetime facilitates experimental investigation of such states, and their quantum characteristic may usually be determined with a higher degree of reliability than in levels with shorter lifetimes. It is just because of this that the investigation of isomeric states is of particularly great significance in nuclear spectroscopy.

It is recalled that the interpretation of isomerism islands in odd nuclei is one of the most important achievements of the shell model, since it has essentially proved that this model is capable of describing the excited states of nuclei.

The more detailed analysis of an isomerism island with ZN = 39-49 and the discovery there of type-E3 isomeric transitions was an important stimulus to the improvement of the shell model, i.e. the construction of the many-particle model. The study of isomers in deformed odd nuclei for the first time showed the validity of the Nilsson scheme of single-particle levels for the description of excited states.

The discovery of isomerism, due to K-forbiddenness, in $_{72}Hf^{180}_{108}$, $_{72}Hf^{178}_{106}$, $_{76}Os^{190}_{114}$ was one of the most important proofs of the soundness of the generalized model.

Isomerism of the nucleus $_{83}Bi^{210}_{127}$ made it possible to explain certain essential characteristics of p−n interaction, which assisted the understanding of the characteristics of multiplets in odd−odd nuclei.

The first examples of more complex states, three-particle, four-particle and so forth, were discovered in the study of isomeric states.

It is possible that future research on isomeric states will result in important discoveries, which will substantially strengthen our ideas on the nucleus. One example of such discoveries which may be mentioned is the discovery of short-lived isomers disintegrating by spontaneous fission.

We shall consider some types of isomeric states in greater detail.

Isomeric states in odd spherical nuclei may be divided into two quite distinct classes.

Most of them are found in nuclei, arranged in isomerism islands with Z, N = 39-49, 63-81, N = 113-125, and as a first approximation are well explained in the framework of the shell model as states of the type: $p_{1/2}$, $(g_{9/2})^{3.57}_{7/2}+$, $g_{9/2}$, $h_{11/2}$, $i_{13/2}$. In two odd nuclei, $_{50}Sn^{113}_{63}$ [1] and $_{52}Tl^{115}_{63}$ [3], where as the result of pair correlations, the level $g_{7/2}$ was found to be closest to the ground state $S_{1/2}$, it naturally became isomeric, and the observed isomeric γ-transition, as was to be expected, had multipole order of M3 type ($I_i = {}^7/_2 + \rightarrow I_f = {}^1/_2{}^+$).

We shall examine in more detail the isomeric states observed in nuclei not included in isomerism islands and having other quantum characteristics.

In 1964-5, in the nuclei $_{21}Sc^{43}$, $_{21}Sc^{45}$, $_{21}Sc^{47}$ low type ${}^3/_2{}^+$ isomeric states were found, which were coupled to type ${}^7/_2{}^-$ ground states by M/2 γ-transitions. Table 1 presents data on the energies and half-lives of these levels, and the degree of hindrance of the γ transitions [4, 5, 7].

The unexpectedness of this discovery, from the shell-model point of view, was caused by the fact that levels with positive parity $d_{3/2}$ in odd isotopes of $_{21}Sc$ could arise through rupture of the proton shell Z = 20 or on transition of an odd proton to the $d_{3/2}$ level of the next shell. It had been assumed that any of these processes could only occur at high excitations (above 1 MeV) and that consequently the $d_{3/2}$ levels ought to be very high.

TABLE 1

Nucleus	E_γ (M2), keV	$T_{1/2}$, sec	$F = \frac{T_{exp}}{T_{o.h.}}$
$_{21}Sc^{43}_{22}$	150±3	$4.4 \cdot 10^{-4}$	190
$_{21}Sc^{45}_{24}$	13±1	0.31	210
$_{21}Sc^{47}_{26}$	774±10	$2.8 \cdot 10^{-7}$	410

Several possible reasons have been suggested for the occurrence of isomerism in odd nuclei.

Bansal and French [8] drew attention to the fact that on excitation of a $d_{3/2}$ hole level, a $(d_{3/2})^2_{I=0}$ pair ought to be formed in the filled shell but a $(f_{7/2})^2_{I=0}$ pair is formed. Since the pairing energy of nucleons in the $f_{7/2}$ level is greater than in the $d_{3/2}$ level, this process is favored, and the $d_{3/2}$ hole level may sometimes be found to be low enough.

Assessment of such factors for isotopes of $_{21}Sc$ led to excitation energies in good agreement with experiment.

Two other viewpoints on the nature of these levels have been advanced by Sheline and his coworkers.

One of them [7] is the suggestion that the nuclei $_{21}Sc^{43}$, $_{21}Sc^{45}$, $_{21}Sc^{47}$ are deformed (prolate $\delta > 0$). In this case, the Nilsson scheme comes into play and the single-particle level with $\Omega = {}^3/_2 + (d_{3/2})$ is situated closer to the ground state. It follows from the Nilsson scheme that the distance between them is a minimum for maximum deformation.

In the framework of this concept, maximum deformation ought to be in $_{21}Sc^{45}_{24}$ which has a half-filled neutron shell. Table 1 shows that in $_{21}Sc^{45}_{24}$, the ${}^3/_2{}^+$ level is indeed the lowest of all.

However, in the Nilsson scheme, the ground states of $_{21}Sc$ ought to have $K = \Omega = {}^1/_2{}^-(f_{7/2})$. To explain the observed spin value $I = {}^7/_2{}^-$, Sheline pointed out that the decoupling parameter for this state, calculated by means of the Nilsson wave functions, was large: $a = (3\text{-}4)^-$. In the rotational band with $K = {}^1/_2$, described by the formula

$$E = \frac{\hbar^2}{2J}\left[I(I+1) + a(-)^{I+1/2}(I + 1/2)\right]$$

the rotational level with $I = {}^7/_2{}^-$ ought, therefore, to be lower than the level with $I = {}^1/_2{}^-$.

If the coupling effect of rotational bands with $\Delta k = 1$ (Coriolis force effect) is taken into account, the level with $I = {}^7/_2{}^-$ is moved still lower and may be lower than the state of the band $\Omega = K = {}^1/_2{}^-$. This provides an explanation for the experimental value of ${}^7/_2{}^-$ for the spin.

Naturally, the soundness of such an interpretation can only be assessed after experimental determination of the spins and parities of other excited states and not only the isomeric state.

Sheline and Wildermuth [6] proposed another possible interpretation for the isomeric level in $_{21}Sc^{43}_{22}$. They draw attention to a certain resemblance between the nuclei $_{21}Sc^{43}_{22}$ and $_{9}F^{19}_{10}$. It is not difficult to see that they both have identical structures: A doubly magic core ($_{20}Ca^{40}_{20}$ and $_{8}O^{16}_{8}$) + p·2n, and very low levels are found in both, having a parity opposite to the parity of the ground state (for $_{9}F^{19}_{10}$, the ground state has $I = {}^1/_2 + 1$, and the 112 keV level has $I = {}^1/_2{}^-$).

This characteristic of ${}_9F^{19}_{10}$ may be explained in the framework of the cluster model on the assumption that in the ground state the nucleus has the structure ${}_8O^{16}_8 + {}_1H^3_2$ with $I = {}^1/_2{}^+$, while in the state with E = 112 keV it has the structure ${}_7N^{15}_8 + {}_2He^4_2$ with $I = {}^1/_2{}^-$.

Similarly, it may be conceived that close to the ground state of ${}_{21}Sc^{43}_{22}$, having the structure ${}_{20}Ca^{40}_{20} + 1\ H^3_2$ with $I = {}^7/_2{}^-$, there ought to be a state with the structure ${}_{19}K^{33}_{20} + {}_2He^4_2$ with $I = {}^3/_2{}^+$.

It is interesting that reasoning in just this way, the authors predicted the existence of this isomer even in 1960, long before the experimental work. They also noted a peculiar similarity between the odd isotopes of ${}_9F^{19}$ and ${}_{21}Sc^{43}$.

In recent years, nine different models from the single-particle model for the spherical nucleus to the asymmetric rotator and cluster model have been used for describing the spectrum of the levels of F^{19}, and they have all led to almost the same good agreement with experiment. We are evidently facing a similar experience in the examination of the isotopes of Sc. This means that the same principles underly all these externally different models, and only the form of their expression differs.

Three-Particle Isomeric States

In some odd nuclei, isomeric states are found which cannot be described in the framework of the shell model like single-particle states. They are all characterized by high excitation energy (more than 800 keV) and very high spin $I > {}^{13}/_2$. Data on such isomers are given in Table 2.

Their formation is due to excitation of the odd–odd core of the nucleus. Isomerism of this kind has therefore been called core isomerism. Owing to the high value of $T_{1/2}$, these states cannot be collective.*

The high spins of many such isomers is evidence of the contribution of a number of unpaired nucleons, and the excitation energy limits this number to three nucleons (at energies of $\lesssim 2$ MeV only one pair of nucleons may be ruptured). All of them, therefore, ought to be three-particle isomers.

What is the specific interpretation of these isomers? Is it possible to establish relationships enabling nuclei to be distinguished in which there are conditions for the existence of these isomers† and can one determine their spin and parity?

Let us examine the possible answers to these questions. The simplest way of finding these answers is the direct calculation of the spectra of the nuclear levels. If these calculations are performed correctly, it will be immediately clear in what nuclei the conditions for isomerism exist. At the present time, however, such a method can be used only for the analysis of spectra of the levels of nuclei which are particularly simple from the viewpoint of shell concepts. The simplest of the nuclei mentioned in Table 2 is ${}_{84}Po^{211}_{127}$, in which, in addition to the filled shells, there are only three nucleons – one neutron and two protons. Since the level with $I = {}^{25}/_2{}^+$, observed in the spectrum of single-particle neutron levels, is absent, it is necessary to calculate the spectrum of the three-particle configuration

*Otherwise, they would be discharged with $T \lesssim 10^{-8}$ sec to lower states of the same rotational or vibrational band.

†Such a condition may be the requirement that all the lower levels have much lower (or higher) spins, and the multipolarity of the isomeric transition is not less than M/2.

TABLE 2. Isomeric States of Odd Nuclei

Nucleus	$T_{1/2}$	E, keV	$I\pi$ experiments	Configuration p	Configuration n	Rule	$I\pi$ theor.	Literature
$_{42}Mo^{93}_{51}$	6,7h	2428	21/2+	$(g_{9/2})^2_{8+}$	$(d_{5/2})^1$	A2	21/2+,	[13]
$_{45}Rh^{103}_{58}$	⩾ 4 years						11/2+	[27]
$_{49}In^{109}_{60}$	0.21 sec	2106	19/2	$(g_{9/2})^{-1}$	$(g_{7/2})^2_{6+}$	A3	19,2+	[30, 31]
$_{55}Cs^{135}_{80}$	53 min	1620	19/2−	$(g_{7/2})^1$	$(h_{11/2})^{-1}$, $(d_{3/2})^{-1}$	A3	19/2−	[14, 15]
$_{57}La^{137}_{80}$	$12\cdot10^{-3}$ sec	⩾1120	>15/2	$(g_{7/2})^1$	$(h_{11/2})^{-1}$, $(d_{3/2})^{-1}$	A3	19/2−	[32]
$_{61}Pm^{141}_{80}$	$2.2\cdot10^{-3}$sec	⩾840	>13/2	$(d_{5/2})^1$	$(h_{11/2})^{-1}$, $(d_{3/2})^{-1}$	A3	17/2−	[16]
$_{71}Lu^{177}_{106}$	155 days	968	23/2−	7/2+[404]	7/2−[514], 9/2+[624]		23/2−	[23, 24, 25]
$_{75}Re^{189}_{114}$	~ 120 days							[26]
$_{83}Bi^{195}_{112}$	33 sec	[800—900]	(1/2)	$(h_{9/2})^1$	$(f_{5/2})^2_{4+}$	B2	1/2−	[17]
$_{83}Bi^{197}_{114}$	8 min	[800—900]	(1/2)	$(h_{9/2})^1$	$(f_{5/2})^2_{4+}$	B2	1/2−	[18]
$_{83}Bi^{199}_{116}$	24.4 min	[800—900]	(1/2)	$(h_{9/2})^1$	$(f_{5/2})^2_{4+}$	B2	1/2−	[18]
$_{83}Bi^{201}_{118}$	52 min	[800—900]	1/2	$(h_{9/2})^1$	$(f_{5/2})^2_{4+}$	B2	1/2−	[18]
$_{83}Bi^{203}_{120}$	~ 720 min	[800—900]	(1/2)	$(h_{9/2})^1$	$(f_{5/2})^2_{4+}$	B2	1/2−	[19, 18]
$_{84}Po^{207}_{123}$	2.8 sec	1390	19/2−	$(h_{9/2})^2_{8+}$	$(f_{5/2})^{-1}$	B3	19/2−	[20]
$_{84}Po^{211}_{127}$	25 sec	1456	25/2+	$(h_{9/2})^2_{8+}$	$(g_{9/2})^1$	B1	25/2+, 7/2+	[21, 22, 34]

It will be seen that the interaction of the nucleons in this configuration amounts to two-particle couplings of p−p and p−n type. In spherical nuclei, the presence of this interaction leads to splitting of the configuration multiplet with given values of j_{p_1}, j_{p_2}, j_n. It follows from analysis of experimental data on the spin of ground states of neighboring odd nuclei that the configuration of the ground state of $_{84}Po^{211}_{127}$ is $[p(h_{9/2})^2, n(g_{9/2})^1]_{9/2^+}$.

Sliv and Kharitonov [9, 35] calculated the spectrum of this multiplet: they used the parameters of the p−p and p−n interactions, obtained from an analysis of the levels of the even−even nucleus $_{84}Po^{210}_{126}$, in which the spectrum of the two-particle configuration $p(h_{9/2})^2$ is determined by the residual p−p interaction, and of $_{83}Bi^{210}_{127}$, the spectrum of the two-particle configuration $p(h_{9/2})^1$, $n(g_{9/2})^1$ of which is determined by residual p−n interaction.

Calculation showed that in the spectrum of the multiplet $p(h_{9/2})^2$, $n(g_{9/2})^1$ there are in fact conditions for the existence of an isomeric level with $I = {}^{25}/_2{}^+$, since the maximum spin of the levels situated below it is $I = {}^{13}/_2{}^+$.

It is clear that disintegration of the isomeric state with emission of γ quanta is strongly forbidden ($\Delta I \geq 6$). This is in agreement with the fact that only α decay has been found in the isomer Po^{211} [21, 22, 34]. The spectrum of another relatively simple nucleus $_{42}Mo^{93}_{51}$ has been similarly calculated; the configuration of its ground state is $[p(g_{9/2})^2, n(d_{5/2})]_{5/2^+}$ [9], like those of the nuclei $_{83}Bi^{195}$–Bi^{203} [11]. Calculation of the spectrum of the three-particle configurations: $[p(h_{9/2})^2, n(g_{9/2})^1]$, $[Po^{211}]$, $p(h_{9/2})^1$, $n(g_{9/2})^2$, $[Bi^{211}]$ [9, 35] and some others has shown that the interaction of two identical nucleons with momentum j may be accounted for effectively by assuming that they always make the maximum contribution I_0^{max} to the spin of the isomeric level.

The configurations of the isomeric states of Po^{211} and Mo^{93} may therefore be written in the form

$$\left[p(h_{9/2})^2_{I_0^{max}=8+},\quad n(g_{9/2})^1\right]_{25/2^+} (Po^{211}),$$

$$\left[p(g_{9/2})^2_{I_0^{max}=8+},\quad n(d_{5/2})^1\right]_{21/2^+} (Mo^{93}).$$

The residual p−n interaction is of decisive value in the formation of the spin I of an isomeric level from the angular momenta of a nucleon pair I_0^{max} and an odd nucleon of another type j.

However, it is this very interaction which leads to the splitting of the multiplet in odd-odd nuclei, for which have been found certain relationships permitting determination of the spins of its lower state.

It is therefore possible that the same relationships determine the spins of the lower states of three-particle multiplets of the type

$$p(jp)^2_{I_0^{max}}, \quad n(j_n)^1 \quad \text{or} \quad p(j_p)^1, \quad n(j_n)^2_{I_0^{max}}.$$

They may be represented in the form of configurations of the type [12]:

particles−particles or holes−holes

1. $j_p = l_p \pm {}^1/_2,\ j_n = l_n \mp {}^1/_2,$

$$I = |I_0^{max} - j_0| \text{ for } A \leqslant 150, \tag{A1}$$

$$I = |I_0^{max} \pm j_0| \text{ for } A \sim 210, \tag{B1}$$

2. $j_p = l_p \pm {}^1/_2,\ j_n = l_n \pm {}^1/_2,$

$$I = |I_0^{max} \pm j_0| \text{ for } A \leqslant 150, \tag{A2}$$

$$I = |I_0^{max} - j_0| \text{ for } A \sim 210 \tag{B2}$$

particles−holes

$$I = |I_0^{max} + j_0 - 1|. \tag{A3-B3}$$

If the configurations of three-particle isomeric states are already known, it may be seen from the last columns of Table 2 that these rules actually describe their spins. At the same time, they describe correctly the spins of the isomeric states of ${}_{49}In^{190}_{60}$ and ${}_{84}Po^{207}_{126}$, the spectra of which have not yet been calculated owing to their complexity.

It is interesting to note that if the selected configuration of In^{109m} is correct, the analogous level of ${}_{49}I^{111}_{62}$ ought to belong to a multiplet of configuration $p(g_{9/2})^{-1};\ n(g_{7/2})^{-2}_{I_0^{max}=6+}$ of A1 type and according to the rules, its spin I = $^3/_2$, i.e., it cannot be isomeric. It is easy to show that in the heavier isotopes $In^{113,115}$, similar excited levels cannot have such high spins as In^{109m}, and consequently core isomerism ought not to be found in them.

At the same time, the three-particle isomer ${}_{49}I^{107m}_{58}$ is possible. The results of recently conducted unsuccessful searches for short-lived isomers in In^{111} and In^{113} are in agreement with these conclusions.

The theoretical analysis of the multiplets of three-particle configurations of the type $[j_p, (j_{n_1}, j_{n_2})I_0^{max}]$ or $[(j_{p_1}, j_{p_2})I_0^{max}, j_n]$ is more complicated, and their spectra have not yet been calculated. An isomeric state of such a type, however, is known in ${}_{55}Cs^{135}_{80}$ $[p(g_{7/2})^1, n(h^{-1}_{11/2}, d^{-1}_{3/2})I_0 = 7^-]$. The experimental value of its spin (see Table 2) is understandable on the assumption that rules A3-B3 also apply to such complex configurations. We shall now consider the methods of determining a configuration [10].

For this, we draw attention to the fact that $\mathbf{I} = \mathbf{I}_0^{max} + \mathbf{j}_0$, where $\mathbf{I}_0^{max}$ is the maximum spin in the multiplet of a configuration of two identical nucleons. It is obvious that if we remove an odd nucleon from the nucleus, an even−even nucleus is left, in which there ought to be the

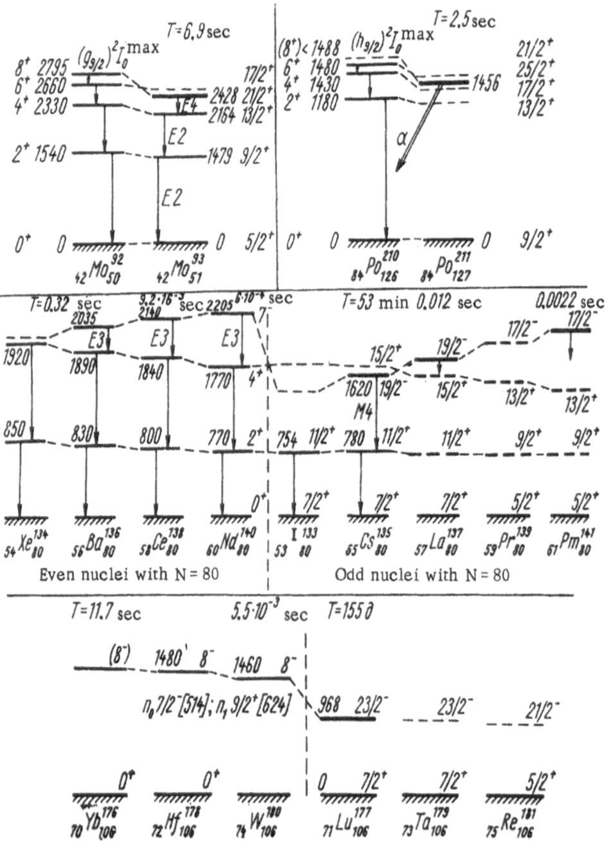

Fig. 1. Connection of three-particle isomeric states in nuclei with odd A and two-particle states in neighboring even–even nuclei.

same two-particle configuration and in the constitution of its multiplet a level with I-I_0^{max}, i.e., it is possible that there ought to correspond to the three-particle isomeric state of an odd nucleus with spin I a real level with I = I_0^{max} in the neighboring even–even nucleus with A = 1.

Although the converse is true in the general case (since the condition of the formation of a three-particle isomer is mainly determined by the p–n interaction), it is possible that when a configuration of the type B1, A2, A3-B3 occurs in an odd nucleus, a three-particle isomeric level of the odd nucleus will often correspond to the two-particle excited level of the even–even nucleus with high I_0^{max}.

Therefore, if a long-lived isomer with high spin (two-particle proton or neutron isomer) is discovered, a three-particle isomer is possible in the neighboring nucleus.

Figure 1 shows that such correspondence of levels of even–even and odd nuclei actually occurs in the presence of adequate experimental data in the case of ${}_{42}Mo_{50}^{92}$ - ${}_{42}MO_{51}^{93}$; ${}_{84}Po_{126}^{210}$ - ${}_{84}Po_{127}^{211}$; ${}_{56}Ba_{80}^{136}$ - ${}_{58}Ce_{80}^{138}$ - ${}_{60}Nd_{80}^{140}$ - Cs_{80}^{135}; ${}_{72}Hf_{106}^{178}$ - ${}_{74}W_{106}^{180}$ - ${}_{76}Os_{106}^{182}$ - ${}_{71}Lu_{106}^{177}$.

It is also clear that a three-particle level is always shifted downwards compared with two-particle levels. A similar analogy is also found between levels with lower spins, through which the γ discharge of two-particle and three-particle levels occurs. Here also, the levels are displaced downwards, the displacement increasing with increase in I_0^-.

If, therefore, the configuration of a two-particle level with $I = I_0^{max}$ in an even–even nucleus is known, it is possible to determine the configuration of a three-particle isomer in an odd nucleus. The configuration of the majority of the isomers cited in Table 2 were determined in just this way [10]. It is important to note that they all proved to be three-particle levels of (2p, n) or (p, 2n) type; so far levels of (3p) or (3n) type have not yet been found.

Figure 1, in particular, shows that such isomers ought to be in the nuclei: $_{53}I^{133}_{80}$, $_{57}La^{132}_{80}$, $_{59}Pr^{139}_{80}$, $_{61}Pm^{141}_{80}$, $_{73}Ta^{179}_{106}$, $_{75}Re^{181}_{106}$. Further, it is possible that the isomer considered in I^{133} like Cs^{135} ought to have a long life (since the isomeric transition is of M4 type), while the isomers in La^{137}, Pr^{189}, Pm^{141} ought to have $T_{1/2} \sim 10^{-3}$ sec (since the isomeric transitions in them are of M2 type).

Recently, in the Khar'kov Physicotechnical Institute isomeric states have been found in $_{57}La^{137}_{180}$ [32] and $_{61}Pm^{141}_{80}$ [16] with $T_{1/2} = 12 \cdot 10^{-3}$ sec and $T_{1/2} = 2.2 \cdot 10^{-3}$ sec, which judging by available data ought to be identical with the expected three-particle isomers.

It seems to us that the γ rays observed in the disintegration of these isomers are produced on the discharge of intermediate levels, while isomeric M2-transitions with an energy of $E_\gamma \leq$ 100 keV have not yet been found.

Since in $_{76}Os^{190}_{114}$ there is a two-particle neutron level with $I_0^{max} = K = 10^-$, there ought to be in the neighboring nucleus $_{75}Re^{189}_{114}$ a three-particle isomer with $I = {}^{25}\!/_2$. This isomer has evidently been discovered and has T = ~ 120 days [26].

It is interesting to nate that levels with $I = 7^-$ are analogous to isomeric levels in nuclei with N = 80 ($_{56}Ba^{136}_{80}$ [28, 29], $_{58}Ce^{138}_{80}$ [16], $_{60}Nd^{140}_{80}$ [16]), found also in nuclei with Z = 80 ($_{80}Hg^{192-198}_{112-118}$, configuration $p(h^{-1}_{11/2}, d^{-1}_{3/2})$; however, these levels are not isomeric since levels with $I = 5^-$ $[p(h^{-1}_{11/2}, s_{1/2})]$ are somewhat below them in nuclei of $_{80}Hg$, and between them there is a fast E2-transition.

It is possible that three-particle levels of odd nuclei of Hg with $I_\pi = {}^{17}\!/_2{}^+, {}^{15}\!/_2{}^+$ also ought to have a very short $T_{1/2}$ and consequently do not appear as isomeric levels.

Four-Quasiparticle Isomeric Levels

If a nucleon is added to the odd nuclei enumerated in the preceding section and occurring in a three-particle isomeric state with high spin, conditions for the existence of isomerism may arise in the resulting four-particle multiplet of the even–even or odd–odd nucleus. Evidently, isomerism is possible if a lower level of this multiplet has the maximum possible spin.*

Unfortunately, the calculation of the levels of such multiplets is very difficult and as yet has been done only for the simplest configurations of type $p(h_{9/2})^2$, $n(g_{9/2})^2$ in $_{84}Po^{212}_{128}$, containing in addition to filled shells only two protons and two neutrons [35, 9]. In configurations

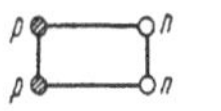

of such a kind, both p–n interaction and interaction of identical nucleons p–p and n–n are essential.

*A less rigid and more general condition for the existence of a four-particle isomeric level is the requirement that all the lower levels of the multiplet have much less spin.

Sliv and his coworkers [35] have shown, however, that the latter may be effectively taken into account on the assumption that pairs of identical nucleons provide the maximum possible contribution to the spin of the isomeric state. The configuration of the isomeric state of ${}_{84}Po^{212}_{128}$, therefore, has the form

$$p\,(h_{9/2})^2_{I_0^{max}=8^+}, \qquad n\,(g_{9/2})^2_{I_0^{max}=8^+}.$$

The character of the splitting of the multiplet, due to p−n interaction, is the same as in the two-particle configurations of odd−odd nuclei and is determined by the parity of the Nordheim number

$$N = j_p + l_p + j_n + l_n.$$

In heavy nuclei (A ~ 210) in the case of an even number $N(j_p = l_p \pm 1/2;\ j_n = l_n \mp 1/2)$, the spin of the lower level of the four-particle multiplet is determined by the expression: $I = |I_0^{max}(2p) \pm I_0^{max}(2n)|$ and is equal to $I = 16^+$ or $I = 0^+$. A state with $I = 0^+$ cannot be isomeric, since it readily decays to the ground state of Po^{212} through an intermediate state with $I = 2^+$ and $I = 1^-$.

As shown by Sliv's calculations, below the level with $I = 16^+$ in Po^{212}, there are only levels with maximum spin $I = 8^+$.

Under these conditions, isomeric transition of the type $I_i = 16^+ \to I_f = 8^+$ are practically impossible, and the isomeric level ought to experience α decay. In 1962, such an isomeric level was found in Po^{212} with E* = 2930 keV, $T_{1/2} = 45$ sec, $I \geq 16$, undergoing α decay [33, 34].

As yet there is no information at all on four-quasiparticle levels in other odd−odd nuclei.

Isomeric State of ${}_{63}Eu^{152}_{89}$ with $I = 0^-$

If two states of the same nucleus have different deformation parameters, it is possible that the probability of the γ transition binding them will be greatly decreased and the lifetime of the isomeric level will be substantially greater than that obtainable on the basis of the usual single particle estimates. The isomeric state of such a type is evidently observed in ${}_{63}Eu^{152}_{89}$ [44]. A ground state with T = 13 years and $I = 3^-$ and an isomeric state with T = 9.3 h and $I = 0^-$ have already long been known in this nucleus. It follows from an analysis of the limit energies of β spectra that the level with T = 9.3 h is higher than the ground state by 50 ± 15 keV [1].

However no isomeric γ-transition of M3 type has yet been observed. Further, it has been shown that it cannot occur in more than 0.002% of the number of disintegrations with T = 9.3 h [42, 43].

Consequently, the considered isomeric transition M3 $0^- \to 3^-$ is very strongly hindered compared with the single-particle estimate $\left(F = \frac{T_{exp}}{T_{\gamma+e}} > 5\cdot 10^8\right)$. At the same time, if the β, ε decays of states of Eu^{152} with T = 13 years ($I = 3^-$) and T = 9.3 h ($I = 0^-$) are compared, it may be seen that the first decays with greater probability to levels of the deformed even−even nucleus ${}_{62}Sm^{152}_{90}$ ($\beta^+ + \varepsilon \sim 80\%$, $\log ft \approx 9.7$; $\beta^- \sim 20\%$, $\log ft \approx 12.5$), while the second decays to levels of the spherical nucleus ${}_{64}Gd^{152}_{88}$ ($\beta^+ + \varepsilon \sim 26\%$, $\log ft \approx 8.7$; $\beta^- \sim 74\%$, $\log ft \approx 7.5$).

Since on passing from nuclei with N = 88 to nuclei with N = 90 there is a sharp variation in the deformation parameter, it may be assumed that in the intermediate nucleus with N = 89 the equilibrium states having strongly differing equilibrium forms are close to each other.

The characteristics of the β decay of both isomeric states and the strongly hindered γ-transition between them which have been considered give reason to assume that they are pre-

TABLE 3. Characteristics of Levels of $_{81}Tl$

Nucleus	$T_{1/2}$, sec	E_γ (E3), keV	E_{level}, keV
$_{81}Tl^{193}_{112}$	126	<25	~390–365
$_{81}Tl^{195}_{114}$	3.6	99	482
$_{81}Tl^{197}_{116}$	0.53	387	607
$_{81}Tl^{199}_{118}$	$2.9 \cdot 10^{-2}$	382	749
$_{81}Tl^{201}_{120}$	$2.1 \cdot 10^{-3}$	590	923
$_{81}Tl^{207}_{126}$	~1	990	1340 ($h_{11/2}$)

cisely representative of states having different equilibrium shapes, the ground state (T = 13 years) being strongly deformed ($\beta \approx 0.28$) and the isomeric state (T = 9.3 h) having a spherical or almost spherical shape [44].

Anomalous Isomerism in the Odd Isotopes $_{81}Tl^{193-201}$

All the odd isotopes of Tl have Z = 81, and according to the shell model are included in the isomerism islands.

According to Meyer's scheme in this Z region the adjacently situated levels $S_{1/2}$, $d_{3/2}$, $h_{11/2}$ and possibly $d_{5/2}$ are filled.

Among the lower excited levels of Tl, therefore, the level $h_{11/2}$ will be isomeric and the isomeric transition may be of the type E5 ($h_{11/2} \rightarrow s_{1/2}$), M4 ($h_{11/2} \rightarrow d_{3/2}$), E3 ($h_{11/2} \rightarrow d_{5/2}$). In complete agreement with these simple predictions, isomeric levels have been found in the odd isotopes $_{81}Tl^{193}$ - $_{81}Tl^{201}$; in the decay of these levels an isomeric γ-transition of E3 type has been observed, as well as other transitions of lower multipolarity. Since the ground states of $_{81}Tl$ are always $S_{1/2}$ (their spins were measured) the disintegration scheme of the isomers may be represented as [1]

$$^{11}/_{2} \xrightarrow{E3} {}^{5}/_{2}{}^{+} \xrightarrow{M1} {}^{3}/_{2}{}^{+} \xrightarrow{M1} {}^{1}/_{2}{}^{+} .$$

Recently, however, Diamond and Stephens [38] showed that γ transition of type E3 did not occur to the known level $^{5}/_{2}{}^{+}$ but to a level with I = $^{3}/_{2}{}^{+}$, and consequently the spin of the isomeric state in these isotopes was not $^{11}/_{2}{}^{-}$ ($h_{11/2}$) but $^{9}/_{2}{}^{-}$, its excitation energy being situated in the range from ~ 380 keV (Tl^{193}) to 920 keV (Tl^{201}) [38-41].

Table 3 gives the characteristics of these levels of $_{81}Tl$. The level with I = $^{9}/_{2}{}^{-}$ is not a single-particle level (there is no such level in this region of the Meyer scheme), and, due to the low energy value, it is not a three-particle level of (p, 2n) type. One possible interpretation is that the $^{9}/_{2}{}^{-}$ level is formed as the result of the combination of a nucleon in the state $h_{11/2}$ (j = $11/2^{-}$) and a quadrupole phonon with $\Lambda = 2$ ($I_0 = 2^{+}$). The scheme* of levels of $_{81}Tl$ produced as the result of such a combination is shown in Fig. 2, from which it will be seen that such a model is in good agreement with experiment. Two interesting features of the interpretation considered may be mentioned.

*In constructing the scheme it was assumed that the known excitation levels with I = $^{3}/_{2}{}^{+}$ and $^{5}/_{2}{}^{+}$ are also produced in consequence of combination of a quadrupole phonon with a proton in the $S_{1/2}$ state.

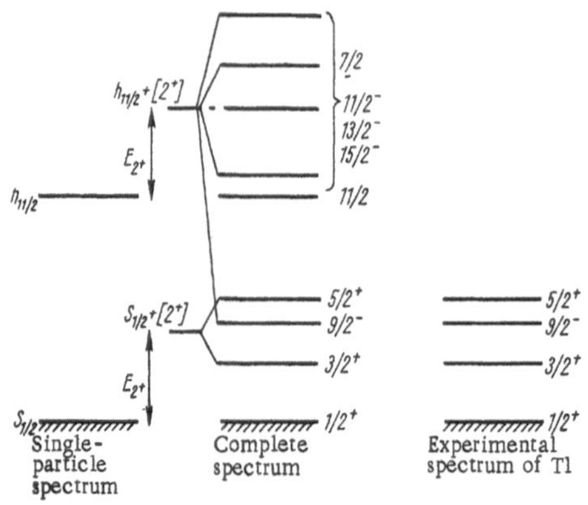

Fig. 2. Spectrum of levels of thallium isotopes.

1. The interaction of the proton $h_{11/2}$ and quadrupole phonon ought to be such that the lower level of the multiplet $\mathbf{j} + \mathbf{I}_0$ has spin $I = j - 1$. This is a hypothesis. In experimentally investigated multiplets of this type, however, in ${}_{27}Co^{59}_{32}$ [$j = {}^7/_2{}^- + (I_0 = 2^+)$] and ${}_{49}In^{115}_{86}$ [$(j = {}^9/_2{}^+) + (I_0 = 2^+)$], it has been shown that the spin of the lower level is also described by the relationship $I = j - 1$.

2. This interaction ought to be so strong that the lower level of the multiplet is displaced downwardly more than the value of the energy of the phonon.

If the interpretation considered is correct, in the odd isotopes of Tl we are dealing for the first time with a vibrational isomeric state.

It is interesting to note that in ${}_{81}Tl^{207}_{126}$, a nucleus with one hole in doubly filled shells, an isomeric level with $T \sim 1$ sec, $E = 1340$ keV, has also been found, but in contrast to the other isotopes, its spin is ${}^{11}/_2{}^-$ ($h_{11/2}$), and not ${}^9/_2{}^-$. It is easy to see that this should have been expected since the energy of the quadrupole phonon in ${}_{80}Hg^{206}_{126}$ - ${}_{82}Pb^{208}_{126}$, nuclei each with a filled neutron-shell, is very high (a level with $I = 2^+$ has not hitherto been observed) and consequently the vibrational level in the odd nucleus ${}_{81}Tl^{207}_{126}$ is high.

LITERATURE CITED

1. Dzhelepov, B. S., et al., Izv. Akad. Nauk SSSR, Ser. Fiz., 1963.
2. Dzhelepov, B.S., and Peker, L. K., Decay Schemes of Radioactive Nuclei with A < 100, Moscow, Nauka, in press.
3. Khudelidze, D. E., et al., Zh. Éksperim. i Teor. Fiz., 47:1167 (1964).
4. Holland, R. E., et al., Phys. Rev. Lett., 13:241 (1964).
5. Yntema, G. L., and Satchler, G. R., Phys. Rev., 134:B976 (1964).
6. Sheline, R. K., and Wildermuth, K., Nucl. Phys., 21:196 (1960).
7. Plendl, H. S., et al., Nucl. Phys., (1965).
8. Bansal, R. K., and French, J. B., Phys. Lett., 11:145 (1964).
9. Kharitonov, Yu. I., Dissertation LFTI (1965); Izv. Akad. Nauk SSSR, Ser. Fiz., 29:315 (1964).
10. Peker, L. K., Izv. Akad. Nauk SSSR, Ser. Fiz., 28:303 (1964).
11. Peker, L. K., and Kharitonov, Yu. I., Izv. Akad. Nauk SSSR, Ser. Fiz., in press (1965).
12. Peker, L. K., Yad. Fiz., in press (1966).
13. Alburger, D. E., and Thulin, S., Phys. Rev., 89:1146 (1953).

14. Wsrhanek, H., Nucl. Phys., 33:639 (1962).
15. Haller, I.B., and Yung, B., Nucl. Phys., 52:524 (1964).
16. Remaev, V. V., et. al., Zh. Éksperim. i Teor. Fiz., 42:408; 43:1649 (1962).
17. Karras, M., et al., Arkiv. Fys., 23:57 (1963).
18. Silvola, A., et al., Nucl. Phys., 52:449.
19. Dunlavey, D.C., and Seaborg, G. T., Phys. Rev., 85:757 (1952).
20. Hargrove, C.K., and Martin, W. M., Can. J. Phys., 40:964 (1962).
21. Spiess, F. N., Phys. Rev., 94:1292 (1954).
22. Jentsche, W. J., et al., Phys. Rev., 96:231 (1953).
23. Jorgensen, M. J., et al., Phys. Rev. Lett., 1:321 (1962).
24. Kristensen, L., et al., Phys. Rev. Lett., 8:56 (1964).
25. Alexander, P., et al., Phys. Rev., 133:B284 (1964).
26. Blinchert-Toft, P. H., Arkiv. Fys., 26:241 (1964).
27. Kalkstein, M. I., Bull. Amer. Phys. Soc., 8:376 (1963).
28. Campbell, E. G., and Fettweis, P. F., Nucl. Phys., 33:272 (1962).
29. Reising, R.R., and Pate, B. D., Nucl. Phys., 65:609 (1965).
30. Demin, A. G., and Kushakevich, Yu. P., Yad. Fiz., Vol. 1, No. 2 (1965).
31. Aleksander, K. F., et al., Paper presented at the Conference on Neutron-Deficit Isotopes of the Rare Earths, Dubna, Joint Institute for Nuclear Research, 1965.
32. Gritsyna, V.T., and Klyucharev, A. P., Paper presented at the Conference on Neutron-Deficit Isotopes of the Rare Earths, Dubna, Joint Institute of Nuclear Research, 1965.
33. Karnaukhov, V. A., Zh. Éksperim. i Teor. Fiz., 42:973 (1963).
34. Perlman, I., et al., Phys. Rev., 127:917 (1962).
35. Sliv, L.A., and Kharitonov, Yu. I., Zh. Éksperim. i Teor. Fiz., 46:811 (1964); 44:247 (1963).
36. Flerov, G.N., and Polikanov, S. M., Paper presented at the Conference on Nuclear Spectroscopy, Minsk (1965); Paper presented at the Conference on Neutron-Deficit Isotopes, Dubna, Joint Institute for Nuclear Research, July, 1965.
37. Zaretskii, D., and Urin, M., Comptes Rendus du Congrès International de Physique Nucléaire, Paris, 1964.
38. Diamond, R. M., and Stephens, F. S., Nucl. Phys., 45:632 (1963).
39. Brandi, K., et al., Nucl. Phys., 59:33 (1964).
40. Grizina, V.T., and Foster, Nucl. Phys. (1965).
41. Demin, A. G., et al., Zh. Éksperim. i Teor. Fiz., 45:1344 (1963).